Microelectronics Processing

ACS SYMPOSIUM SERIES **295**

Microelectronics Processing: Inorganic Materials Characterization

Lawrence A. Casper, EDITOR
Honeywell Inc.

Developed from a symposium sponsored by
the Division of Industrial and Engineering Chemistry
of the American Chemical Society

American Chemical Society, Washington, DC 1986

Library of Congress Cataloging-in-Publication Data

Microelectronics processing.
  (ACS symposium series, ISSN 0097-6156; 295)

  Includes bibliographies and indexes.

  1. Microelectronics—Materials—Congresses.

  I. Casper, L. A. II. American Chemical Society. Division of Industrial and Engineering Chemistry. III. Series.

TK7874.M486   1986   621.381'7   85-30648
ISBN 0-8412-0934-0

Copyright © 1986

American Chemical Society

All Rights Reserved. The appearance of the code at the bottom of the first page of each chapter in this volume indicates the copyright owner's consent that reprographic copies of the chapter may be made for personal or internal use or for the personal or internal use of specific clients. This consent is given on the condition, however, that the copier pay the stated per copy fee through the Copyright Clearance Center, Inc., 27 Congress Street, Salem, MA 01970, for copying beyond that permitted by Sections 107 or 108 of the U.S. Copyright Law. This consent does not extend to copying or transmission by any means—graphic or electronic—for any other purpose, such as for general distribution, for advertising or promotional purposes, for creating a new collective work, for resale, or for information storage and retrieval systems. The copying fee for each chapter is indicated in the code at the bottom of the first page of the chapter.

The citation of trade names and/or names of manufacturers in this publication is not to be construed as an endorsement or as approval by ACS of the commercial products or services referenced herein; nor should the mere reference herein to any drawing, specification, chemical process, or other data be regarded as a license or as a conveyance of any right or permission, to the holder, reader, or any other person or corporation, to manufacture, reproduce, use, or sell any patented invention or copyrighted work that may in any way be related thereto. Registered names, trademarks, etc., used in this publication, even without specific indication thereof, are not to be considered unprotected by law.

PRINTED IN THE UNITED STATES OF AMERICA

ACS Symposium Series

M. Joan Comstock, *Series Editor*

Advisory Board

Harvey W. Blanch
University of California—Berkeley

Alan Elzerman
Clemson University

John W. Finley
Nabisco Brands, Inc.

Marye Anne Fox
The University of Texas—Austin

Martin L. Gorbaty
Exxon Research and Engineering Co.

Roland F. Hirsch
U.S. Department of Energy

Rudolph J. Marcus
Consultant, Computers &
  Chemistry Research

Vincent D. McGinniss
Battelle Columbus Laboratories

Donald E. Moreland
USDA, Agricultural Research Service

W. H. Norton
J. T. Baker Chemical Company

James C. Randall
Exxon Chemical Company

W. D. Shults
Oak Ridge National Laboratory

Geoffrey K. Smith
Rohm & Haas Co.

Charles S. Tuesday
General Motors Research Laboratory

Douglas B. Walters
National Institute of
  Environmental Health

C. Grant Willson
IBM Research Department

FOREWORD

The ACS SYMPOSIUM SERIES was founded in 1974 to provide a medium for publishing symposia quickly in book form. The format of the Series parallels that of the continuing ADVANCES IN CHEMISTRY SERIES except that, in order to save time, the papers are not typeset but are reproduced as they are submitted by the authors in camera-ready form. Papers are reviewed under the supervision of the Editors with the assistance of the Series Advisory Board and are selected to maintain the integrity of the symposia; however, verbatim reproductions of previously published papers are not accepted. Both reviews and reports of research are acceptable, because symposia may embrace both types of presentation.

CONTENTS

Preface ix

1. **Analytical Approaches and Expert Systems in the Characterization of Microelectronic Devices** 1
   D. E. Passoja, Lawrence A. Casper, and A. J. Scharman

2. **Electrical Characterization of Semiconductor Materials and Devices** 18
   Dieter K. Schroder

3. **Dopant Profiles by the Spreading Resistance Technique** 34
   Robert G. Mazur

4. **Scanning Electron Microscopic Techniques for Characterization of Semiconductor Materials** 49
   Rodney A. Young and Ronald V. Kalin

5. **Semiconductor Materials Defect Diagnostics for Submicrometer Very Large Scale Integration Technology** 75
   G. A. Rozgonyi and D. K. Sadana

6. **Applications of Secondary Ion Mass Spectroscopy to Characterization of Microelectronic Materials** 96
   Mary Ryan–Hotchkiss

7. **Applications of Auger Electron Spectroscopy in Microelectronics** 118
   Paul A. Lindfors, Ronald W. Kee, and Douglas L. Jones

8. **X-ray Photoelectron Spectroscopy Applied to Microelectronic Materials** ... 144
   William F. Stickle and Kenneth D. Bomben

9. **Application of Neutron Depth Profiling to Microelectronic Materials Processing** 163
   R. G. Downing, J. T. Maki, and R. F. Fleming

10. **Thermal-Wave Measurement of Thin-Film Thickness** 181
    Allan Rosencwaig

11. **Characterization of Materials, Thin Films, and Interfaces by Optical Reflectance and Ellipsometric Techniques** 192
    D. E. Aspnes

12. **Measurement of the Oxygen and Carbon Content of Silicon Wafers by Fourier Transform IR Spectrophotometry** 208
    Aslan Baghdadi

13. **Application of the Raman Microprobe to Analytical Problems of Microelectronics** 230
    Fran Adar

14. **Characterization of Gallium Arsenide by Magneto-optical Photoluminescent Spectroscopy** 240
    D. C. Reynolds

15. Thermal-Wave Imaging in a Scanning Electron Microscope..............253
    Allan Rosencwaig

16. Fourier Transform Mass Spectrometry in the Microelectronics Service
    Laboratory..267
    W. H. Penzel

17. Materials Characterization Using Elemental and Isotopic Analyses
    by Inductively Coupled Plasma Mass Spectrometry....................284
    B. Shushan, E. S. K. Quan, A. Boorn, D. J. Douglas, and
    G. Rosenblatt

18. Activation Analysis of Electronics Materials..........................294
    Richard M. Lindstrom

19. Trace Element Survey Analyses by Spark Source Mass Spectrography.....308
    Fredric D. Leipziger and Richard J. Guidoboni

20. Characterization of Components in Plasma Phosphorus-Doped Oxides....320
    Jana Houskova, Kim-Khanh N. Ho, and Marjorie K. Balazs

21. Process Control of Vacuum-Deposited Nickel–Chromium
    for the Fabrication of Reproducible Thin-Film Resistors................333
    Vineet S. Dharmadhikari

22. Characterization of Spin-On Glass Films as a Planarizing Dielectric.......349
    Satish K. Gupta and Roland L. Chin

23. Effects of Various Chemistries on Silicon-Wafer Cleaning...............366
    D. Scott Becker, William R. Schmidt, Charlie A. Peterson, and
    Don C. Burkman

24. Monitoring of Particles in Gases with a Laser Counter..................377
    C. E. Nowakowski and J. V. Martinez de Pinillos

25. Microelectronics Processing Problem Solving: The Synergism
    of Complementary Techniques.....................................398
    J. N. Ramsey

INDEXES

Author Index...429

Subject Index..429

PREFACE

INTEGRATED CIRCUIT (IC) TECHNOLOGY has moved steadily toward increasing circuit density and improved performance during its brief history. These improvements have been achieved through the development of the microlithography process, which now permits production of devices with feature sizes of 1 μm. Submicrometer devices are actively being developed, and production is expected later in this decade.

This continuous reduction in size has been paralleled by advances in thin-film materials technology. In certain devices, materials structures that are only a few hundred nanometers in dimension are being used. Processing of semiconductor, conductor, and dielectric materials to such dimensions requires that the materials and processes be characterized to an extent unsurpassed in any other technology. Exceedingly small concentrations of impurity may result in significant degradation of the operating characteristics of the final products. Further, the economic viability of advanced IC production is directly linked to the achievement of high yield through the identification and elimination of yield inhibitors.

The role of chemical analysis and materials characterization has become a central one in advanced IC fabrication. Starting materials, the raw semiconductor wafer, as well as each gas, liquid, and solid used in the processing sequence are coming under increasing scrutiny to assure the highest standards of purity. As a result, analytical methods for bulk chemicals and materials are often used at their lowest attainable detection limits, and techniques are being developed to meet new demands.

Surface analytical tools have increased in use as device dimensions have decreased into the "near-surface" realm. High spatial resolution and high analytical sensitivity are basic requirements for the characterization of impurities and defects in near and submicrometer devices. Electron, ion, and photon beam techniques are widely employed to characterize IC fabrication at each processing step and to analyze failures as they occur in processing as well as in the field.

The purpose of the symposium from which this book developed was to provide a forum for the discussion of advanced analytical techniques and their application to all aspects of IC processing. This task is rather large because nearly every major analytical technique finds application in some aspect of IC processing. The focus was narrowed by concentrating on the inorganic side of analysis; the analysis of organics and polymers is yet another subject.

Some important techniques are not discussed such as Rutherford backscatter and other accelerator techniques, whereas other topics overlap to a certain degree. Nevertheless, the chapters presented here provide a useful overview of current analytical practice and supplement material published in related books and in reports and proceedings that have been issued periodically by several technical societies.

LAWRENCE A. CASPER
Solid State Electronics Division
Honeywell Inc.
Plymouth, MN 55441

July 1985

Analytical Approaches and Expert Systems in the Characterization of Microelectronic Devices

D. E. Passoja[1], Lawrence A. Casper[2], and A. J. Scharman[3]

[1] Linde Division, Union Carbide Corporation, Tarrytown, NY 10591
[2] Solid State Electronics Division, Honeywell Inc., Plymouth, MN 55441
[3] Central Scientific Laboratory, Union Carbide Corporation, Tarrytown, NY 10591

With further device miniaturization anticipated for microelectronics in the future, more stringent requirements will be placed on contaminants and impurities in the chemicals, materials and gases that are supplied to the industry. It is suggested that absolute determination of the role of a particular contaminant during device processing will be difficult. It is proposed that the present analytical "characterization" of a device is an image composed of texels - discrete picture elements - defined in a space of magnification and sensitivity. By viewing the microanalytical problem from a broader perspective based upon the geometric organization of the microscopic world, it is also proposed that by the use of "expert systems", more structured analytical algorithms and some simple scaling laws it should be possible to provide more efficient problem solving methods for the analyst.

In the past 10 years the driving force of miniaturization in the microelectronics industry has resulted in a consistent decrease in the feature sizes of integrated circuits. Commensurate with this miniaturization there has been an increase in the complexity of the technology base (Figures 1-3). Whereas the early solid state technology was based primarily on knowledge of systems and circuits, materials science and chemistry, had a significant but limited role in it. With the advent of the integrated circuit, however, the technology base expanded and to include a broad range of disciplines that is now a continuum of overlapping technologies merging to produce the final outcome: the integrated circuit.

In progressing from discrete to integrated devices at the levels of small and medium scale integration, it became necessary to incorporate device physics into the technology base. As device structures were made progressively smaller, they behaved neither as "bulk" three dimensional devices nor as strict two dimensional

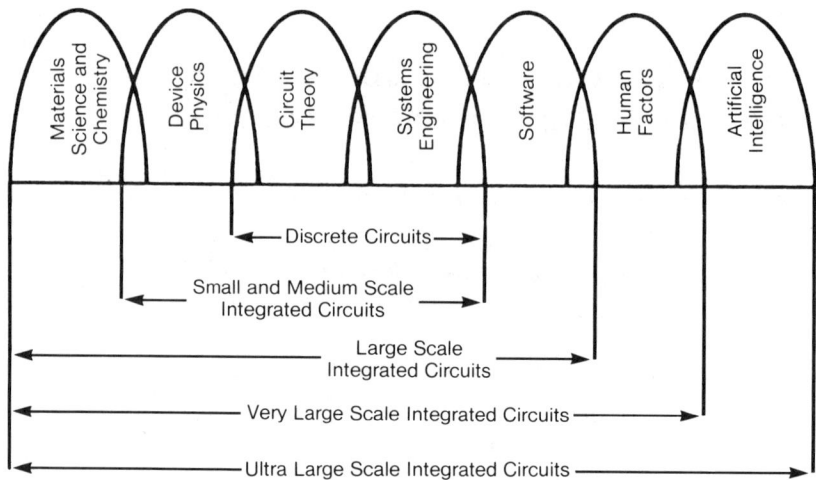

Figure 1. Science and engineering disciplines used in the integrated circuit technology.

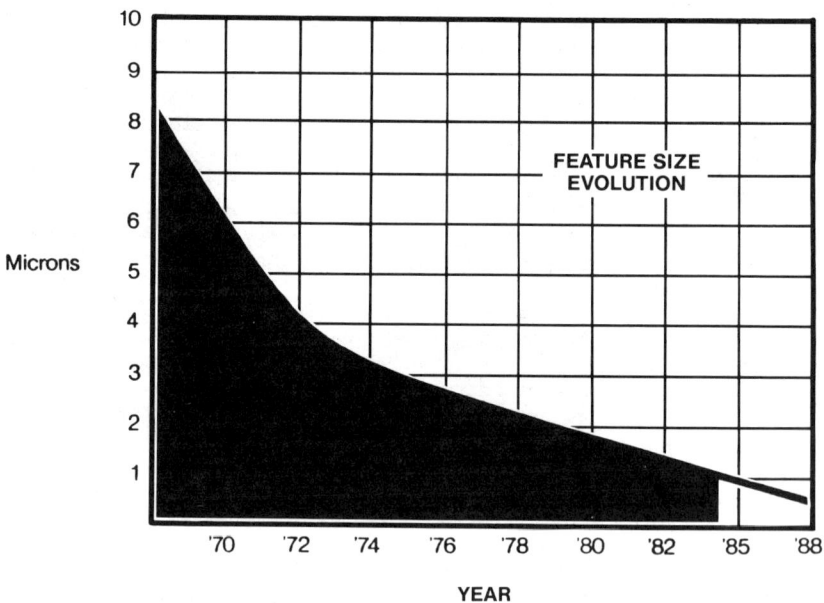

Figure 2. A plot showing a chronology of minimum device feature size achievable.

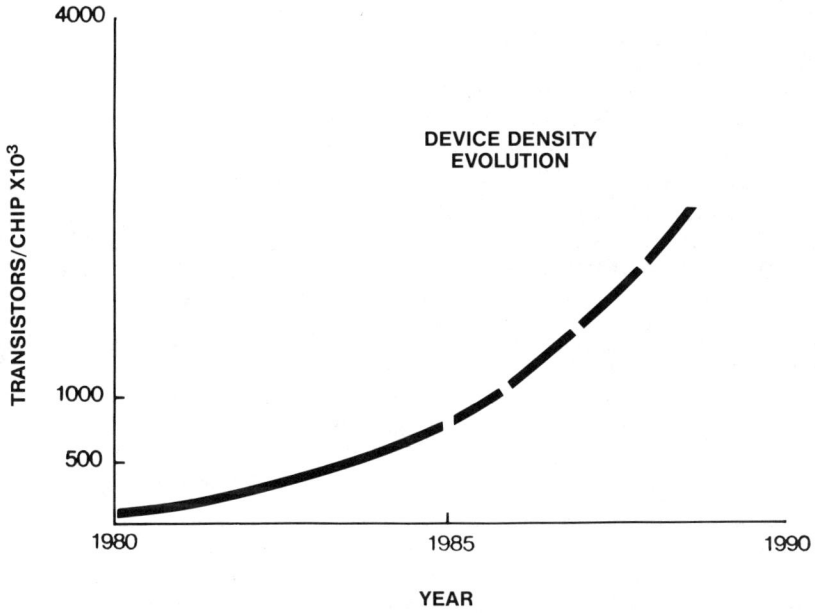

Figure 3. A plot showing the progress of increasing device density.

devices. At these levels of miniaturization, defects and
contaminants are not minor constituents since failure of only a
small fraction of circuits on the device level results in a
substantial decrease in process yields. As miniaturization
progresses further, defects and contaminants will become even more
influential on device yield, performance and reliability.

As the semiconductor industry progresses into the latter part
of this decade, submicron technology will become a reality. From
scaling laws we know that thin films are not thin enough. For
example in metal-oxide-semiconductor (MOS) technology, gate oxides
will be 200 in thickness or less. Such layers should probably
not be thought of as thin films since they are more accurately
described as "surface region" or using the terminology of surface
chemistry, "interphase" region.

The high sensitivity of VLSI technology to defects and
contamination has necessitated a substantial investment in new
processing technology. Contamination and defect control is now a
basic requirement in all aspects of processing. Accompanying this
new ultra-clean technology is the need to characterize the
materials and chemicals, the unit processes and the ambients to
which the wafers are exposed.

At a more fundamental level, VLSI devices must be studied to
determine the complex relationships that exist between a device's
structure, composition and electrical characteristics. It is
presently uneconomic to achieve process control by determining the
detailed changes which occur in a device's structure and chemistry
once processing begins. Process control is usually achieved
instead by monitoring critical parameters such as film thickness
and electrical characteristics thereby eliminating (or reworking)
wafers that are not in the control range.

The latitude of a process and the sensitivity of certain
designs to yield losses are, in part, a direct reflection of the
evolution of microscopic structural imperfections that develop
during processing. Such details as a defect's scaling behavior
need to be studied in order to determine the relationship between
their spatial evolution, formation mechanisms and frequency of
occurrence. This behavior is quite complex and will certainly
include more surface related issues as device dimensional scaling
continues. It would be expected that at some level of
miniaturization the synergism between the accumulated process
transients and randomness inherent at the atomic level might
represent a fundamental physical limitation of current technology
and make further miniaturization difficult if not impossible.

As surface scientists, material scientists and chemists it
would be useful at this time to step back and reflect on what
impact further miniaturization will have on us. Of particular
interest to us is the analysis of process performance and device
yields using surface analysis techniques. With further
miniaturization will it be possible to make meaningful analyses on
device failures in an attempt to answer such questions as: why
did this device fail? or will it soon be necessary to speak in
more general terms and develop answers in terms of probabilities.
For example, if one die is characterized in great detail should it
be considered as a unique observation or should it be considered

as part of an ensemble of dice in the process? If the latter
consideration is the more appropriate one then the task of
characterization is to reconstruct a probability distribution (for
a set of dice in the process) from incompletely sampled data.
This is a familiar problem which can be analyzed by using the "
maximum entropy method",(1-4) a technique used extensively
elsewhere in the field of image analysis.

More specifically, an important question to be answered is:
in failure analysis of microelectronic devices what are the
consequences of miniaturization and what impact will further
miniaturization have on the way an analysis is performed? Simple
questions such as - what technique should be used? and what is a
defect, would I know one if I saw one? appear to have answers at
one level of miniaturization but at another level the answers
become more elusive and expensive to obtain directly by direct
microanalytical methods.

As we look to the future expansion of this technology base
the analytical problems indeed look challenging and more
difficult. It is anticipated that with further miniaturization
the use of machine intelligence could assist the analyst in
problem solving. We have assembled some necessary information
that we hope will foster future developments in this important
area.

Problem Solving - A Look at Expert Systems

In the last 15 years or so, several research groups working in
artificial intelligence (AI) have been able to make some formal
statements concerning problem solving. These methods have been
used to develop some effective, specialized computer based methods
to analyze problems such as the identification of molecular
structure from mass spectral data (DENDRAL) and diagnosis of
cardio-pulmonary diseases (PUFF)(5). These machine implemented
"expert systems" are just beginning to be applied to selected
problems with notable success (for example, PROSPECTOR'S
identification of a mineral deposit potentially worth $100
million)(5).

The meaning of knowledge is difficult to define yet we are
aware of its usefulness in our everyday experiences. "Knowledge
engineering" has been developed even though a precise definition
of knowledge isn't available. This has occurred because a
pragmatic, working definition for knowledge has been adopted by
the AI community. Knowledge can be stated in general terms as:(5)

Knowledge = Facts + Beliefs + Heuristics

This statement doesn't facilitate an understanding of knowledge in
detail but instead defines its boundaries in rather broad terms.

Knowledge also contains certain features and although they
are often expressed in broad terms, they can be programmed to
define a knowledge base. Descriptions using the knowledge base
seem like sentences in a language whose elementary components
consist of primitive features or concepts. Certain descriptions
within this system define unique relationships that can be used to
develop and analyze other relationships. Two major themes of
knowledge representation are thus:

A representation system must specify how to
represent any chunk of knowledge in a domain.
A representation system must supply algorithms for
manipulating its data structures and determining
that some expressions "follow from" others.(5)

An explanation of a solution to a problem can be developed by
defining a path from a problem state to a solution state. An
"expert system" is thus a machine of accessible knowledge
representation that can be used to find a solution to a problem by
selecting an appropriate path from a problem state to a solution
state from a large number of possible alternatives(5). In this
endeavor an analyst must be successful which is defined as:
Success = finding a good enough answer with the available
resources.

In the course of assembling expert opinion and technical
resources for this paper, it became apparent that certain
guidelines were being used by the microanalysis community to solve
microelectronic process and failure problems. Not everyone was
aware of these guidelines, not everyone was aware that any such
rules existed nor was their application uniform, since not every
problem was solved by the analysts. Typical comments were made
such as:(12)

" people who know how to do these problems well
seem to be hard wired or born with the knowledge..."

" people who don't know the rules don't do microanalysis because they're just not efficient at it..."

It was generally agreed that guidelines were indeed being used and
they were considered to be "intuitively obvious" to those who used
them.

The authors felt that since these guidelines seem almost
trivial to some analysts, but to others (perhaps the less
successful ones) they are unknown, we believed that it would be
helpful to state a few of them explicitly. Our purpose here is to
begin assembling shared knowledge so that we can develop more
formalized procedures and protocols with hope to construct an
expert system for solving process related problems in microcircuit
manufacturing.

We state below some of the rules that experts used for
solving microanalysis problems. The list is not meant to be
exhaustive, nor is it complete. It is stated merely as an attempt
at assembling some collective knowledge from the materials
analysis community on problem solving.

Some Expert's Opinions

It is recognized that any problem solution is probably goal
oriented and that answers must be obtained efficiently. As
experts we should try to replace knowledge for search whenever
possible since it is more efficient: learning must occur quickly
from observations that are made during any search. Errors in
understanding and/or in learning should be avoided since mistakes
can propagate and grow in the course of our search. If this
occurs our problem solving will become inefficient as it
progresses.

Some Guidelines for Problem Solving(6)

- Use low magnification first and work to higher and higher magnifications. Part of the basis for this approach appears to be inherent in the definintion of any microscopy problem. Low magnification defines a spatial average of some characteristic of interest. Higher magnification observations usually reveal fluctuations (higher moments of the distribution) and "structures within structures".
- Try to develop an understanding (a working hypothesis) at each magnification that is used by asking questions such as: Is this observation significant? How can this influence the observed failure? Are there more of these defects? How could these defects interact with their environment? Are there physical models that are available that would help to explain the role of this defect?
- Read the literature to make the search for an answer more efficient.
- Use information from more than one source (when possible) to interpret the information obtained from the microanalysis.
- Recognize that there are certain types of problems and that each has a particular structure. There are not an unlimited number of problem types. The unique structures usually depend upon statistics that underlie the phenomena that caused the defect in the first place. If possible, a complete description of a defect's distribution function(s) would probably be helpful, but this is usually too expensive to measure or the problem is too poorly understood to spend much time on its measurement.
- Information collected from a hierarchial level is more efficient than that collected from a more fundamental level. This is J. Ramsay's "molecular level information" observation (in this symposium's proceedings). This can also be stated in a different way as: "By searching an abstracted representation of a space, the combinatorimetrics can be reduced. The search of the abstract space is quicker because it is smaller; single steps in the abstract space correspond to big steps in the ground search space."(5)
- Try to relate observations made at one magnification to those made at another magnification. What scales are independent of one another?
- Use solid scientific information in any facts that are developed in the search, don't try to invent new science. The chances are small that it is possible to do this without making mistakes.

- Don't ask questions indescriminately. If this is done the problem will grow uncontrollably since each question spawns new ones. Be conservative and thoughtful in developing critical inquiry.
- Measurements are real, speculations about reality are not. When in doubt get some measurements!
- After an analysis has been done there is a tendency to believe it, this should be resisted if the results cannot be supported by some physical reason or correlated with other measurements.
- Quantitation or transformation of the problem into a more abstract one can be helpful since new modes of thinking can appear. This should be done with some caution since mistakes can be made.
- Each analytical technique has its limitations that must be fully understood in the context of each application. Don't push analytical methods and techniques beyond their limits for any applications.
- Avoid using a "single technique" approach, it is usually inefficient.
Background information is useful since it places the problem in perspective; this perspective is valuable since it forms a "solution space" that defines the context of the problem.

Some Features of the Analytical Image Space

A general task that microanalysts often encounter is to collect information concerning the structure and composition of failed devices or to characterize some aspect of a microelectronic manufacturing process. Usually a number of analytical techniques must be used in order "to perform a characterization". We suggest that the meaning of "to characterize", as understood by the analyst, can be developed further by borrowing some concepts used in AI and image analysis.

Figure 4 describes an elemental unit constructed from a set of orthogonal coordinates of sensitivity and magnification. (In an interesting historical note, Weiner(7) considered magnification to be a transformation of spatial coordinates, a potentially far more significant concept than is usually ascribed to magnification nowadays. These coordinates are quite familiar to every analyst as they directly impact the cost of an analysis. Four points are described in this space in familiar terminology as shown in Figure 4.

In order "to characterize" something, the analyst usually defines some point in this space by the instrument that he chooses and develops geometric and chemical information about the object of interest.

The space in the vicinity of the chosen point can be extended somewhat by using a series of overlapping techniques and thus defining movement in the space. The extension is realized once spatial variations in composition are measured in space-time determining an object's "texture" and thus the object is "characterized". The analyst essentially decides how to build an image of the object in this space by using texture primitives or "texels" similar to picture primitives or pixels.

Expressing this in a less abstract manner; the analyst must describe the object with as few magnification steps as are necessary in order to span a given range. These depend upon the structure under study and must be adjusted accordingly. At each magnification, the object under investigation is scrutinized and its structure and composition are determined as a function of the spatial coordinates. The analysts dilemma can be best described by the following:

Magnification and Sensitivity Considerations

Increasing Complexity and Costs
- A. Low magnficiation -
  Low sensitivity, "a little bit everywhere"
- B. Low magnification -
  High sensitivity, "a sensitivity window"
- C. High magnification -
  Low sensitivity, "a typical microscopy problem"
- D. High magnification -
  High sensitivity, "fools rush in..."
  the paradox of no positive results after commencing.

In the following section we touch on issues of magnification and sensitivity by considering some specific examples. The observation of space and the microstructural heterogeneities embedded in it has a <u>physical</u> basis that is derived from probability distributions. As Prigogine(<u>8-9</u>) has suggested, macroscopic irreversible phenomena are associated with probability distributions and their evolution in time. Partitioning a system introduces an "internal" time to a system. The introduction of partitions into a system merely defines a subset that may be observed by increasing the magnification. We follow Prigogine's basic line of reasoning in the examples that follow by considering probability distributions as a function of magnification (instead of time). Our discussion is not exhaustive but illustrates what meaning particular spatial defect distributions can have for the analyst when a characterization is performed.

Geometric Elements and Physical Constraints

Figure 5 describes the geometric building blocks or independent geometric features of which microelectronic devices are composed. The list is ordered to reveal smaller and smaller geometric features from the top to the bottom list. By applying Rule 1 above it is possible to first recognize the boule's structure, then the wafer's etc. At higher and higher magnification finer and finer structural details can be observed until finally atoms can be seen. At each level of magnification there is a limited field of view and a limited resolution that defines a texel size; we observe finite sets of defects (generic) having given structures. The information that is collected at each magnification is a result of knowing more and more about less and

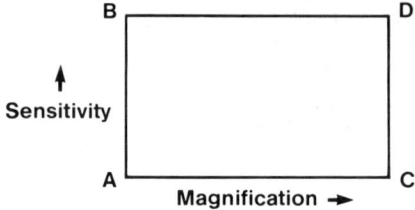

Figure 4. A schematic representation showing sensitivity and magnification space used for describing the virtues and limitations of an analytical technique.

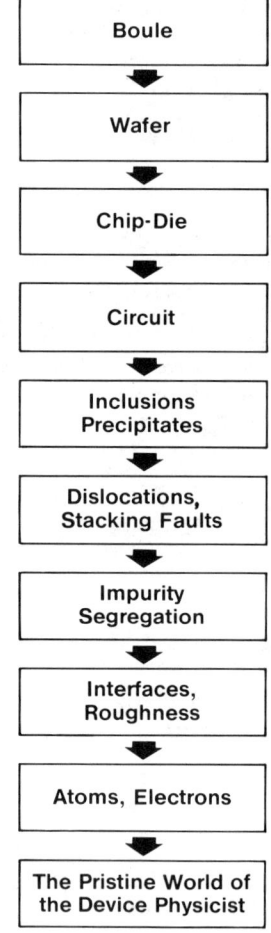

Figure 5. A block diagram showing just some of the building blocks that microelectronic devices are comprised of.

less of the structure. Observations made at higher and higher magnification result in more specialized knowledge about the microstructure. In order to describe the structure of the device, more area must be scanned and a cost is associated with this activity. (A similar problem has been discussed by A. Wheeler(11) and is known as the traveling salesman problem. Here the problem is to optimize the time that the salesman is to visit remote points in a search space.)

Probability Distributions

Spatial probability distribution functions play an essential role in determining what is observed and how observations are made in microstructural hierarchies. For purposes of illustration, in Figure 6 we consider a random two dimensional Gaussian distribution (a Raleigh distribution) of small spherical particles of density N_s in the plane with radius R. At low magnification we cannot observe any particles as the resolution of our instrument is too low. The particles soon become observable as the magnification is increased, but as it is increased further, the spacing between the particles increases also. Figure 6 shows how the particle distribution influences the time that is required to make a given number of observations (assuming that the observations are made at constant rate). A minimum occurs at the optimum magnification range; the shape of the minimum depends upon the particle size distribution function. (See appendix for a further development of this point.)

Different distributions of microstructural hierarchies result in different types of observations. For example, Figure 7 shows the relation between the number of defects per cm^3 in silicon vs length of the defect(10). This relationship is quite different than the one described above. As higher and higher magnifications are used more and more defects are observed and the spacings between the defects increase but at a much slower rate than in the above example. This behavior probably arises from the agglomeration of point defects that occur during processing of the silicon. The essential point to be made here is that probability distributions determine how observations are to be made.

Defects in an Oxide Film

Figure 8 shows some of the microstructural relationships that are observed at different magnifications in an oxide film on silicon. This example is merely a more detailed extension of the fundamental geometric features shown in Figure 4 as they apply to an oxide film.

Different aspects of the microstructure can be observed at each magnification shown in the figure. Each magnification reveals added information but the lower magnification images generally reveal the microstructural hierarchies more easily. The interface represents a discontinuity in the properties and therefore it is expected that this area would be of great importance as indeed it is. The operation of an MOS device relies upon the process used to fabricate it as being able to produce a

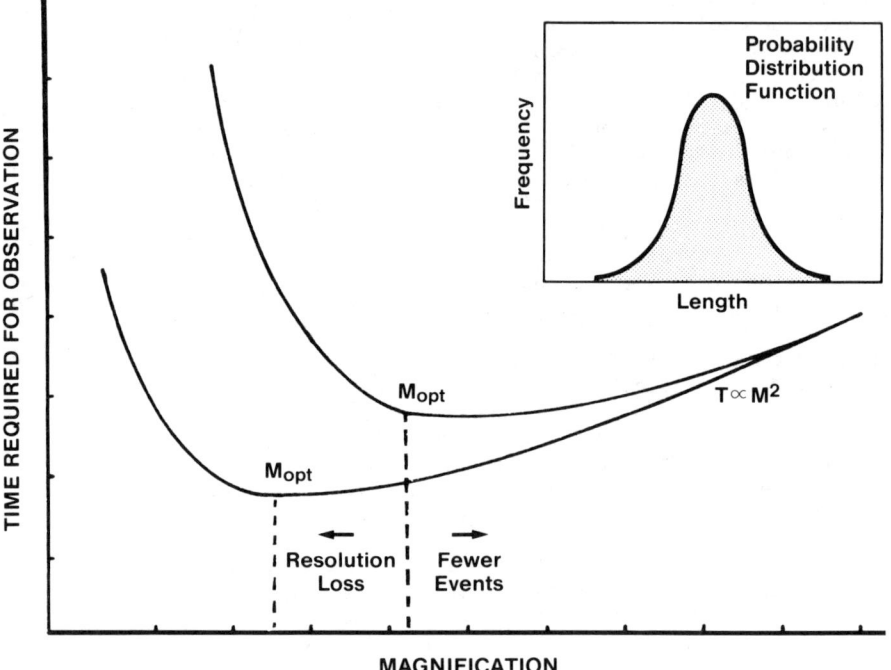

Figure 6. A plot of magnification vs. observation time showing how an optimal magnification range could be determined.

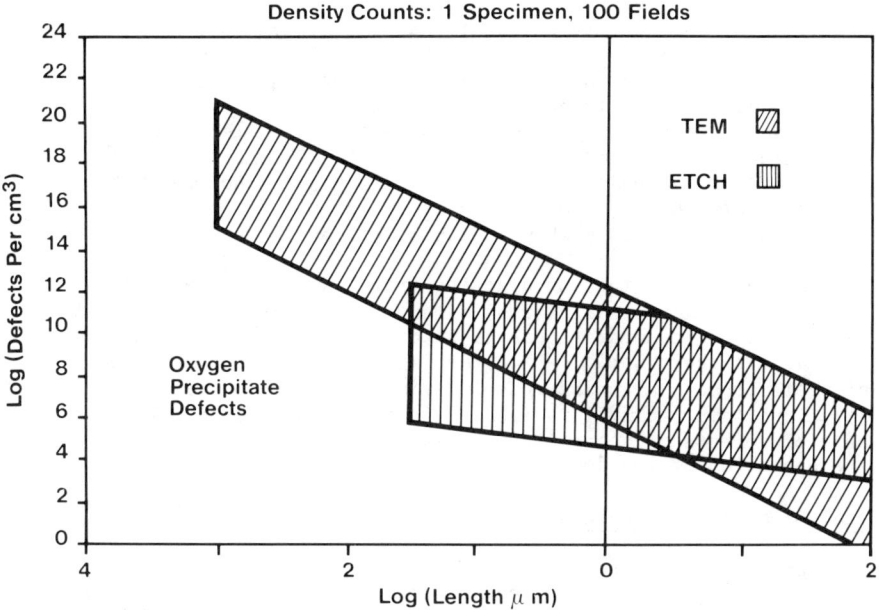

Figure 7. The size distribution of defects in bulk silicon. Reproduced with permission from Ref. 10. Copyright 1984 Electrochemical Society Inc.'

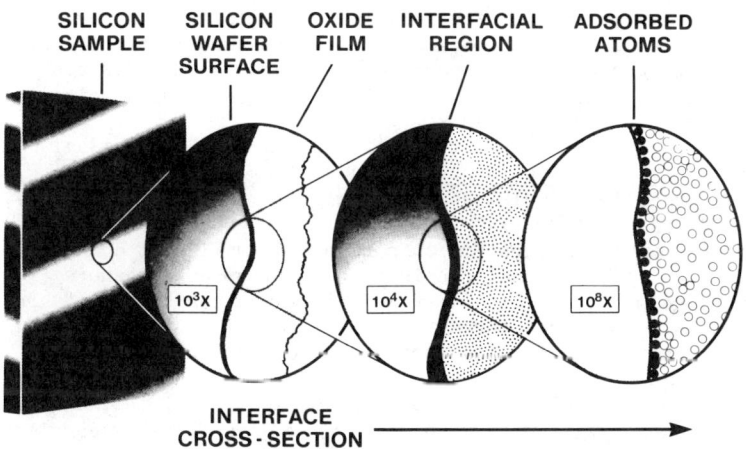

Figure 8. A schematic diagram showing the effects of magnification on the type of information from the Si-SiO$_2$ interface revealed by microscopy.

well defined interface with a consistent charge distribution.
Measurement of a C-V relationship would yield microscopic
information, whereas a cross sectional TEM study would yield 1
dimensional information about the interfacial structure - more
microscopic information. SIMS, on the other hand, would yield
more two dimensional information about the interface at a
sensitivity range that is lower than that of the electrical
measurements i.e. sodium levels that would influence the device
would not be measured by this method. To "solve the problem" the
analyst must characterize the structure, relate it to the
microchemistry and then show how this influences the operation of
the device.

At each magnification starting from low to high we observe
more and more detailed information until atoms are observed. It
is possible to describe a hierarchy of defects since they are
embedded in the material - they are merely members of a set of
discrete objects which represent deviations from the ideal
structure. The analyst is faced with the task of describing the
sets of defects and conditions under which they are detrimental to
the operation of the device.

Analytical Requirements

The analytical requirements in IC processing equal or surpass
those of any other industry and present a formidable problem to
the analyst. The fabrication of a semiconductor device involves
interaction of materials, processes and environment to produce a
complex, geometrical configuration of many materials. When
processed through to the package level a typical IC is a composite
of semiconductors, metals, ceramics and polymers containing over a
dozen distinct materials.

In the process of fabrication these materials have been
exposed to a wide range of materials or chemicals, ranging from
plasmas to aggressive chemical etchants. Environmental effects
may also be produced as a result of exposure to ambients such as
vacuum and high temperature, as well as air laden with moisture
and a variety of contaminants.

In the face of this complexity there are few analytical
techniques which do not apply! An advanced IC process will
utilize traditional bulk chemical analysis, i.e. chromatography,
spectroscopy, titrimetry, etc., as well as the array of ion and
electron beam techniques for thin film and small spot analysis.
In addition, process analysis is becoming more common where
plasmas, gases, distilled water and other process "streams" are
continuously monitored for a variety of constituents. The goal of
such extensive analytical support is to establish and maintain the
process "window".

Philosophies differ on the need for analytical instrumentation
in IC characterization. These range from exceptionally complete
(and expensive) support facilities to bare-bone operations which
rely on the more sophisticated techniques only when catastrophic
failure occurs. The latter approach is Edisonian in nature, i.e.
highly empirical. The problem arises that through a lack of
analytical information the process window is not well defined.

Excursions outside these process limits may produce a yield crash or, worse, a longer-term reliability problem. The well-characterized process is less prone to such problems and when they do occur the capability is at hand to support failure analysis.

The type and sophistication of materials characterization instrumentation made available for a given IC facility will depend on several factors: cost/benefit, quality perceptions, process complexity, maturity of the process. Commercial support labs are becoming more readily available to meet the VLSI challenge. The mix of in-house versus external laboratory support then becomes a matter of choice based on considerations of sample turn-around time, maintaining security of proprietary devices or processes, and outside versus inside expenditures.

Regardless of the analytical instrumentation mix, the art of problem-solving remains. A person or team must make the decisions about analysis: what, when, and how. There must be an analytical "expert" be it man or computer, who can make choices of appropriate technique.

The choice of technique and the interpretation of resulting data constitutes the art and science of analysis. The science is the consideration of the many factors in choosing an analytical technique, among them:
- Spatial resolution
- Analytical sensitivity
- Detection limit
- Interferences
- Sample preparation
- Analytical volume
- Sample consumption or destruction
- Elemental or molecular information

Such complex information is known by experts in each analytical area. The problem for most practitioners is to obtain the needed information or knowledge and proceed with an analytical scheme. The extent to which this knowledge becomes available in computerized "expert systems" the easier and more efficient this task will become. This creates more freedom for the art, or the intuitive/creative side of analysis. The critical role of the analyst is then the building of scientifically consistent mental picture, a concept or a model, to fit the accumulated data.

Conclusions

The problem solving guidelines used by various experts appear to have a fairly well defined structure. The rules themselves seem to be self evident and quite general but their general application is not uniform since problem solving is presently an art usually based on the analyst's intuition and individual problem solving ability. We suspect that with further miniaturization this approach will not suffice but will need to be replaced by more consistent, systematic methods that are machine based. This should standardize the methods and make them more efficient.

New manufacturing technology will probably increase the need for machine oriented problem solving methods. Any movement of the technology toward variable point-to-point microfabrication such as laser processing will only accelerate the need for protocols and more standardized problem solving methods. With refinement of our manufacturing processes, material changes will be produced at higher resolution thereby increasing the information content/volume of our manufactured items. The need for more efficient microscopic information collection and analysis is therefore expected to accompany these trends.

The advances in technology that have driven the miniaturization trends have created a complex situation for microanalysis that can only become more complex in the future. Even though technological advances might improve the resolution of some of our instrumentation, miniaturization itself will only complicate our ability to characterize devices in the future. Artificial intelligence and expert systems appear to have an excellent potential for enhancing our problem solving ability. It is expected that with proper development, this tool could become an essential item in the microanalyst's repertoire of techniques in tomorrow's technology.

Literature Cited

1. Skilling, J. Nature, Vol. 309, June 28, 1984, p. 748.
2. Tikochinsky, Y.; Tishby, N.Z., Levine, R. D. Phys. Rev Letters, Vol. 52, No. 6, April 16, 1984, p. 1357-1360.
3. Jaynes, E. T. Phys Rev, Vol. 106, No. 4, May 1957, p. 620-630.
4. Frieden, R. B. Jrnl Optical Society of America, Vol. 62, No. 4, April 1972, p. 511-518.
5. Fredrick, H.; Waterman, D. A.; Lenat, D. B. eds., Building Expert Systems, Addison-Wesley, Reading, Mass. 1983, p. 8-29.
6. Ramsay, J.; Linfors, P; Balazs, M.; Passoja, D. E., Casper; L.S. Private communication April, 1984
7. Weiner, N. Cybernetics, MIT press. Cambridge, Mass., 1965, p. 50.
8. Prigogine, I.; Stengers, I, Order out of Chaos, Bantam Books, NY, NY, 1984, p. 272.
9. Prigogine, I. From Being to Becoming, Freeman and Sons, San Francisco, CA. 1982.
10. Craven, R.A.; Shimura, F.; Hockett, R. S.; Shive, L. W.; Faundrot, T. B.; Keefe-Frandort, G. "Characterization Techniques for VLSI Silicon", Proceedings of 2nd International Symposium on VLSI Science and Technology, Electrochemical Society Inc., Princeton, NJ 1984, p. 20-35.
11. Wheeler, J. A. Americal Journal of Physics, Vol. 51, No. 5, May 1983, p. 398-404; Armour, R. S.; Wheeler, J. A. op cit, p. 405-406.
12. Linfors, P. Physical Electronics - Private Communication, April, 1984.

Appendix

It is suggested that the cost/magnification curve shown in Figure 6 is fundamental. The particular one shown is derived from man as an observer of either the microscopic or the macroscopic world. Man's time frame is dictated by his own internal clockwork that has evolved and has survival value to him - this is translated into cost in our contemporary society. Magnification is dictated by the resolution of his measuring device which, for a scanned microanalytical device is 0.2mm or what his eye can observe on a CRT in the diffraction limit. Taking the particle size distribution to be Gaussian and the two dimensional distribution function to be Raleigh it is possible to write:

Size distribution: $P(r) = A \exp(-N_s r^2)$
Planar spacing: $P(l) = 2N_s l \exp(2N_s r)$

It is evident that by increasing the magnification it becomes possible to observe the particles in the plane of observation and that the probability of observing them approaches 1 as their magnified size becomes greater than 0.2mm. It is possible to observe all particles >0.2mm or the probability of observation is the integral of the pdf up to the limit Mr : $P(r)dr = \text{erf}(N_s r^2)$. On the other hand the spacing grows more rapidly than this since the number of particles on the <u>plane of observation</u> decreases with magnification. The shape of the curve is dictated by three influences: the increasing probability of observing a particle, the standard deviation of the particle pdf and the decreasing density of particles on the plane of observation.

RECEIVED July 8, 1985

2

Electrical Characterization of Semiconductor Materials and Devices

Dieter K. Schroder

Center for Solid State Electronics Research, Arizona State University, Tempe, AZ 85287

> A few selected techniques that are representative of recent advances are described as examples of the much broader field of semiconductor electrical characterization. In particular, resistivity, carrier concentration, junction depth, generation/recombination lifetime, deep level transient spectroscopy and MOSFET mobility measurements are discussed. The importance of non-contacting methods is stressed and recent trends in this direction are outlined. This paper serves as an introduction to some of the following papers in this volume.

Most semiconductor devices operate with electrical input and output signals. Some, such as light emitting diodes, use electrical inputs to give optical outputs, while others, like photodetectors, use optical inputs to give electrical outputs. The electrical characteristics of all of these devices are therefore important, and electrical characterization is the ultimate test. There are, of course, non-electrical characterization techniques in use. They are, however, largely employed as various types of process monitors during device fabrication or as characterization tools during failure analysis.

Characterization techniques fall into three main categories

- process monitors
- failure analysis
- device performance indicators.

Non-electrical methods fall, by and large, into the first two categories, while electrical methods fall into all three.

The basic mechanisms in the material or device that are utilized by the various characterization techniques are generally well known. Much progress has been made in the last few years in the resolution and sensitivity of the equipment used. The driving force has been an increasing complexity of circuits and devices, more quality control, as well as computer automation of test equipment, formerly under manual control. This has generally meant more expensive equipment, but the higher cost of material growth, circuit design and manufacturing operations has justified the investments in expensive characterization equipment.

I will limit myself here to a discussion of some recent trends in electrical characterization by using examples of a few techniques. Non-electrical methods are discussed in substantial detail in some of the following chapters.

Device Characteristics

One of the easiest ways to visualize the material and device characteristics that need to be measured is to consider a semiconductor device. For this I have chosen in Figure 1 a metal-oxide-semiconductor field-effect transistor (MOSFET) as representative of a typical semiconductor device. Indicated on it are the important material and device parameters that need to be measured. Only some of them are addressed in this chapter. Other devices could have been chosen, but the MOSFET incorporates most of the parameters of interest and is the most common integrated circuit device.

The characteristics are:

- Resistivity
  - Substrate
  - Source and drain
  - Gate (poly-silicon, silicide, metal)
- Carrier Concentration
  - Source and Drain
  - Implanted Channel, Channel Stop
- Insulator
  - Charge Density
  - Interface State Density
- Mobility
  - Channel
- Resistance
  - Source and Drain
  - Contact
- Impurities
  - Deep Level Impurities
  - Oxygen and Carbon
  - Generation/Recombination Lifetime
  - Structural Imperfections (dislocations, stacking faults)
- Physical Dimensions
  - Junction Depth
  - Epitaxial Film Thickness
  - Channel Length and Width
  - Oxide Thickness

For other devices there will be minor variations in these parameters. A detailed discussion of all of these factors is beyond the scope of this article. I will use some examples to illustrate recent trends in electrical characterization.

Some of these trends are:

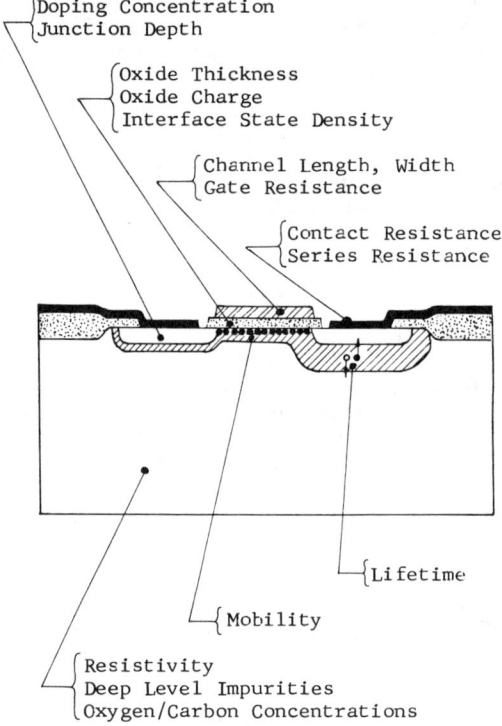

Figure 1. A MOSFET used here as a representative semiconductor device. Indicated are the various parameters that need to be characterized.

- Measurement of increasingly smaller features
- Instrumentation with increasing sensitivity
- Non-contacting measurement techniques
- Computer data acquisition and display.

Non-Contacting Measurements

Most electrical characterization techniques require physical contacts between the wafer and the measuring instrument. They can be non-permanent contacts (e.g. four-point probe) or permanent contacts (e.g. evaporated metal). For some applications such permanent contacts are not permissible. They may, for example, create damage or leave residues that are deleterious during subsequent processing. Non-contacting methods allow complete inspection of all wafers because no physical contact is made.

Various principles are utilized. Resistivity is measured by inserting the wafer between two coils connected to an oscillator. The time varying magnetic fields, set up by the oscillator, induce eddy currents in the wafer. These eddy currents dissipate energy supplied by the oscillator. A measure of the power required to stabilize the oscillator is an accurate measure of the resistivity.

By combining this technique with capacitive coupling or ultrasound reflection, wafer thickness and wafer flatness information is also obtained. A further step is to wafer-map the data. Using optical scanning, surface defect maps are generated (1) and insulator thickness variations are measured ellipsometrically and displayed. As discussed further on, recombination lifetime maps can also be generated by non-contacting methods.

The trend in non-contacting characterization is its extension to material or device parameters, measured by contacting techniques in the past. Furthermore, it is being combined more and more with two- or three-dimensional data display.

Resistivity

The four-point probe is commonly used for resistivity measurements (2). It is well understood and, until recently, was largely manually operated with a limited number of data points per wafer. Computer advances have added a dimension to this technique that gives very useful information.

By placing the mechanical probe stepping as well as the data acquisition under computer control, it is easy to gather many data points per wafer and display the data in a way for certain trends to become obvious. For example, in Figure 2 such data points are plotted as contours of equal resistance (3). These plots clearly show non-uniformities like gas flow patterns during epitaxial growth or diffusion. They have also been successfully used to characterize implant patterns of ion implanters (1). A systematic variation of the implanted dose across a wafer, which may be the result of the implant beam or substrate holder motion of the implanter, gives rise to threshold voltage variations of MOSFETs across that wafer. Doping uniformity mapping techniques are now being accepted by implanter vendors for the characterization and qualification of their systems.

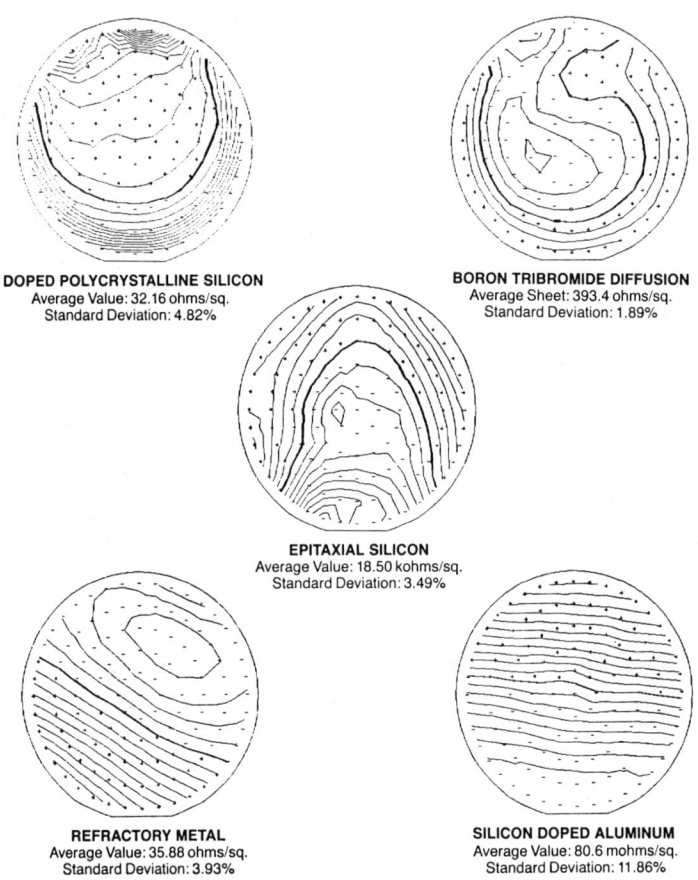

Figure 2. Examples of resistance plots using wafer mapping. Courtesy of D.S. Perloff.

Doping and Carrier Concentrations

The doping concentration is often confused with the carrier concentration. In uniformly and moderately doped substrates the two are virtually identical at room temperature. This is no longer true when the substrate is heavily doped or when a diffused or ion-implanted layer is measured. Even if all the dopant atoms are electrically active, the carrier concentration in heavily-doped material is lower than the doping concentration, as described by Fermi-Dirac statistics (4). This is further aggravated if, for example, the ion-implanted layer is not wholly activated or if there is a steep doping gradient. It is important that one is aware of these difficulties.

Various techniques are used for these measurements. The most popular are capacitance-voltage (C-V) profiling, spreading resistance and secondary ion mass spectroscopy (SIMS). SIMS is not a strictly electrical characterization technique, but is included here because it is routinely used to measure the dopant atom distribution. The basis for these three techniques are very different and I will briefly describe them.

C-V Profiling

C-V profiling requires a reverse-biased space-charge region (scr). It can be implemented with any one of these devices:

- Schottky barrier contact
  - permanent metal contact
  - temporary mercury contact
  - temporary liquid contact
- pn junction
- MOS capacitor or transistor

In the conventional approach, the carrier concentration is determined by measuring the capacitance as a function of reverse-biased voltage (5). The carrier concentration p and the depth W are given by

$$p = -2/(qK_s\epsilon_o A^2 d(1/C^2)/dV) \qquad (1)$$

$$W = K_s\epsilon_o A/C \qquad (2)$$

where $q = 1.6 \times 10^{-19}$ coul, $K_s\epsilon_o$ is the semiconductor dielectric constant, A is the device area, C is the measured capacitance and V is the applied voltage.

By varying the applied voltage, the scr width is increased and a plot of p vs. W is obtained. There are two limits (i) how close to the surface, and (ii) how deep into the wafer can the device be profiled. Limit (i) is several Debye lengths from the surface and limit (ii) is set by the voltage breakdown of the device (6).

Frequently it is desirable to profile deeper than voltage breakdown allows. This is especially important for heavily doped layers where the breakdown limit allows only an extremely shallow profile measurement. A clever solution is the electrochemical profiler shown in Figure 3 (7). The Schottky barrier is created by a liquid-semiconductor contact, eliminating device processing and making the measurement very easy. A fixed voltage coupled with a capacitance measurement gives one datum point on a p-W plot. The application of a current to the electrolyte causes the semiconductor to be etched, with the etch depth controlled by the duration of the current flow. Both p and W are remeasured for the next point and so on. An entire p-W plot is generated in this manner with no depth limitation.

An example of such a plot is shown in Figure 3, where the increment between experimental data points is 35Å, making this an extremely high resolution instrument. It is preferable to a conventional C-V profiler because it has no depth limitation and uses wafers without the need to fabricate devices. It is destructive, however, by leaving an approximately 1 mm diameter hole in the wafer.

Spreading Resistance Profiling

In the spreading resistance profiler (SRP), two metal probes contact the semiconductor (8). A small voltage of typically a few mV, is applied between the contacts and the resistance is measured. It is the sum of several components, with the dominant one being the spreading resistance under the probes. The success of this method rests on two chief features: (i) a mechanical arrangement that allows the probes to contact the semiconductor surface in a well controlled manner, and (ii) a calibration of the measured spreading resistance to calibrated standard wafers. If these standards are calibrated in doping rather than carrier concentrations, the SRP actually gives doping concentration as a function of distance.

For shallow implanted or diffused layers, the semiconductor is angle-bevelled and the probe moved along the bevelled surface. A shallow angle coupled with short probe steps, allows equivalent shallow depth steps. Examples are given in (8) with depth resolution of 200Å. The accuracy is limited by the correction factors in the algorithm used in the data reduction.

Secondary Ion Mass Spectroscopy

In the SIMS technique, an oxygen or cesium ion beam incident on the sample, sputters atoms from the surface. Either negatively or positively charged ions are mass analyzed and their density displayed as a function of sputter time. By using calibration standards, the density is calibrated as concentration/cm^3, and by measuring the sputter crater depth/ the time axis is converted to a distance axis, giving a dopant concentration vs. depth plot.

This technique gives the total dopant concentration, not just the electrically active portion. We mention it here, even though it is not an electrical characterization technique in the sense that the others are, because it is routinely used to characterize the dopant concentration and depth of ion-implanted and diffused layers. When the dopant atoms are electrically active, then SIMS is found to give results very close to those obtained from spreading resistance measurements (9). When electrical activation is not complete, then there will be significant deviations between SIMS and SRP or C-V data.

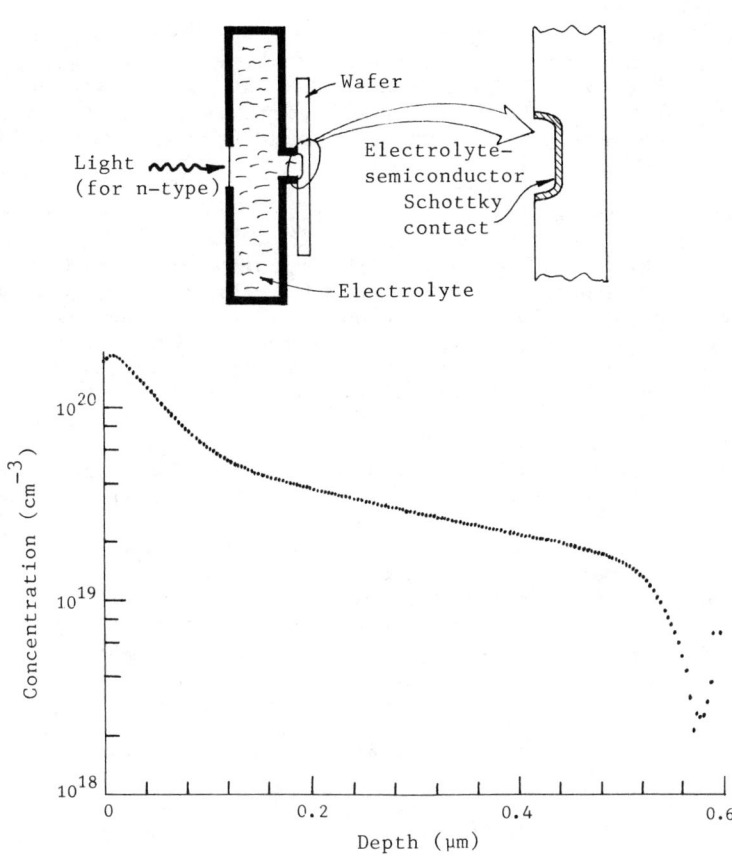

Figure 3. A schematic of the electrochemical profiler and a carrier concentration vs. depth plot. The example is a Zn diffusion into GaAs. Courtesy of R.J. Roedel.

Junction Depth

All of the three preceding techniques are used to measure junction depth. There is general agreement between the three techniques for deep junctions with junction depths of the order of 1-2 μm or greater. However, for shallow junctions it is believed that SIMS gives the most accurate results. This question is under active investigation at present because junctions with depths of ≤ 0.5 μm are not easy to measure precisely.

The trend is for finer resolution in these measurements so that increasingly shallower junctions can be unambiguously determined. A better understanding of just exactly what is measured is also being addressed. For example, in SRP the algorithms to extract doping concentration data from the measured spreading resistance are being improved.

Deep Level Impurities

Impurities that perturb the single crystal periodicity of a semiconductor, generally introduce energy levels into the energy gap. These impurities are divided into shallow level impurities whose energies lie close to either the conduction or valence band edge and deep level impurities with energy levels deeper in the bandgap. There is no precise definition, but those impurities whose energy levels lie within about 0.1 eV of the band edges are generally considered to be shallow.

Optical techniques like photoluminescence (10) and infrared photothermal spectroscopy (11) work well for the characterization of shallow level impurities, while electrical techniques work well for deep level impurities. There are a number of methods that have been used for electrical characterization. I will only discuss deep level transient spectroscopy (DLTS), however, because it has become the most popular and gives a fairly complete characterization.

Before discussing DLTS, it is appropriate to talk about some device parameters that are affected by these impurities. The main parameters are the recombination and generation lifetimes because they affect junction leakage current, device switching speed, light emitting diode efficiency and a number of other device performance indicators.

Recombination-Generation Lifetime

A semiconductor in the non-equilibrium state either has excess carriers, due to an electrical forward bias or optical excitation, or it has a depletion of carriers, due to an electrical reverse bias. In the former the excess carriers recombine in a characteristic time - the recombination lifetime, τ_r. In the latter carriers are generated in a characteristic time - the generation lifetime, τ_g.

The recombination lifetime for electrons is given by

$$\tau_r = (\sigma_n v_{th} N_T)^{-1} \qquad (3)$$

where σ_n is the capture cross-section, v_{th} the thermal velocity and N_T the deep level impurity concentration. τ_r depends on the capture cross-section and impurity concentration.

The generation lifetime is (12)

$$\tau_g \approx \tau_r \exp[(E_T-E_i)/kT] \tag{4}$$

where E_T and E_i are the impurity and intrinsic energy levels, respectively. Clearly, the generation lifetime is very sensitively dependent on the energy level, while the recombination lifetime is not. A comparison of Equations 3 and 4 shows that τ_g can be much higher than τ_r, which is generally observed experimentally (12).

There are numerous techniques to measure the recombination lifetime. Some of the better known are photoconductive decay (13), diode reverse recovery (14), diode open circuit voltage decay (15), surface photovoltage (16) and forward-biased pn junction I-V characteristic (17). I will describe one particular photoconductive decay method, because it is a relatively new, non-contact method that requires no junctions. This makes it very suitable for a large number of measurements as for a process sequence characterization tool.

It is the microwave reflection method of Figure 4 (18). Light pulses incident on the wafer generate electron-hole (e-h) pairs, modulating its conductivity. These non-equilibrium e-h pairs recombine causing a conductivity change. A microwave signal is directed onto the wafer, reflected from it, detected, amplified and displayed. The microwave reflectivity is dependent on the wafer conductivity. Hence the displayed signal is a representation of the time dependent excess carrier density. The temporal behavior of this signal is determined by the recombination lifetime. The advantage of this technique over conventional photoconductive decay is the absence of contacts to the wafer.

Wafer mapping is easily accomplished. A representative two-dimensional τ_r map is shown in Figure 4 (19). The data are coded through the number of horizontal lines for each location. The more lines, the higher the τ_r. The numerical values are also available and two cross-sections through the two-dimensional display are also shown.

The lower limit for short lifetimes in this technique is determined by the optical excitation source turn-off time to about 0.1 μs. For shorter lifetimes steady-state diffusion length measurements are more suitable. The diffusion length is related to the recombination lifetime by the equation $L = \sqrt{(D\tau_r)}$. Suitable techniques are surface photovoltage and scanning electron microscope electron beam induced current. They lend themselves to lifetimes down to the nano-second range.

In contrast to τ_r measurements, in which the decay of excess carriers is monitored, the generation lifetime is determined from the reverse-biased pn junction leakage current or from the pulsed MOS capacitor (20). In the latter and the more popular of the two, an MOS-C is pulsed into deep depletion and the capacitance is monitored as a function of time. An appropriate analysis of the C-t response yields τ_g.

An important point in τ_g and τ_r measurements is that τ_g is measured in the scr of a reverse-biased device. The thickness of the scr is typically a few microns and is located near the surface. It is furthermore determined by the applied gate voltage and therefore

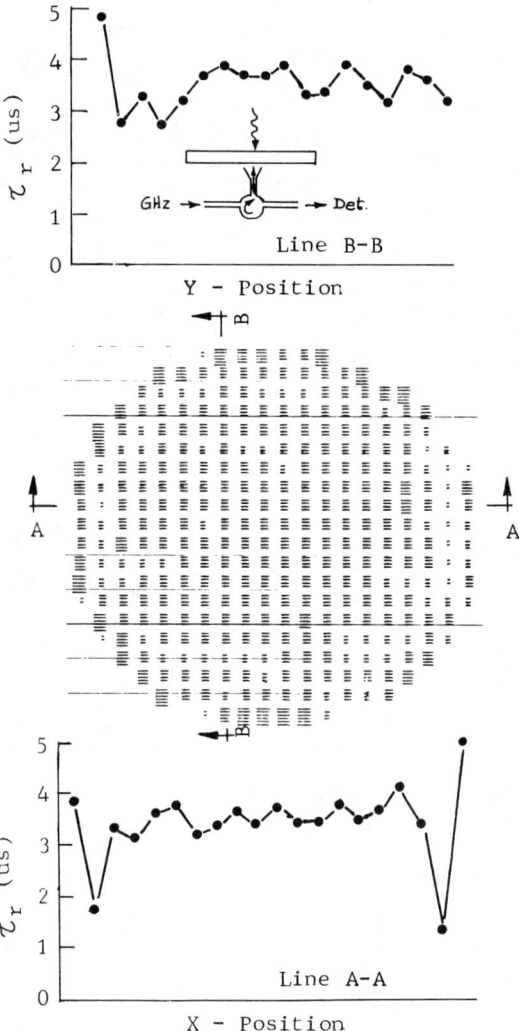

Figure 4. A recombination lifetime wafer map taken on a 3 inch diameter silicon wafer. Two sections through the data are shown. Also shown is a schematic of the microwave reflection measurement technique. Courtesy of K. Hunter.

under the experimenter's control. The distance over which τ_r is measured is the minority carrier diffusion length, which is typically around 100 μm for n-type Si of 10 μs recombination lifetime. It is a property of a particular material and therefore not under the experimenter's control. So the generation lifetime is measured within a few microns from the surface, while the recombination lifetime is measured over a much larger distance. This has important implications.

Equation 4 states that for a wafer with uniformly distributed recombination/generation centers (N_T), τ_g is generally higher than τ_r. If N_T is non-uniformly distributed, then this discrepancy is further enhanced. An example of this is found in modern Si integrated circuit processing. The high oxygen concentration of Czochralski-grown silicon has led to a process sequence in which the oxygen is diffused out of a region near the surface, sometimes called the denuded zone. This zone is typically around 20 μm wide. In the wafer interior the oxygen is made to precipitate, causing a high density of structural defects such as dislocations and stacking faults. These defects act as gettering sites removing impurities from the denuded zone. Consequently, the denuded zone is of very high perfection for high yield devices with a correspondingly high generation lifetime. The recombination lifetime, determined partially by the denuded zone but largely by the precipitated interior, is likely to be very low. Measurements, using the pulsed MOS-C for both τ_g and τ_r, have given (21) $\tau_g \approx 2.7$ ms and $\tau_r \approx 1.8$ μs. The ratio is over 1000 and is clearly the result of differing regions of the wafer having significantly different properties.

The trend here is for simpler lifetime measurement methods and a better understanding of the mechanisms within the device or material, giving rise to the measured values. For example, for the denuded zone/precipitated interior intrinsic gettering mechanism, a high τ_g and low τ_r is clearly an indicator of a successful process.

Deep Level Transient Spectroscopy

The lifetimes of the previous section depend on the concentration, capture cross-section and energy level of the impurities. Lifetime measurements, however, cannot easily be used for these determinations. DLTS is the technique most frequently used instead. It is based on the concept in Figure 5 (22). First consider the n-type Schottky barrier diode of Figure 5 (a) with no deep level impurities. A reverse bias $-V_1$ creates a scr of width W. When the bias is reduced to zero, the scr is also reduced. The scr capacitance, being inversely proportional to W, is small and equal for cases A, C and D and large for case B. If the voltage is pulsed between zero and $-V_1$, the capacitance follows almost instantaneously and no time-dependent capacitance is observed.

A different behavior is observed when deep level impurities are present, as indicated in Figure 5(b). This figure is a composite showing the device and its scr, but superimposed on it is the energy level corresponding to the impurity, N_T. It is merely shown for convenience. Initially, the device has been reverse biased for some time and N_T is empty of electrons (A). A short pulse brings it to

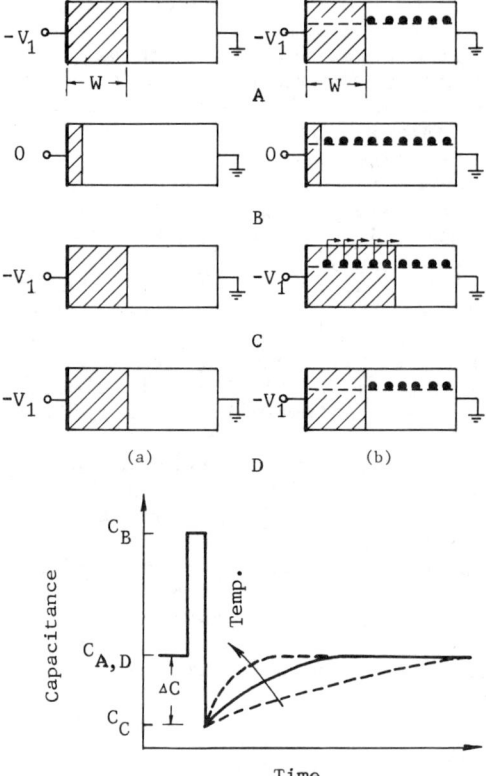

Figure 5. A schematic representation of DLTS and the corresponding C-t response.

zero bias (B) and then back to $-V_1$. During the zero bias pulse, N_T is filled with electrons and when the bias is returned to $-V_1$, those N_T in the scr are initially filled with electrons (C). Since the capacitance is given by

$$C = \sqrt{K_s \epsilon_0 Q/2(V_{bi}-V)} \qquad (5)$$

the space-charge $Q = q(N_D^+ - N_T^-)$ at (C). Electrons are then thermally emitted, indicated by arrows in (C), and as time progresses $Q \to qN_D^+$ (D) i.e. when all electrons have been emitted, the charge in the scr is identical to Figure 5(a). Hence the capacitance, which is initially C_C rises to C_D, with a time constant

$$\tau_e = (\sigma_n v_{th} N_c)^{-1} \exp(E_T/kT) \qquad (6)$$

where N_c is the effective density of states in the conduction band. Note that τ_e contains both σ_n and E_T. It can be shown (22) that the capacitance change ΔC in Figure 5(c) is related to N_T by

$$N_T = 2N_D \Delta C/C_A \qquad (7)$$

We see then that the main parameters that characterize deep impurity levels can be extracted from DLTS measurements. The technique is implemented by repeatedly pulsing the sequence in Figure 5 and detecting the C-t transient with a lock-in amplifier or boxcar integrator (22). The measurement is usually performed as a function of temperature.

Although only the impurities in the scr can be detected by DLTS, it is always possible to increase the reverse-bias voltage to increase the scr width. It is difficult, however, to pulse to high voltages. Therefore, the width is ultimately limited by voltage breakdown considerations. Then, one can etch the wafer repeatedly, form Schottky barrier devices each time and measure to profile the N_T concentration.

The DLTS method has been well accepted and is used by many laboratories. The trend is to place both the data acquisition and data interpretation under computer control. In addition to making the measurement easier, it allows certain signal processing of the experimental data that simplifies parameter extraction.

<u>Mobility</u>

For the final example, I have chosen the MOSFET mobility measurement because depending on how it is measured, different values are obtained if the data interpretation is done incorrectly.

The drain current-drain voltage relationship of a MOSFET is (23)

$$I_D = (Z/L)\bar{\mu}C_{ox}[V_G - 2\phi_F - V_D/2)V_D - K(V_D+2\phi_F)^{1.5} - (2\phi_F)^{1.5}] \qquad (8)$$

where K is a constant that depends on doping and oxide capacitance and the other symbols have their usual meaning. From the drain conductance $g_D = \partial I_D/\partial V_D$ at constant V_G we derive for $V_D \ll 2\phi_F$ an effective mobility as

$$\mu_{eff} = Lg_d/(ZC_{ox}(V_G-V_T)) \qquad (9)$$

Here ϕ_F is the Fermi potential and V_T is the threshold voltage.
A field-effect mobility is sometimes defined, using the transconductance $g_m = \partial I_D/\partial V_G$ at constant V_D, as

$$\mu_{FE} = Lg_m/(ZC_{ox}V_D) \qquad (10)$$

When these two measured mobilities are compared, generally μ_{FE} is of smaller value. The reason for this is the mobility variation with gate voltage. It decreases with increasing gate voltage, and that variation is not considered in Equation 10. If it is considered, then the equation should be written as

$$\mu_{FE} = Lg_m/(ZC_{ox}V_D) - (V_G-V_T)\partial\mu_{eff}/\partial V_G \qquad (11)$$

The difficulty with this expression is that generally $\partial\mu_{eff}/\partial V_G$ is not known a priori and for that reason it is Equation 9 that should be used. Representative data to illustrate this point can be found in (24). I merely wish to point out this fact which is usually overlooked.

A further complication for mobility measurements is the device series resistance R_s. Both drain and transconductance are degraded by R_s according to

$$g_D' = g_D/(1+g_D R_s) \quad ; \quad g_m' = g_m/(1+g_m R_s) \qquad (12)$$

R_s values of several hundred ohms are sometimes found in MOSFETs. They can degrade the mobility from its intrinsic value by 50% or more (25).

There is no particular trend here, just a note of caution. Be aware what is measured and be certain that extraneous factors like resistance do not degrade the mobility.

<u>Summary</u>

Semiconductor material and device characterization is a very comprehensive subject and worthy of a book by itself. In this chapter, I have limited myself to a few selected topics. I have chosen them because they are relatively new and perhaps not well-known. In some cases, the technique is old, but by combining it with modern computer-automated data-acquisition and display, it is significantly enhanced.

The trend in characterization is toward higher sensitivity, non-contacting wherever possible and computer-automation. For example, ion-implantation is capable of sheet resistance uniformity to better than 1% across a wafer. It is only with computer-driven four-point probe measurements, that this type of uniformity can be meaningfully displayed.

Lifetime characterization is infrequently utilized for MOSFETs. However, with denuded surface zone/precipitated bulk wafer fabrication, both generation and recombination lifetimes are extremely useful process descriptors. Their meaning should be better understood than it is today. They are also very useful as a check on process cleanliness. This chapter addresses these concepts.

Acknowledgments

This work was partially supported by NSF grant ECS-82-12336. I wish to thank Chrina Darrington for typing the paper.

Literature Cited

1. Burggraaf, P. Semicond. Internat. 1984, 7, 52-57.
2. Runyan, W.R. "Semiconductor Measurements and Instrumentation"; McGraw-Hill: New York, 1975; pp. 69-75.
3. Perloff, D.S. Personal Communication.
4. Blakemore, J.S. "Semiconductor Statistics"; Pergamon Press: New York, 1962; p. 122.
5. Johnson, W.C.; Panousis, P.T. IEEE Trans. 1971, ED-18, 965-973.
6. Nicollian, E.H.; Brews, J.R. "MOS Physics and Technology"; Wiley: New York, 1982, p. 382.
7. Ambridge, T.; Faktor, M.M. J. Appl. Electrochem. 1975, 5, 319-328.
8. Mazur, R.G. This Volume.
9. Ehrstein, J.R.; Downing, R.G.; Stallard, B.R.; Simons, D.S.; Fleming, R.F. "Semiconductor Processing"; ASTM: Philadelphia, 1984; pp. 409-425.
10. Kolbesen, B.O. Appl. Phys. Lett. 1975, 27, 353-355.
11. Kogan, S.M.; Lifshits, T.M. Phys. Stat. Sol., 1977, 39a, 11-39.
12. Schroder, D.K. IEEE Trans. 1982, ED-29, 1336-1338.
13. Runyan, W.R. "Semiconductor Measurements and Instrumentation"; McGraw-Hill: New York, 1975; pp. 107-113.
14. Kuno, H.J. IEEE Trans. 1964, ED-11, 8-14.
15. Wilson, P.G. Solid-State Electron. 1967, 10, 145-154.
16. Goodman, A.M. J. Appl. Phys. 1961, 32, 2550-2552.
17. Neugroschel, A., Lindholm, F.A., Sah, C.T. IEEE Trans. 1977, ED-24, 662-671.
18. Mada, Y. Jap. J. Appl. Phys. 1979, 18, 2171-2172.
19. Hunter, K. Personal Communication.
20. Schroder, D.K.; Guldberg, J. Solid-State Electron. 1971, 14, 1285-1297.
21. Schroder, D.K.; Whitfield, J.D.; Varker, C.J. IEEE Trans. 1984, ED-31, 462-467.
22. Miller, G.L.; Lang, D.V.; Kimerling, L.C. Ann. Rev. Mat. Sc. 1977, 7, 377-448.
23. Sze, S.M. "Physics of Semiconductor Devices"; Wiley: New York, 1981; p. 440.
24. Sun, S.C.; Plummer, J.D. IEEE Trans. 1980, ED-27, 1497-1508.
25. Duvvury, C.; Baglee, D.; Duane, M.; Hyslop, A.; Smayling, M.; Maekawa, M. Solid-State Electron. 1984, 27, 89-96.

RECEIVED July 29, 1985

3

Dopant Profiles by the Spreading Resistance Technique

Robert G. Mazur

Solid State Measurements, Inc., Monroeville, PA 15146

Spreading resistance profiles are made by stepping a pair of specially-conditioned "point"-contact probes across the bevelled surface of a sample. The near-zero-bias resistance between the probes is measured at each point. Measurement accuracy depends very strongly on properly calibrating the probes with known-resistivity samples. In addition, a theoretically-derived correction factor must be calculated and applied to each raw data point to account for the effects of PN junctions and other boundaries in the sample. At its current level of development, the spreading resistance technique can provide detailed dopant density profiles on essentially all silicon structures of interest, including as-grown crystals and diffused, ion-implanted, and epitaxial wafers. For silicon, there are no limits on dopant density and essentially no limits on the depth resolution. A spatial resolution as low as 1 nanometer per point has been reported and layers with a thickness as little as 20 nanometers have been profiled. This paper details the current state-of-the-art of the spreading resistance technique and presents a number of typical examples.

Dieter Schroder has just presented a comprehensive review of the electrical techniques used in characterizing semiconductor materials (1). One point that I'd like to develop further is the need for high spatial resolution dopant profiling. As you know, in the world of microelectronics, the emphasis is always on the "micro"; everything is fantastically small. This is because both the operating speed and the manufacturing yield of integrated circuits improve as individual device elements are made smaller and more chips are put on a wafer. This miniaturization process generates a need for high spatial resolution dopant profiling, because making devices smaller means making the doped layers of which they're composed thinner. At the present time, many devices use

0097-6156/86/0295-0034$06.00/0
© 1986 American Chemical Society

ion-implanted, diffused, or epitaxial layers only a few tens or hundreds of nanometers thick. To get the dopant concentration profiles needed in process development and control of these thin layers, we must section and evaluate them with a spatial resolution on the order of one nanometer.

One method that has the required spatial resolution is the spreading resistance technique (2,3). This technique is based on measuring the contact resistance of specially-prepared point-contact diodes. In this paper, I will first describe the spreading resistance technique as it exists today and present some typical examples. Then I'll try to give you a perspective on spreading resistance by discussing it in relation to SIMS. I'll finish by indicating those situations in which the spreading resistance technique is most effectively used for characterizing semiconductor materials.

The Spreading Resistance Technique

Figure 1 illustrates the experimental procedure used in making spreading resistance measurements. Two probes are carefully aligned and then stepped across the bevelled surface of a semiconductor sample; at each point, the probes are lowered onto the sample surface and the resistance between the two probes is measured and plotted. The technique is referred to as the spreading resistance technique because the dominant resistance of a point contact diode occurs in a very small volume beneath the probe, where the current rapidly spreads out into the sample. Spreading resistance profiles are usually computer-processed to yield resistivity or dopant concentration profiles.

Figure 2 shows a typical automatic spreading resistance system. It consists of a mechanical apparatus to operate the probes and step them across a bevelled test chip and an electronics sub-system to acquire, process and plot the data.

The spreading resistance technique is characterized by four major features:
1) special probes and mechanical apparatus to make the contacts;
2) the use of a very low applied voltage during the resistance measurements;
3) a calibration procedure using samples of known resistivity;
4) a multi-layer correction procedure to correct for boundary effects.

Probes. The most important parts of a spreading resistance system are the probes and the mechanical apparatus that operates them. Figure 3 is a close-up view of a pair of spreading resistance probes mounted on gravity-loaded probe arm assemblies. Note that the probe arms are mechanically massive, thus providing a stable and rigid mounting for the probe pins. Each probe arm is supported by a kinematic-design bearing system with a total of five point contacts. This construction gives the probe arm only one degree-of-freedom, a rotation around a horizontal axis. When the probe arm is pivoted to

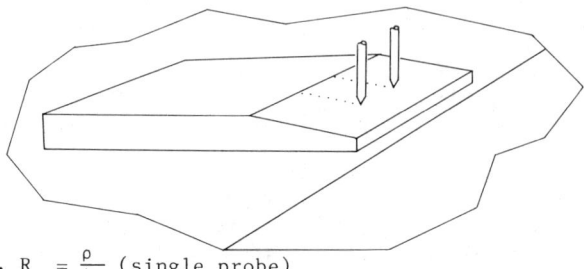

1. $R_m = \frac{\rho}{4a}$ (single probe)

2. $R_m = f(\rho,$ radius conductivity type, orientation, surface finish, etc.)

Figure 1. A schematic illustration of dopant profiling with the spreading resistance technique. Reproduced with permission from Ref. 5. Copyright 1984 American Society for Testing and Materials.

Figure 2. An ASR-100C/2 Automatic Spreading Resistance Probe.

bring the probe into contact with the test sample, the probe pin itself makes the sixth point contact between the probe arm assembly and the rest of the apparatus, eliminating all six degrees-of-freedom. This is important, because it means that the probe doesn't move laterally (i.e.,"scrub") when it makes contact with the sample. Because the probe doesn't "scrub," probe tip wear is essentially eliminated and damage to the semiconductor surface is minimized.

Figure 4 is a scale drawing of a spreading resistance probe tip and the nominal contact size for a 10 gram probe load. The probe tips are a hard tungsten-osmium alloy; they have a precisely-controlled shape and are massive in relation to the contact spot diameter. Because of their relative size and hardness, the probes do not undergo gross deformation during measurements. When set down, they deform only elastically; thus they make very reproducible mechanical contacts.

Electrically, we have to "condition" the probes in a special way to get a good contact. This is necessary because all silicon samples are covered with a tough native oxide some 2 to 3 nanometers thick. Since the probes don't "scrub" on contact, they cannot break through the oxide layer the way that other probes would. Instead, we achieve an electrical contact with the right characteristics by controlling the micro-topography of the probe tips such that the contact area consists of a large number of microscopic protrusions, or "asperities." These asperities are sufficiently small that they fracture the oxide by pressure alone, generating a micro-contact under each asperity. Because of the controlled shape of the probe tip, the resultant cluster of micro-contacts is closely enough grouped that it acts electrically as a single contact. In recent years we've learned to condition probes so that we get a large number of micro-contacts, thus reducing the pressure at each of them. This minimizes probe penetration and produces good profiles, even on the extremely thin layers now common in silicon technology. As an example of current capability, one spreading resistance probe user recently reported profiling an ion-implanted layer just 20 nanometers thick.

The improvement in spreading resistance profiles obtained through controlling probe penetration is illustrated in Figure 5. The structure is an NPNN$^+$ transistor. The lowest plot was done several years ago with 20 gram penetrating probes; the middle plot shows the improvement obtained at that time by using a Gorey-Schneider probe grinder to control probe penetration (4). The uppermost profile was measured recently, using controlled low penetration 10 gram probes, also conditioned with a Gorey-Schneider probe grinder, but done according to our current procedures and criteria (5).

A further example of the depth resolution possible with properly conditioned spreading resistance probes is shown in Figure 6. This is a narrow-base NPN transistor, profiled with probes loaded to about 5 grams and with a very shallow bevel angle to obtain a depth increment of just 2.6 nanometers per point (Figure 6a). Note that it is possible to achieve even finer resolution by simply going to a smaller horizontal step on the bevel surface,

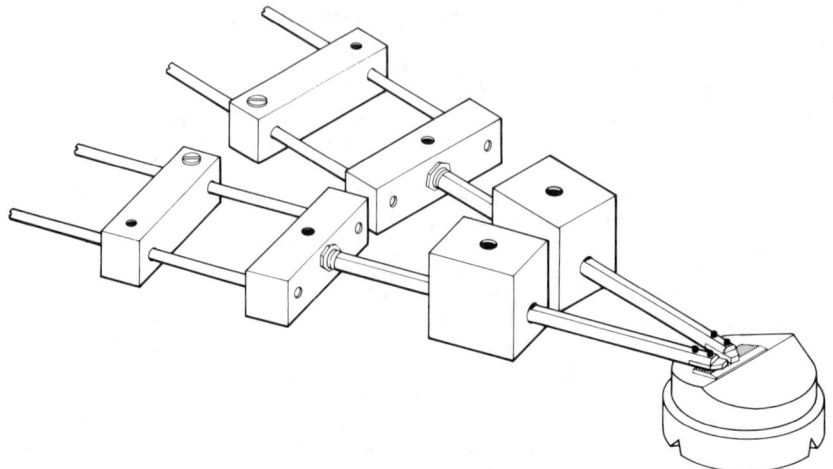

Figure 3. A pair of standard spreading resistance probe arm assemblies with probe pins attached, shown in measurement position on a bevelled test sample.

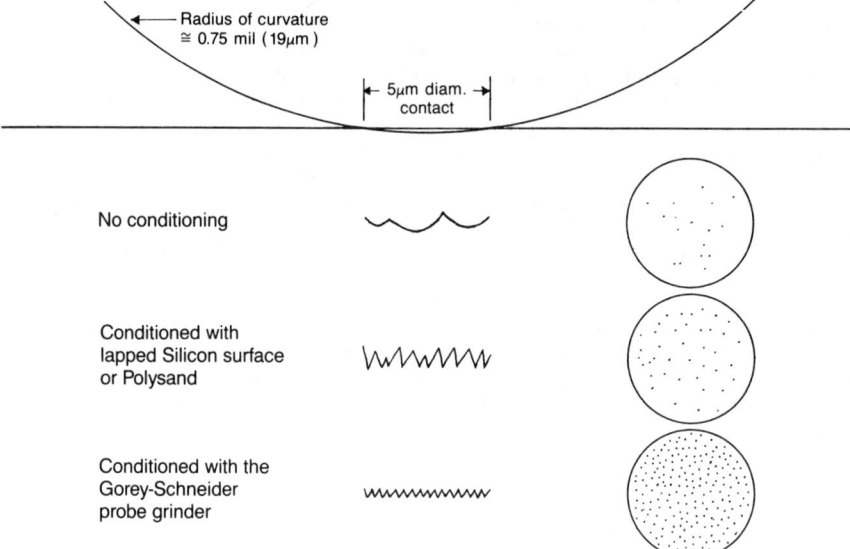

Figure 4. Scale drawing of a spreading resistance probe tip and typical microcontact clusters produced at various degrees of probe conditioning. Reproduced with permission from Ref. 5. Copyright 1984 American Society for Testing and Materials.

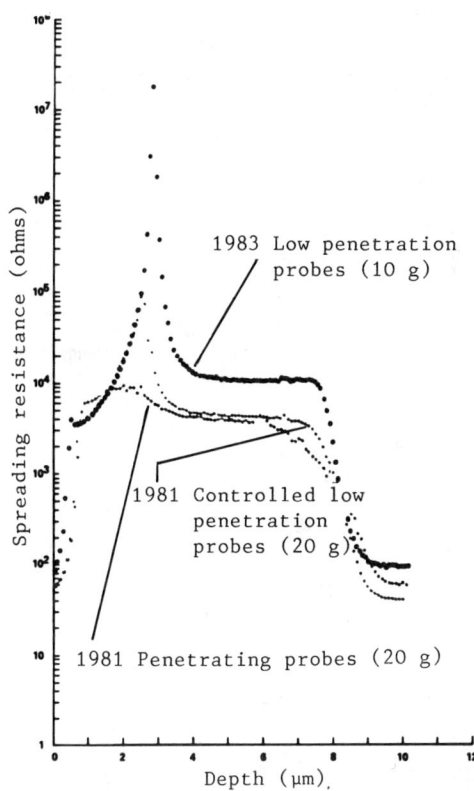

Figure 5. Spreading resistance profiles of an NPNN⁺ transistor structure, as measured with penetrating probes and with controlled low penetration probes. Reproduced with permission from Ref. 5. Copyright 1984 American Society for Testing and Materials.

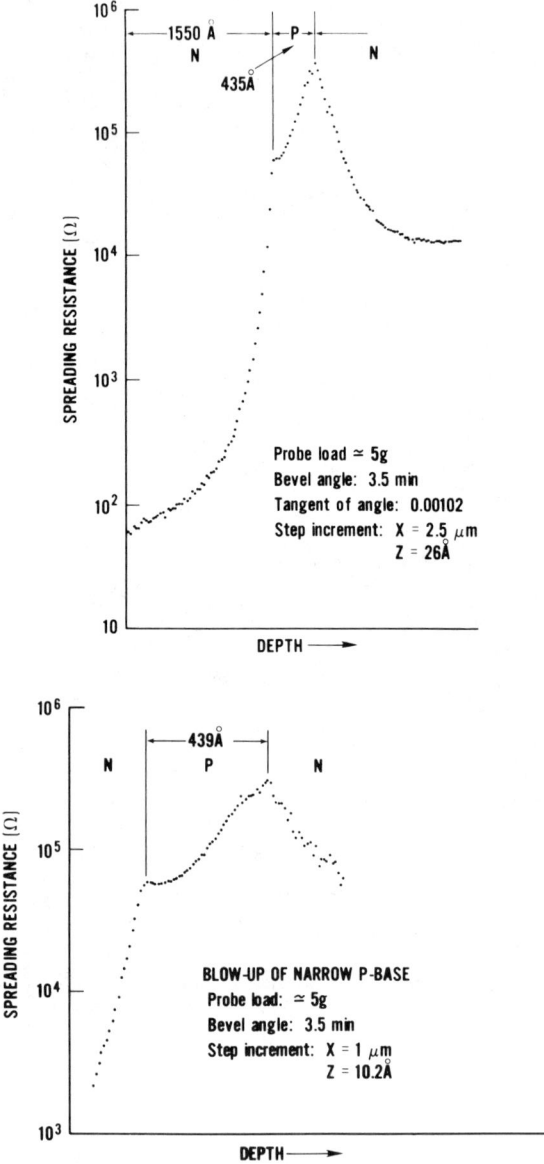

Figure 6. Spreading resistance profiles of a narrow base NPN transistor.

yielding a point-to-point depth increment of just 1.02 nanometers (Figure 6b).

This discussion of the mechanical and electrical properties of spreading resistance contacts makes it clear that the probes used differ from any others previously used in either point-contact diode work or in semiconductor electrical measurements.

System Electronics. In addition to using specialized contacts and mechanical apparatus, spreading resistance profiling also requires specialized electronics. In particular, we use an ohmmeter with a very low applied voltage of 5 mV (2.5 mV across each probe). This low voltage minimizes several effects normally associated with point contact diodes; e.g., excess carrier injection. The ohmmeter's single logarithmic range spans resistances from 1 ohm to 10^8 ohms, allowing measurements on material with a resistivity of from less than 10^{-3} ohm-cm on up to the intrinsic level.

Calibration. The third key feature of the spreading resistance technique is its use of calibration curves. Because measurements are made on real silicon surfaces, the relationship expected on the basis of simple theory doesn't hold (Figure 1, equation 1). Instead, the measured resistance, R_m, depends not only on the sample resistivity ρ and the contact radius a, but also on the sample's conductivity type, crystallographic orientation, and surface finish (see Figure 1, equation 2).

Therefore, we make resistivity measurements by first generating calibration curves on known-resistivity samples of the same type, orientation, and surface finish as the test specimens to be profiled. Calibration curves are generated for a particular pair of probes at a particular time, using known-resistivity samples of the highest quality available. This calibration procedure is a particularly noteworthy characteristic of the spreading resistance technique. It means that spreading resistance is a comparison method, and that its ultimate accuracy is therefore limited only by the calibration material available. Fortunately, it's now possible to obtain complete sets of calibration samples from the National Bureau of Standards.

Multilayer Corrections. Finally, the spreading resistance technique, as we use it today, is also characterized by the use of a multilayer correction procedure. Corrections to the raw data are necessary because of the effects of boundaries, such as PN or low-high junctions, in the vicinity of the probes. These corrections are made using a method based on a point-by-point solution of the Laplace equation, treating each point on the spreading resistance profile as a separate sub-layer in the structure; hence the term "multilayer." A considerable amount of work has been done on these corrections over the last fifteen years, beginning with the original multilayer calculations of Schumann and Gardner at IBM (6), and continued by D'Avanzo, Rung, and Dutton at Stanford University (7) and by S. C. Choo and his co-workers at the

University of Singapore (8). As a result of this work, we now have programs that enable us to do a rather effective job of making spreading resistance corrections.

More recently, additional work to improve the accuracy of calculated correction factors has been done by Berkowitz and Lux at Fort Monmouth (9), by Piessens, Vandervorst, and Maes at the University of Leuven in Belgium (10), and, again, by Choo and his group in Singapore (11). These newer procedures and programs are now being implemented; they offer the promise of giving us even better corrections in the future.

Figure 7 is an example of a spreading resistance profile corrected with a multilayer procedure. It shows a shallow boron ion implant into a P-type substrate, along with the carrier concentration profile derived from the spreading resistance data, and a theoretical curve (solid line), calculated with the program SUPREM II. Notwithstanding the fact that we don't know whether the experimental profile or the theoretical calculation is the more correct, the relatively good agreement between them gives us some idea of the accuracy of the spreading resistance-derived concentration profile.

Figure 8 is a more recent spreading resistance and carrier concentration profile of a low dose boron ion-implant into an N-type substrate. The peak in the spreading resistance profile at about 9×10^6 ohms indicates the position of the PN junction. The calculated carrier concentration rises from a value of 10^{17} cm^{-3} at the surface to a maximum of about 2×10^{17} cm^{-3} before falling to a low value at the PN junction at a depth of 1.2 μm. This profile was done with 10 gram, controlled low penetration probes, using a nominal 17' bevel angle and a step increment along the bevel surface of 5 μm to give a point-to-point depth increment of 210Å or 21 nanometers.

Spreading Resistance Profile Accuracy

An obvious question at this point is "how accurate are the concentration profiles derived from spreading resistance measurements?"

In our day-to-day profiling at SSM, we depend primarily on comparisons with theory (as in Figure 7), or on comparisons between the known ion implant dosage and a dosage calculated by integrating the dopant profile. For example, in Figure 8, the known implant fluence was 2×10^{13} cm^{-2} while the dosage calculated from the carrier concentration profile was 1.5×10^{13} cm^{-2}. We also frequently compare the measured sheet resistance of a layer with a ρ_s value calculated from the measured resistivity profile.

We expect to get a better idea of spreading resistance profile accuracy from a "round robin" test now in progress in the ASTM F-1 committee. This test is being done in various laboratories throughout the world and will result in an experimental determination of the precision of the spreading resistance technique, in support of a published ASTM standard on spreading resistance profiling (12).

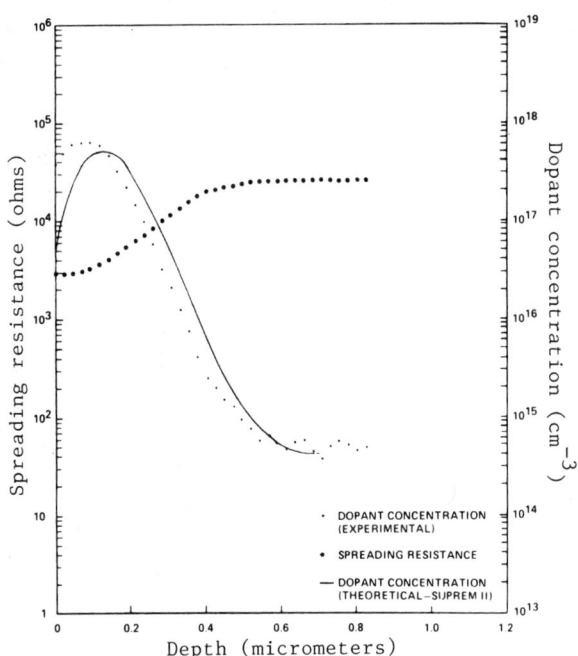

Figure 7. Spreading resistance, carrier concentration and SUPREM II profiles of a boron ion-implant. Reproduced with permission from Ref. 4. Copyright 1981 Solid State Technology.

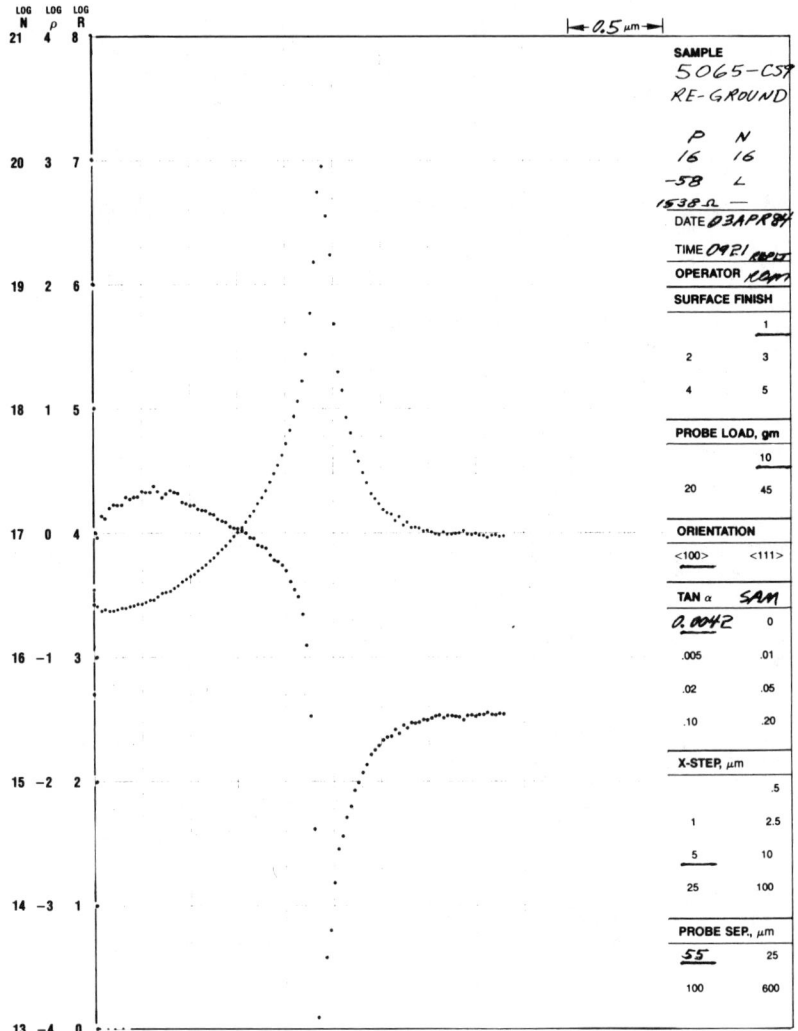

Figure 8. Spreading resistance and carrier concentration profiles of a boron ion-implant into an N-type substrate as measured with controlled low penetration, 10 gram probes.

In addition, a number of authors have recently published comparisons of spreading resistance and SIMS profiles. For example, Marek Pawlik of the GEC Research Laboratories in England and Jim Ehrstein of the National Bureau of Standards both gave papers comparing spreading resistance and SIMS at the February, 1984 Symposium on Semiconductor Processing in San Jose (13,14). Jim Ehrstein's paper was actually a three-way comparison of boron implants using spreading resistance, SIMS, and a Neutron Depth Profiling (NDP) method which is now under investigation at the NBS and which will be described in a paper to be given later at this meeting by R. G. Downing of the National Bureau of Standards.

Comparison of Spreading Resistance and SIMS. Figure 9 is taken from Jim Ehrstein's paper at the San Jose Symposium. It shows the dopant concentration profiles of three ion implants of ^{10}B into an N-type substrate, as measured after annealing by SIMS, NDP, and spreading resistance. The implants were done at 70 keV, with fluences of 1×10^{15}, 4×10^{15}, and 1×10^{16} cm^{-2}.

These profiles show three major differences between the two physical profiling methods and the spreading resistance technique.

First, SIMS and NDP both show an increase in boron concentration at the surface, due to a redistribution of boron into a thin oxide grown during the annealing process. This boron is not electrically active and is therefore not seen by the spreading resistance probe.

Second, there is a hump in the boron concentration at about 0.2 µm in both the SIMS and NDP profiles which is not seen in the spreading resistance profile because it is due to undissolved boron which is also not electrically active.

Finally, there is a significant difference in the concentration of boron measured by spreading resistance and by SIMS at boron concentrations below about 10^{17} cm^{-3}. At the present time, we do not have a logical mechanism to explain this discrepancy. Ehrstein et al suggest that the discrepancy may be caused by a "spill-over" of carriers from the PN junction. However, they also report that the junctions were copper-stained and that the stain agreed closely with the junction position indicated by the spreading resistance profiles. Thus, it is not clear whether we're seeing an artifact of the spreading resistance technique, whether there is still some problem with SIMS measurements at low concentration, or whether the discrepancy is materials-related due to some as-yet unknown mechanism.

Since the other discrepancies between SIMS and spreading resistance are clearly related to the fact that the spreading resistance technique measures only electrically-active impurities, it is tempting to suppose that the cause of the low concentration discrepancy is the same. This hypothesis gets some support from Marek Pawlik's paper on ion-implanted silicon-on-sapphire (14), in which he reports a similar discrepancy between SIMS and spreading resistance at low boron concentrations. Pawlik ascribes the discrepancy to the interstitial diffusion of boron, due to defects in the deposited SOS film. This suggests the possibility that, in

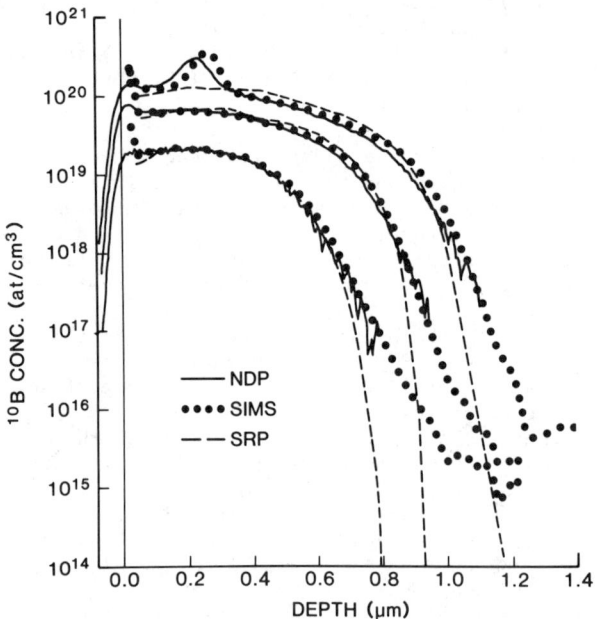

Figure 9. Spreading resistance, SIMS, and NDP (Neutron Depth Profiling) profiles of a ^{10}B ion-implant into an N-type substrate. Reproduced with permission from Ref. 13. Copyright 1984 American Society for Testing and Materials.

Ehrstein's work, the ^{10}B may also have diffused interstitially, perhaps as a result of defects remaining from the ion-implantation damage.

The end results of these comparisons between SIMS and spreading resistance are that the two techniques agree well at high concentrations, but that there is a significant, unresolved discrepancy at low concentrations. Furthermore, there is a developing consensus that the two techniques are complementary rather than competitive. This is due to the primary difference between them; that is, SIMS is species selective, whereas spreading resistance sees only the <u>net carrier</u> concentration, regardless of the chemical nature of the impurities contributing the carriers. Clearly, when used in combination, SIMS and spreading resistance give more useful information than either does by itself.

Spreading Resistance Applications

I can suggest some situations in which spreading resistance measurements are particularly useful. For instance, if you need the ultimate in range of application, you'll find that the spreading resistance technique is virtually unlimited, at least for silicon. Spreading resistance measurements are routinely made on material with dopant concentrations ranging from greater than 10^{21} cm^{-3} to near intrinsic material ($<10^{11}$cm^{-3}). Also, just about any device structure can be profiled, without regard to the number of different layers or to the layer conductivity types or thicknesses. Even polycrystalline silicon is easily and accurately profiled.

The spreading resistance technique will also serve you well when you need very high spatial resolution. Spreading resistance profiles are done with a resolution as little as one nanometer per point—and this is not a fundamental limit. The technique is still being developed, with smaller, lighter, and less-penetrating probes a definite possibility. Also, since spreading resistance probes can be spaced as close together as about 15 to 20 μm, very small patterned structures or devices can be profiled.

Spreading resistance profiling is currently the most accurate method available for depth measurements of PN junctions or other features in silicon, because the probe arm assemblies and the test specimen are mounted on precision micro-positioners. Also, bevel angles can be measured to within $\pm 2\%$ with an optical device fitted to the microscope used in spreading resistance measurements. This combination of precise angle measurements and the lateral position accuracy provided by the micrometers means that spreading resistance profiles can give junction depths accurate to about 2 or 3%.

Spreading resistance profiles are generated quickly, so the technique is useful when speed is essential. Most silicon structures can be bevelled and measured and the raw data then processed and plotted as a resistivity or carrier concentration profile in less than thirty minutes.

Finally, spreading resistance is relatively inexpensive. The capital equipment involved costs well under $100,000, and many systems are satisfactorily operated by well-trained technicians or hourly operators.

The spreading resistance technique does take a bit of learning, practice, and patience. However, when it's done right, there is really no match to the sensitivity and detail in the resultant profiles. That's why the spreading resistance technique has made and will continue to make a significant contribution to the activity that we're all concerned with--the never-ending task of characterizing semiconductor materials.

Acknowledgments

The author would like to thank Dr. James R. Ehrstein of the National Bureau of Standards for providing Figures 6a and 6b and Dr. Marek Pawlik of the GEC Research Laboratories for supplying the boron implant profiled in Figure 8.

Literature Cited

1. Schroder, D. K. Symposium on Materials Characterization in Microelectronics Processing, American Chemical Society, 1984.
2. Mazur, R. G.; Dickey, D. H. J. Electrochem. Soc. 1966, 113, 255-9.
3. Ehrstein, J. E. "Non-destructive Characterization of Semiconductor Materials"; Zemel, J., Ed.; NATO Advanced Institute, Plenum Press: 1978; Chap. 1.
4. Mazur, R. G.; Gruber, G. A. Solid State Technology 1981, 24, 69.
5. Mazur, R. G. In "Semiconductor Processing"; Gupta, D. C., Ed.; ASTM STP 850, American Society for Testing and Materials: Philadelphia, 1984.
6. Schumann, P. A.; Gardner, E. E. J. Electrochem. Soc. 1969, 116, 87.
7. D'Avanzo, D. C.; Rung, R. D.; Dutton, R. W. Stanford Electronics Laboratories; Technical Report No. 5013-2; 1977.
8. Choo, S. C.; Leong, M. S.; Hong, H. L.; Li, L.; Tan, L. S. Solid-State Electronics 1978, 21, 769-74.
9. Berkowitz, H. L.; Lux, R. A. J. Electrochem. Soc. 1981, 128, 1137-41.
10. Piessens, R.; Vandervorst, W. B.; Maes, H. E. J. Electrochem. Soc. 1983, 130, 468-74.
11. Choo, S. C.; Leong, M. S.; Sim, J. H. Solid-State Electronics, 1983, 26, 723-30.
12. ASTM Standard F 672-80, 1982 Annual Book of ASTM Standards, Part 43, American Society for Testing and Materials.
13. Ehrstein, J. R.; Downing, R. G.; Stallard, B. R.; Simons, D. S.; Fleming, R. F. In "Semiconductor Processing"; Gupta, D. C., Ed.; ASTM STP 850, American Society for Testing and Materials: Philadelphia, 1984.
14. Pawlik, M. In "Semiconductor Processing"; Gupta, D. C., Ed.; ASTM STP 850, American Society for Testing and Materials: Philadelphia, 1984.

RECEIVED August 12, 1985

4

Scanning Electron Microscopic Techniques for Characterization of Semiconductor Materials

Rodney A. Young and Ronald V. Kalin

Solid State Electronics Division, Honeywell Inc., Plymouth, MN 55441

The Scanning Electron Microscope (SEM), a powerful and versatile analytical tool, has proven to be indispensable for characterizing semiconductor devices and materials. This paper focuses attention on the working principle of the SEM and the signals available for analysis. Techniques for applying these signals for characterization of semiconductor devices and materials are grouped into three categories: imaging, x-ray microanalysis, and electrical. Imaging techniques vary from standard quality control inspection to in-depth cross-sectional analysis which is required for technlogy development. The advantages of having energy dispersive (EDS) and wavelength dispersive (WDS) x-ray spectrometers for x-ray microanalysis are also discussed. Voltage contrast and Electron Beam Induced Current (EBIC), two electrical techniques usually used for circuit analysis, have been proven useful for semiconductor materials characterization. The SEM has been integrated into semiconductor technology development and will continue to evolve to meet the challenges of Very Large Scale Integration (VLSI) processing.

The scanning electron microscope (SEM) is the most versatile analytical instrument available for the characterization of semiconductor devices and materials. "In no field of physical sciences has the SEM been more productive than in the semiconductor industry; without the SEM the industry would not be nearly as advanced as it is -- conversely, without the enormous economic impetus of the semiconductor industry, the sophisticated advancements in SEM instrumentation would not have come as quickly or in great a profusion as they have" ([1]). As very large scale integration (VLSI) technologies approach near-micron geometries, the importance of the SEM is becoming even greater as the usefulness of the optical microscope diminishes. The spatial resolution of the modern SEM (typically 40 Å) approaches the resolving power of a transmission electron microscope (TEM), yet is without the difficult and time-consuming sample preparation required for TEM analysis. The typical magnification range of 10 to 200,000x allows entire semiconductor chips or 60 nanometer thin films to be viewed in the SEM. The main advantage of the SEM over the optical or transmission electron microscope for defining surface morphology is the greater depth of field of the SEM image. Because objects viewed in the secondary electron image mode appear three-dimensional, interpretation of the SEM image is quickly assimulated even by a casual observer. The only sample preparation required is to coat insulating surfaces with a thin conducting layer. As the low beam energy performance of modern SEMs continue to improve, the SEM can be operated in a "non-loading" condition which makes coating of nonconducting layers unnecessary.

This review paper focuses attention on the working principle of the SEM and the signals available for analysis. Techniques for applying these signals for characterization of semiconductor devices and materials are presented along with a summary of how the SEM has evolved to meet the challenges of VLSI processing.

Working Principle
A simplified diagram of a SEM is shown in Figure 1. An electron gun emits electrons (by thermionic or field emission) which are accelerated down the column by a large potential energy (typically 1-30 keV). Electromagnetic lenses and mechanical apertures are used to demagnify and focus the electrons to form an electron probe of small diameter and high current density. The lenses are adjusted to change the focal length of the primary electron beam to focus the image, and to change the electron beam diameter and current density. The ability to change the focal length of the electronic lenses and size of the apertures allows the electron microscopist to optimize the spatial resolution and depth of field to obtain the desired image or signal. The scanning coils deflect the electron beam over the surface of the specimen in a raster pattern. The scanning circuits allow the operator to vary the size of the area scanned (to change magnification), and to scan at slow speeds for photography or at T.V. scan rates (15 KHz) for observation purposes. The video amplifier synchronizes the raster scan of the SEM cathode ray tube (CRT) with the raster scan of the electron probe striking the sample. This ensures a one-to-one correspondence between the signal collected from any particular point on the specimen surface and the brightness of the analogous point on the CRT screen. The SEM is usually equipped with one large viewing CRT, and one high resolution CRT (typically 2500 lines) for photography. Most SEMs are equipped with video signal processing circuits such as Y-modulation, differentiation, or gamma processing to aid in discriminating surface detail or images with wide ranges in gray (contrast) levels.

One important component of the SEM not indicated in Figure 1 is the specimen stage. The stage must allow precise, fine movements in tilt, rotation, and all three axes to allow observing large samples in all possible orientations. Stages can be adapted to provide heating/cooling, tensile stress, and electrical interfacing to IC packages or wafers to match the needs of the

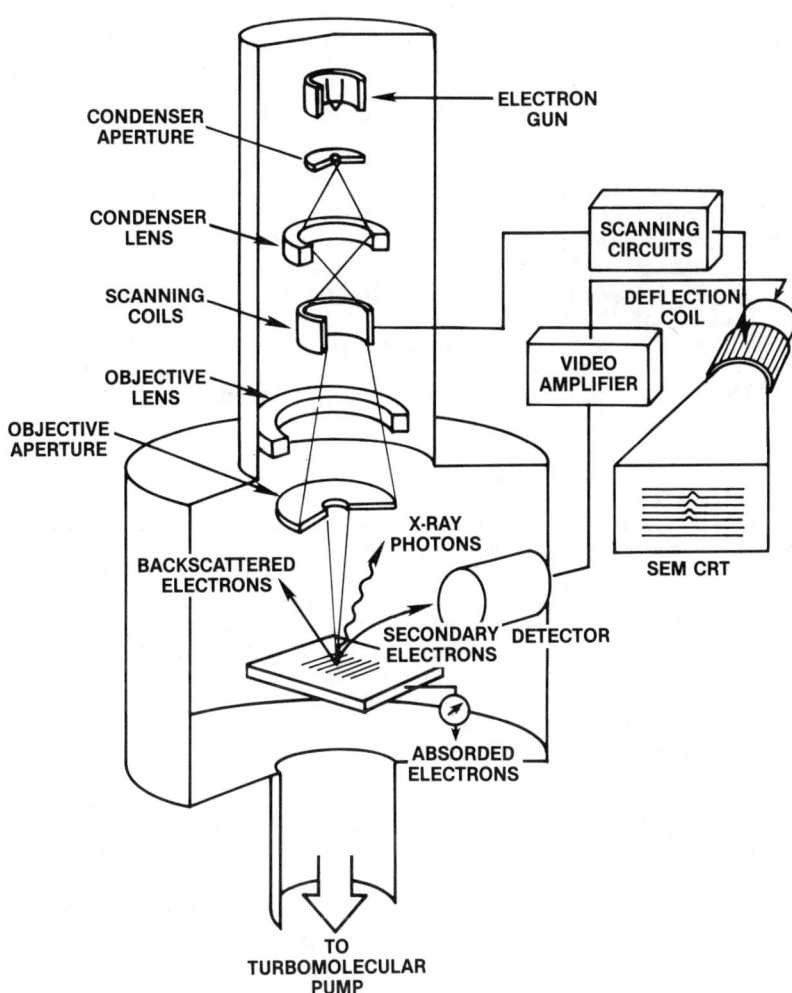

Figure 1. Diagram of SEM and signals available.

analysis. The electron column and specimen stages are both mounted on a specimen chamber which must be large enough to allow the introduction of large samples without compromising the optimum positioning of the various signal detectors. The specimen chamber usually has numerous removeable ports which allow various detectors or other accessories to be added to the instrument. The specimen chamber and electron column are maintained at a high vacuum ($\sim 10^{-3}$ Pa) by a high speed vacuum system. To maintain a vacuum relatively free of hydrocarbon contamination, the final, or high vacuum, pump is usually a liquid nitrogen trapped oil diffusion pump or turbomolecular pump.

Beam-Specimen Interactions
The primary electrons of the electron beam interact with the specimen producing a large number of detectable signals. Multiple scattering of the incident or primary electrons as they enter the specimen causes them to follow numerous trajectories. This diffusion of the electron beam is mainly dependent on the atomic number of the specimen and also on the energy of the primary electrons. A useful equation derived by Everhart and Hoff (2) describes the electron beam energy dissipation range for low atomic number elements as:

$$R_G = 4.0 E_B^{1.75}$$

where R_G = range in mass thickness in $\mu g/cm^2$ (density x range or ρR) and E_B = electron beam energy in KeV. This equation is valid for atomic numbers $10 \leq Z \leq 15$ and the beam energies $5 < E_B < 25$ KeV. Figure 2 shows the depth of signal generation and the spatial resolution of the signals as determined by Goldstein (3). Thus a 10 KeV primary electron beam striking pure silicon results in an energy dissipation range of approximately 1 micrometer. Elastic collisions of the primary electrons with atoms in the specimen result in the generation of backscattered electrons. Since backscattered electrons may lose very little energy they can be generated from deep within the specimen and still escape the surface. Inelastic collisions result in gradual energy loss of the incident electrons until they are absorbed by the specimen (detected as absorbed electrons). Most of the incident electron energy loss results in the generation of heat. However, incident electron-specimen interactions also create x-rays, secondary and Auger electrons, and visible and infrared light (4).

The secondary electrons are low energy (typically < 50 eV) and therefore escape only when created close to the surface of the specimen; because of their limited diffusion inside the specimen they do not degrade the spatial resolution of the probe. For these reasons secondary electrons mainly yield information about surface morphology. Like the backscattered electrons, the x-ray photons can approach the energy of the incident electrons and therefore are generated over a large volume. The spatial resolution of x-ray analysis is generally on the order of one micrometer. The crystal, thermal, and magnetic properties of specimen can also be examined in the SEM by using specialized techniques (5). Table I lists the signals available in a SEM and their application to characterization of semiconductor materials.

Imaging Techniques

Optical microscopy is a very important part of routine inspection of semiconductor devices and materials. The problems encountered with optical microscopes include limited magnification and depth to field due to the wavelength of visible light. The highest useable magnification is approximately 1500X. To achieve this, very high quality optics are required and a very stable platform is needed for the microscope. An example is shown in Figure 3, which is an optical photograph of a VLSI bipolar transistor; note the edges are poorly defined and the depth of field is very shallow. With continued scaling of semiconductor device features and thin films, it is evident that optical microscopy has limited use for inspection and characterization.

By using the secondary electron emission mode of the SEM the resolution and the depth of field can be increased by a factor of 100. Because of their low energy the mean free path of the secondary electrons in the specimen is only 10-100Å, so only those within a thin layer close to the surface leave the specimen. The majority of the secondary electrons are emitted from an area with a diameter comparable to that of the primary electron beam. The secondary electrons are accelerated to the detector by applying a positive voltage (the collector voltage) to a faraday cage or ring which surrounds the detector. This positive voltage can vary from a negative 50 volts to a positive ≈ 250 volts. The detector normally consists of a scintillator and a photomultiplier. Once near the detector the secondary electrons are further accelerated by a positive 10-12 kilovolt potential on the scintillator. The photons emitted as a result of the secondary electrons striking the scintillator are collected by the photomultiplier which feeds the corresponding amplified signal to the video amplifier of the SEM. The high collection efficiency of the detector combined with the small emission site of the secondary electrons results in micrographs with a surprising similarity to optical photographs but with dramatic improvements in resolution and depth of field. A good SEM micrograph is shown in Figure 4 where the same bipolar transistor as shown in Figure 3 has been magnified 15,000 times.

Sometimes the surface being analyzed will have little or no texture; such an image is difficult to interpret. When this is the case Y-modulation can be employed. This is accomplished by vertically deflecting the beam on the viewing CRT where the amount of deflection is proportional to signal intensity. The surface detail of the specimen will be accentuated by this image processing technique. An example of this is shown in Figure 5. The biggest drawback of this technique is that topography can be generated which is not characteristic of the sample surface.

Another imaging technique used to inspect and analyze semiconductors is the use of backscattered electrons. Backscattered electrons travel in relatively straight lines due to their high

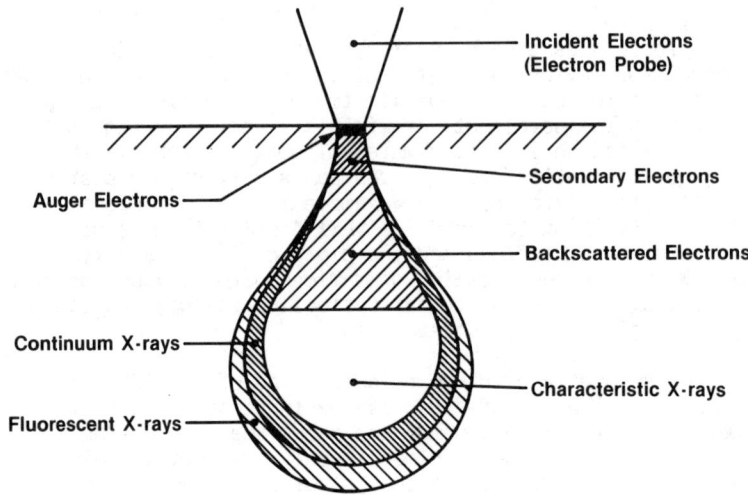

Figure 2. Depth and spatial resolution of signals. (Reproduced with permission. Copyright 1974 ASTM.)

Table I. Material Information Provided by SEM Images

Type of Image	Signal Used	Material Information
Secondary electron	Same	Surface topography
Stereo pair	Secondary electron	Surface topography
Voltage contrast	Secondary electron	Surface potential composition
Magnetic domain	Secondary and backscattered electrons	Magnetic properties
Backscattered electron	Same	Composition, topography
Electron channeling pattern	Secondary and backscattered electrons	Crystalline state
Absorbed electron	Same	Composition, topography
Cathodluminescence	Visible or IR light	Compositon, structure
X-ray dot map or line scan	X-rays fluoresced by E beam	Elemental distribution
EBIC	Electron-hole pairs	Structure, crystalline defects
Thermal wave	Thermally generated acoustic waves	Subsurface structure, thermal properties

Figure 3.   Optical photograph of VLSI bipolar transistor 1500X.

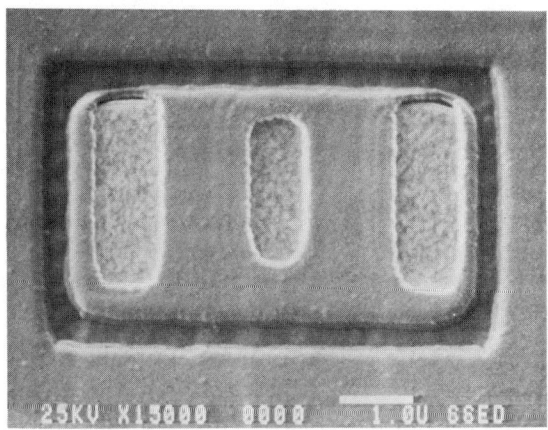

Figure 4.   Secondary electron SEM photograph of same transistor 15,000X.

energy and are not affected by local low level electrical or magnetic fields on the surface and can be used to obtain high contrast images of the surface. Backscattered electron yield is relative to atomic number concentration, with higher yields corresponding to increasing atomic number. Using this technique it is possible to identify materials which differ by < 1 mean atomic number (6).

A secondary electron detector can be employed to detect backscattered electrons if the high voltage on the detector scintillator is turned off. However, since the secondary electron detector is usually positioned to prevent a direct "line of sight" from the sample to the detector, the backscattered electron image may have a shadowing problem. Another detector which is used for backscattered electron detection is a silicon P-N junction, 25 mm or greater in size. When backscattered electrons enter the semiconductor, electron-hole pairs are generated. The current generated by this interaction can be amplified and interfaced to a video amplifier for image generation. The detector can be mounted axially, directly below the pole piece of the final lens, resulting in a high resolution, shadowless backscattered electron image.

Figure 6 shows a backscattered electron image of the bipolar transistor shown in the previous two figures. The bright areas are metal contacts that have $PdSi_2$ on the surface. They appear brighter because more backscattered electrons are being emitted in that area. The gray background is silicon dioxide and the dark ring is an isolation groove. The isolation groove appears dark due to the steep edges and the rough texture of the oxide at the bottom which causes greater backscattered electron scattering.

Cross-Section Analysis
By sectioning a multilayered semiconductor structure and staining it with a selective etch, a great deal can be learned about design and process characteristics. The die or a small piece of wafer is placed into an exacto knife holder and the area of interest is located using an optical microscope. Various grits of paper and polishing compound are used to grind down to the area of interest. The edge is polished and cleaned using deionized water or alcohol, and stained using a selective etch to highlight the features of interest. There are several types of etches which stain various materials; reference is made to papers on this subject (7-9). The SEM micrograph shown in Figure 7 illustrates various layers which make-up a N-channel Metal-Oxide-Silicon (NMOS) transistor. Note the similarities between the designers drawing (Figure 7a.) and the stained cross-section (Figure 7b).

X-Ray Micro Analysis
Of all the signals created by the interaction of the primary electron beam with the specimen in the SEM, x-rays are probably the most commonly used for material characterization. X-ray photons are generated when an electron from the incident beam

Figure 5. Y-modulation SEM photograph of same transistor 15,000X.

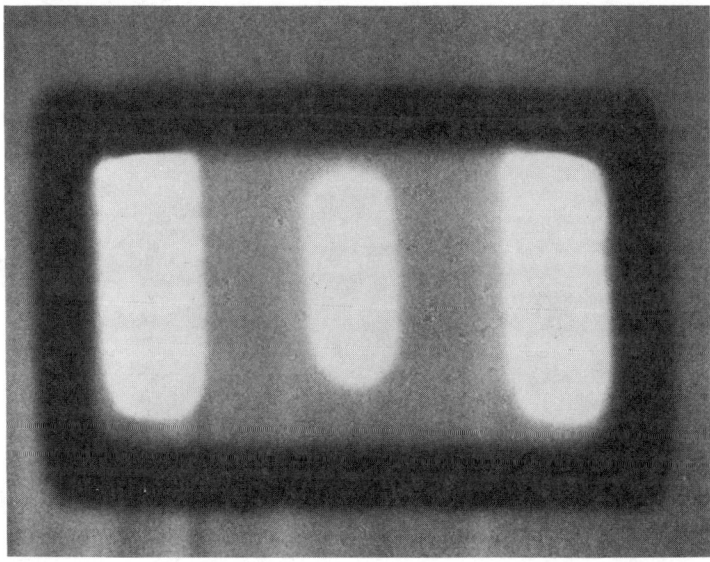

Figure 6. Backscatter electron SEM photograph of same transistor 15,000X.

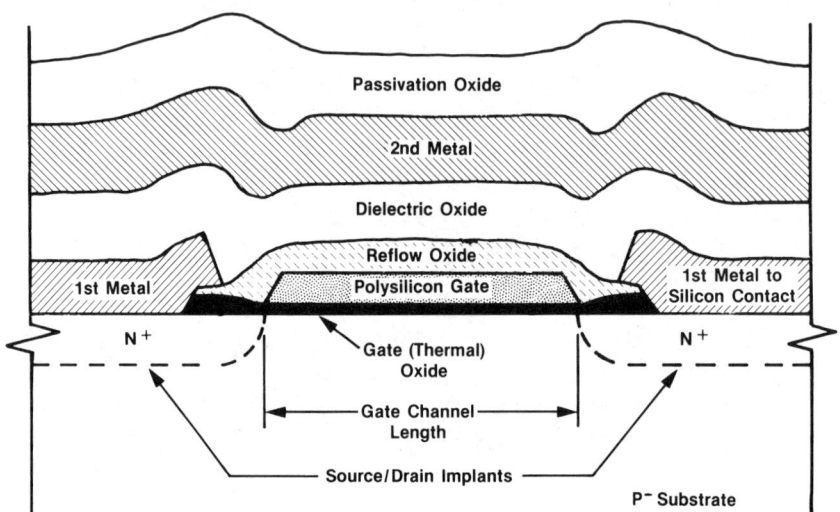

Figure 7a.   Cross sectional diagram of NMOS transistor

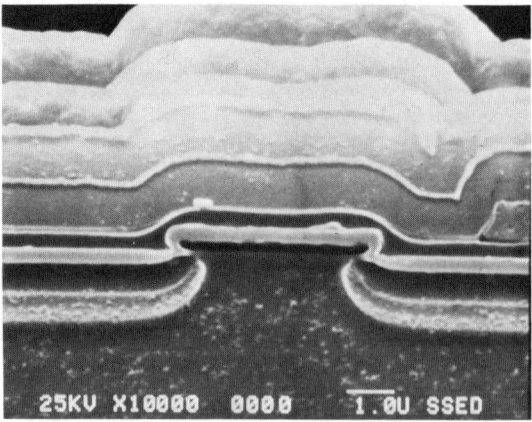

Figure 7b.   Secondary electron SEM photograph of NMOS transistor 7500X.

ejects an inner shell electron from an atom, allowing an electron from another shell to fill the vacancy. The discrete energy released by this electron transition creates a characteristic x-ray photon. Every element has a specific atomic structure and therefore a unique set of x-rays. Two techniques are used to analyze the energy and wavelength of x-ray radiation, energy dispersive x-ray spectroscopy (EDS) and wavelength dispersive x-ray spectroscopy (WDS). The components of EDS and WDS x-ray spectrometers are illustrated in Figure 8. Each spectrometer has unique performance characteristics which complement each other, making it desirable to have both on one SEM. Table II summarizes the performance characteristics of these two x-ray spectrometers.

The solid state detector most often used in EDS passes x-rays through a thin beryllium window which strike a biased p-i-n (p-type, intrisic, n-type) lithium-drifted silicon crystal. The interaction of the x-rays in the Si(Li) detector create electron-hole pairs, where the number of electron-hole pairs created is related to the energy of the incoming x-ray (higher energy x-rays create more electron-hole pairs). The electron-hole pairs are collected and converted to a voltage pulse, which is amplified before being sent to a multi-channel analyzer (MCA). The MCA sorts the pulses by assigning them to memory locations corresponding to preestablished voltage ranges calibrated to correspond to the energy of the x-ray. The contents of the memory locations comprise the x-ray spectrum, which is displayed by a CRT or X-Y recorder. High energy backscattered electrons leaving the sample create continuum x-rays which are not characteristic of the material and show up as background noise (see Figure 9).

EDS has a greater collection and quantum efficiency than WDS, and since all x-rays between 1-20 KeV can be seen simultaneously the time it takes to acquire enough data for qualitative analysis is greatly reduced. The energy range of this detector allows elements from sodium to uranium to be detected. EDS quantitative analysis can be used routinely to monitor thin film metal and oxide processes. The data in Table 3 compares EDS to wet chemical techniques (10) for analysis of phosphorus doped silicon dioxide. The average relative error is only 2.6 percent. The EDS quantitative analysis is standardless, utilizing the "ZAF" algorithm (11). The "ZAF" algorithm improves the accuracy by making corrections for atomic number (Z), x-ray absorption (A), and x-ray fluorescence (F) effects.

Wavelength detectors use a crystal to diffract x-rays into a detector. Only x-rays which strike the crystal at the proper angle and with a defined wavelength are diffracted into the x-ray detector by the analyzing crystal. This is expressed by Bragg's Law:

$$n\lambda = 2d \sin \theta$$

where n = order of diffraction, λ = wavelength of diffraction x-rays, d = interplanar spacing of the analyzing crystal and θ =

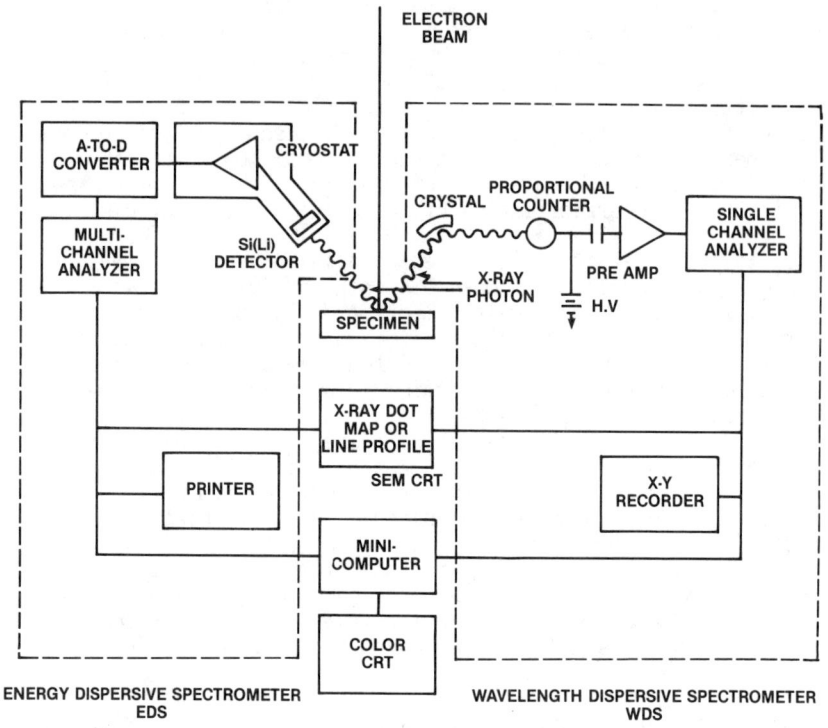

Figure 8.   Diagram of EDS and WDS detector systems.

Table II.   Performance Characteristics of EDS and WDS

Characteristics	WDS Crystal Diffraction	EDS Si(Li) Detector
Quantum efficiency	Variable < 30%	~100% for 2–16KeV
X-ray acceptance range	Spectrometer resolution ~10eV	Entire useful energy range
Data collection time	Minutes to hours	Minutes
Elements detected	All $\geq$ B	All $\geq$ NA
Energy resolution	Crystal dependent ~5eV	Energy dependent ~150eV
Sensitivity	0.1–0.01%	1–0.1%

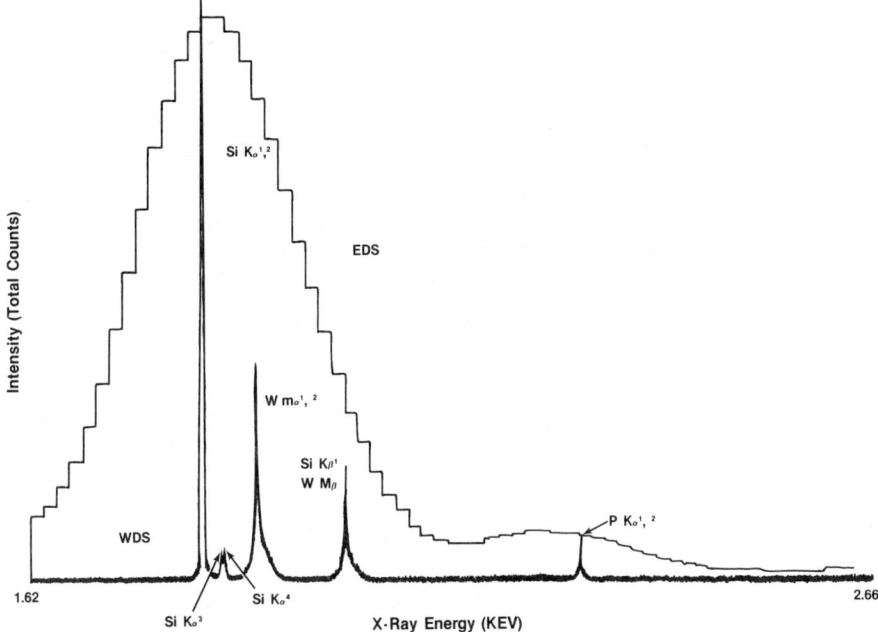

Figure 9.   EDS and WDS x-ray spectra comparison.

Table III.   Quantitative Analysis of Phosphorus-Doped Oxide Films by EDS

Sample Actual Concentration (in wt %)	EDS Computed Concentration	Absolute Error	Relative Error (%)
1.26	1.21	.05	4.0
1.56	1.51	.05	3.2
2.50	2.44	.06	2.4
3.55	3.55	.00	0.0
3.63	3.75	.12	3.3
3.97	3.81	.16	4.0
4.50	4.24	.26	5.8
5.55	5.64	.09	1.6
5.76	5.76	.00	0.0
7.26	7.32	.06	0.8
7.75	7.93	.18	2.3
8.88	8.68	.20	2.3
9.63	9.39	.24	2.5
10.25	10.16	.09	0.9
11.32	10.72	.60	5.3
11.84	11.38	.56	3.9

the angle between crystal surface and the incident and diffracted x-rays. Typically, several different crystals are used in a WDS spectrometer. The interatomic spacing varies with each crystal, allowing element x-ray lines from Boron to Uranium to be detected. Once the x-rays are diffracted they enter a gas flow proportional counter tube through a thin polyproplyene window. The counter tube body acts as a cathode, and inside a fine wire with a positive bias of approximately 1700 volts acts as an anode. As an x-ray enters the body it will ionize the Argon in the filler gas (90% Argon - 10% methane) creating a photoelectron cascade. The positive voltage on the anode results in secondary ionizations, increasing the total charge collected by several orders of magnitude. The signal is amplified further, processed by a single channel analyzer (SCA), and sent to the inputs of an X-Y plotter or CRT.

The advantages of WDS are the improved energy resolution which allows better element resolution, greater peak-to-background ratios, and higher count rates for individual elements. This combination results in better detection sensitivity by approximately 1-2 orders of magnitude. Figure 9 clearly shows the better peak-to-background ratios and energy resolution of WDS compared to EDS.

The location of elements on a sample can be displayed using two techniques, x-ray dot mapping or x-ray line profiles. X-ray maps are generated by placing a "dot" on the CRT of the SEM at each location (of the area scanned) where the element of interest is detected by the x-ray spectrometer. The x-ray line profile is created by slowly scanning one line across the area of interest on the specimen. As in Y-modulation, the deflection in the vertical orientation (Y-axis) of the SEM CRT is dependent upon the x-ray signal intensity at that location. Figures 10a and b show an x-ray dot map and line profile of a metal conductor which opened during a reliability electromigration experiment. The two x-ray imaging techniques clearly show the spatial distribution location of chromium in the metal stripe.

<u>Electrical Techniques</u>
Voltage contrast and Electron Beam Induced Current (EBIC), two electrical techniques usually used for circuit analysis, have proven useful for semiconductor materials characterization. For both techniques the device to be tested is placed in the chamber of the SEM, and electrical connections are made to the device.

Figure 11 illustrates how the voltage contrast image is formed. The electron beam is raster scanned over the device in the usual manner. Since the energy of the secondary electrons is very low their emission (and collection) is affected by local electric fields on the chip surface. In Figure 11, section "A" has a positive bias creating a local retarding electric field which suppresses the emission of secondary electrons causing section "A" to appear darker than section "B". On a typical SEM the voltage resolution is approximately 1 voltage. Quantitative

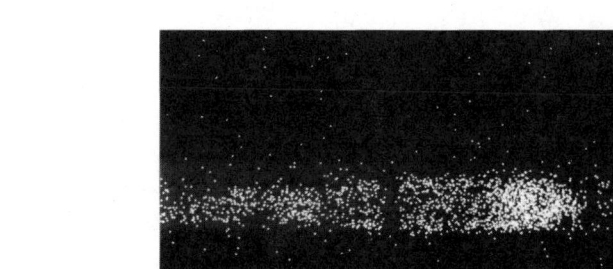

Figure 10a.   Chromium X-ray dot map 2000X.

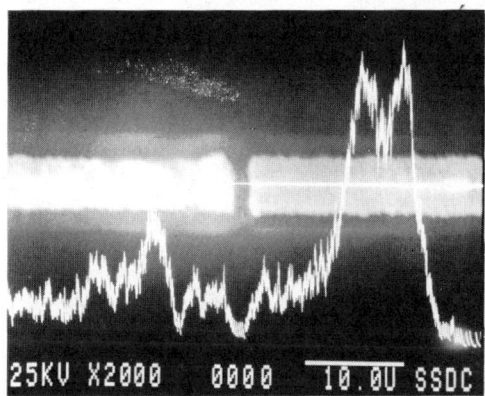

Figure 10b.   Chromium X-ray line profile 2000X.

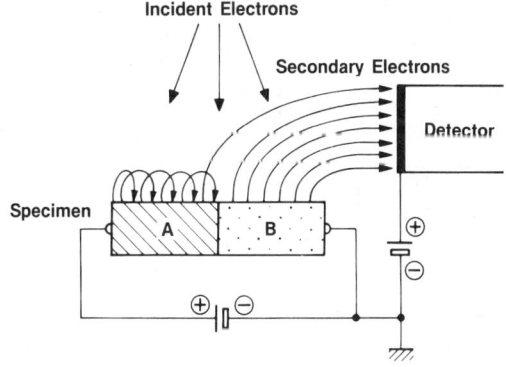

Figure 11.   Voltage contrast mechanism.

voltage measurements are possible only if the secondary electron emission remains the same for different measuring points on the IC surface. Factors which alter secondary electron emission and thereby degrade the voltage resolution include topography, differences in material and their surface condition (12). If a secondary electron analyzing scheme is added and the detrimental effects mentioned above can be minimized, the voltage resolution can be improved to resolve voltage differences on the order of 1 millivolt.

An example of the application of voltage contrast is in the analysis of integrated circuit test structures. Such structures are used to provide empirical information about defect densities to estimate product yield. A typical metal defect test structure has three layers of metal to test for within-layer opens and shorts plus layer-to-layer isolation leakage. Electrical tests can verify these parameters, but cannot always classify the cause of the defect. Figure 12a shows a secondary electron image of this structure with a metal shorting problem. No obvious processing defects, such as a photomask defect or mechanical damage, are present which could cause a metal short. By applying a negative bias or voltage to the left metal stripe and a positive bias to the right hand metal stripe, small metal "streamers" become clearly visible between the metal stripes. This is the cause of the shorting problem. Due to their location and size, these metal streamers could not be identified by high resolution secondary or backscattered electron images.

Electron beam induced current (EBIC) has been used to characterize many electrical and physical properties of semiconductors. Electrical properties defined include the minority carrier life-time and diffusion length, (13) electrical junction location and depletion width (14). Physical properties such as oxide defects, metalization integrity (15) and bulk crystal defects (16) have been quantified by using EBIC.

When an electron beam strikes a semiconductor material, electron-hole pairs are generated; a 15 KeV electron can generate over 3,000 electron-hole pairs in silicon. If the carrier generation takes place a few diffusion lengths away from any junction space charge region, the excessive carriers will recombine within the semiconductor. If these mobile carriers reach or are created within the depletion region of a p-n junction they are swept out by the existing junction potential giving rise to an external reverse current. This electron beam induced current may be amplified to produce an image and thereby gain information about electrical and structural characteristics of the junction. Figure 13 shows the EBIC experimental set-up for a p-n junction device and resultant signal intensity which is proportional to the field intensity in the depletion region.

One of the yield limitations in an advanced bipolar technology is the collector-emitter leakage due to "pipes" which are associated with material defects (17), and generally the pipe density will increase with shallower junctions required for VLSI technologies.

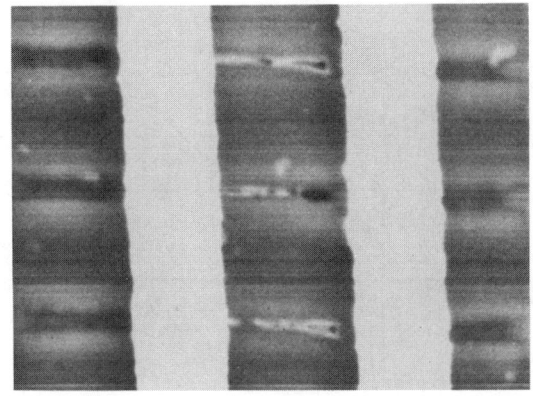

Figure 12a. Secondary electron SEM photograph of shorted metal lines 10,000X.

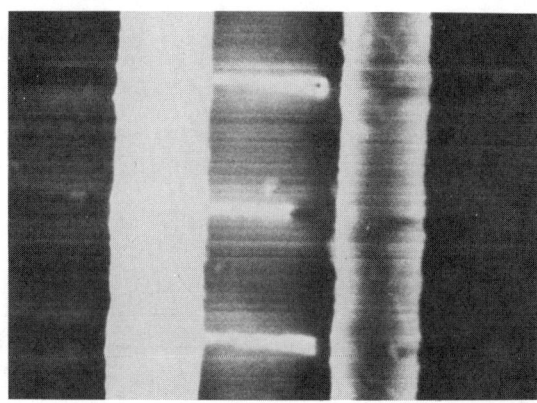

Figure 12b. Voltage contrast SEM photograph of same metal lines 10,000X.

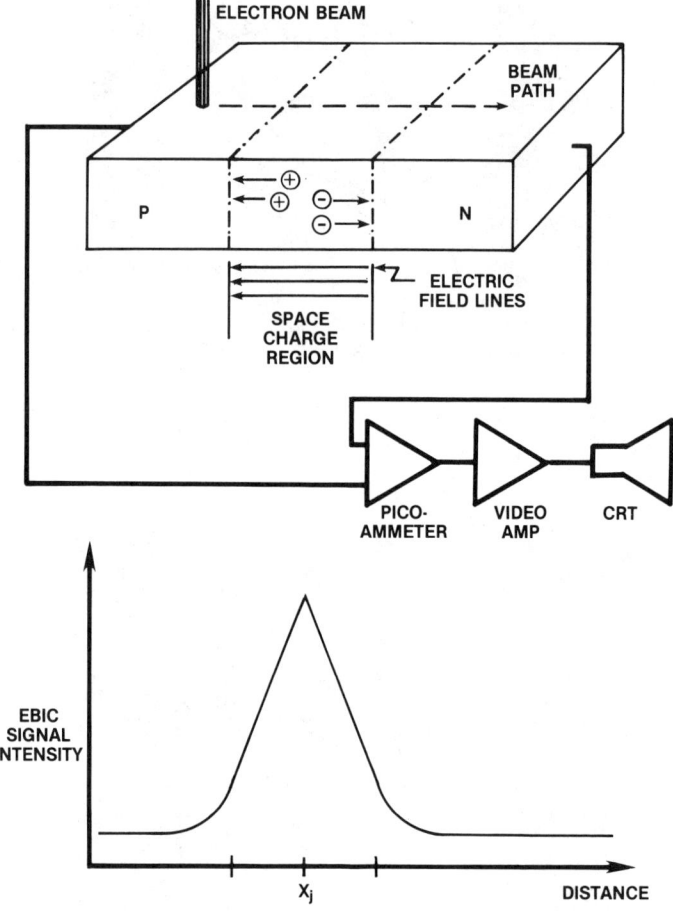

Figure 13. EBIC signal formation and intensity in P-N diode structure.

A typical pipe detector test structure consists of 1,000 transistors connected in parallel. Although the presence of pipes can be verified electrically, in the past the relative density and location of pipes was verified by a destructive technique involving a selective chemical etch to decorate the pipes (18). EBIC can be used to nondestructively determine the location and density of piped transistors. Figure 14a shows secondary electron image of three VLSI bipolar transistors which have no obvious process related defects which would indicate the location of a pipe. Figure 14b shows an EBIC image of the same three transistors. The center transistor is a functional transistor with the emitter-base appearing bright, while a reduced signal level on the two outside transistors across the emitter-base junction indicated the resistive effect of a pipe defect. The dark base-collector junction in the piped transistors is also an indication of abnormal charge collection caused by the pipe.

SEM's Role in VLSI Processing

Because of the numerous applications of the SEM to semiconductor materials characterization and technology development, numerous changes have been made to the instrument to meet the needs of the semiconductor industry. Table IV summarizes the differences between a general purpose SEM and a SEM designed for VLSI semiconductor process development. As mentioned previously, verifying the proper geometries and size of near micron structures is no longer practical with an optical microscope. Before the SEM could be used for in-line wafer inspection and critical dimension (CD) measurements, changes had to be made to the instrument to ensure SEM analysis would have no detrimental effects on in-line process wafers. For SEM analysis to be nondestructive, two issues need to be addressed: electron beam radiation effects and contamination of the wafer surface.

The electron beam of the SEM can create trapped charge in silicon dioxide layers and generate additional interface states between the silicon dioxide-silicon interface which can adversely affect the performance of integrated circuits (19). Developed photoresist patterns are sensitive to electron beam exposure and can be altered in shape and other physical properties. To avoid these effects the electron beam energy and dose (electrons/cm^2/sec) need to be carefully controlled. If the primary electron beam energy is adjusted such that the absorbed current is equal to zero (i.e. primary electrons = backscattered + secondary electrons) no surface charging occurs and the electron dose is less critical. Primary electron beam energies in the range of .6 - 3 KeV are desirable to avoid charging problems and minimize irradiation damage (20).

The potential problems of wafer contamination as a result of being exposed to the environment of the SEM chamber could adversely effect the performance of the devices on the wafer by introducing mobile ions or metallic contamination. Even though the vacuum in the SEM is typically down in the 10^{-3} Pa range, the electron beam can polymerize and/or ionize residual hydrocarbons in the chamber which can then be electrostatically attracted to the wafer surface.

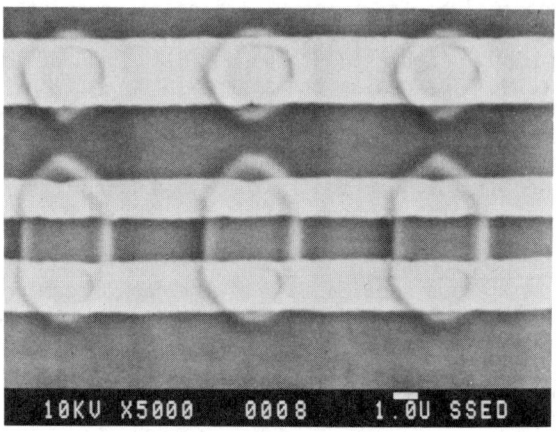

Figure 14a. Secondary electron SEM photograph of VLSI bipolar transistor 5,000X.

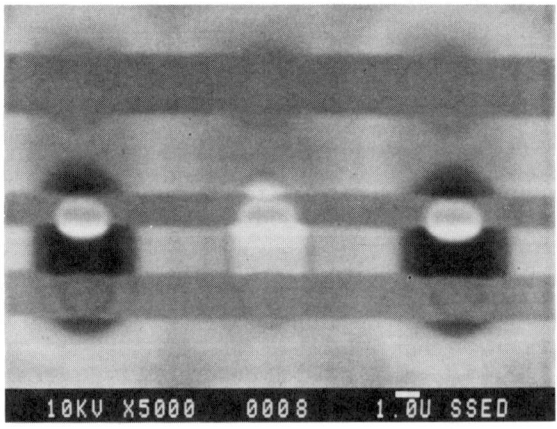

Figure 14b. EBIC photograph of same transistors 5,000X.

Table IV. General Purpose Versus "Semiconductor SEM" Comparison

Parameter	General Purpose SEM	Semiconductor SEM
Electron Gun	Tungsten (Thermionic)	LaB_6 (Therminonic) Or Field Emission Gun
Electron Optics	Designed For Wide Acc. Voltage Operation	Optimized For Low (.5-3KV) Acc. Voltage
Electronic Circuits	Analog	Digital (Microprocessor Controlled)
Stage	Manual, Limited Movements	Motor Driven, 100mm Travel
Chamber	Designed For Small Samples	Designed For Analyzing Wafers
Vacuum System	Oil Diffusion High Vacuum Pump	Oil-Free Vacuum System
Image Processing	Analog	Digital (Computerized)
Operation	Manual	Semi-Automatic Operation

SEMs used for nondestructive inspection of process wafers have been optimized for low (< 3 KeV) beam energy operation and a cleaner vacuum. In the past the increased beam spot size associated with low beam energies resulted in severe spatial resolution degradation. To maintain smaller beam diameters brighter electron guns are being used. A lanthanum hexaboride thermionic electron emitter is typically 10 times brighter than a tungsten hairpin filament, while a field emission gun is 1000 times brighter. The electron optics have been optimized for low beam energy operation by reducing lens aberrations and using higher lens currents to focus the beam. The position and sensitivity of the secondary electron detector can be critical factors in obtaining high resolution images at low beam currents. Some SEM manufacturers are positioning the secondary electron detector above the objective lens of the instrument to improve the collection of secondary electrons. Improved scintillating materials in the typical Everhart-Thornley secondary electron detector improve the signal-to-noise ratio and therefore improve the low beam energy performance. Figure 15 shows a low beam energy SEM image of an uncoated photoresist pattern on a process wafer. Lower beam energy images result in less specimen charging and also show more surface detail than a high beam energy image. A cleaner vacuum is obtained by avoiding the use of oil diffusion high vacuum pumps and rotary roughing pumps. "Dry" pumps previously used on ultra-high vacuum (UHV) analytical instruments are now being used on SEMs. Turbomoleclar pumps have been the most popular, although cryogenic, ion, and sublimation pumps are also being used to maintain a hydrocarbon-free chamber.

The specimen stage and chamber have undergone significant changes to meet the needs of wafer inspection. The stage travel in X and Y-axes is typically 100mm to permit viewing of a 4-inch wafer edge to edge without rotation. The stage usually allows tilting a wafer to a minimum of 60° to allow viewing feature profiles. A eucentric tilt axis speeds wafer inspection by keeping the wafer at the same working distance (therefore in focus). Tilting wafers to high angles increases the working distance (distance from the final lens to the point of focus on the wafer) which degrades spatial resolution. Integrating a conical-shaped pole piece (final lens) with the stage and chamber allow wafer tilting while maintaining a reasonable working distance. To further speed wafer inspection the stage movements can also be motor driven and computer controlled.

SEMs are beginning to benefit from computer control to improve and simplify the operation of the instrument. Computer control of operator dependent parameters such as image contrast, brightness, and focus, not only make the instrument easier to operate but also shortens the analysis time. Having the instrument under microprocessor control allows the SEM image to be digitized for two important applications: (1) critical dimension (CD) measurements and (2) image analysis. A minicomputer can store video intensity profiles by sampling the image signal at

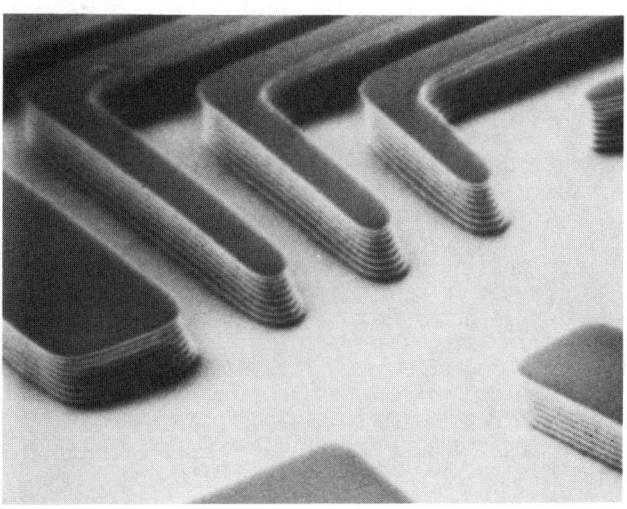

Figure 15. Low beam energy SEM image of photoresist pattern. (Micrograph taken on a Nanometrics CWIKSCAN III FESEM.)

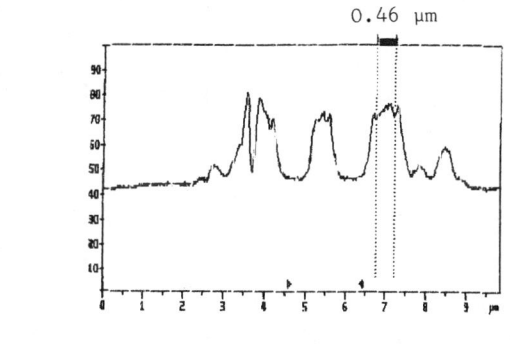

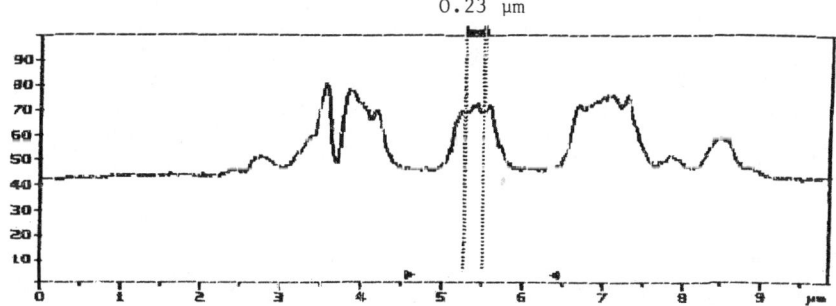

Figure 16. CD measurement via digitized SEM signal profile.

numerous points across the field of view. The stored data points have a spatial resolution calculated from parameters established by the microprocessor. Critical dimension measurements can then be made from the digitized image profile eliminating inaccuracies inherent in visual evaluation of gray levels, CRT non-linearities or distortions. Figure 16 shows a digitized image profile of submicron electron beam lithography features on a device.
Digital image processing and analysis are playing an increasingly active role in SEM analysis. Most x-ray microanalysis systems now offer digital imaging of any signal produced by the instrument. Digital processing allows images to be stored and/or enhanced by a computer to yield different kinds of information. Even with advanced electron optics, low beam energy SEM images of semiconductor wafers often are low in contrast or high in noise due to surface charging effects. "With image processing, the two most attractive features of standard electron microscope (EM) methods can be combined: a fast scan rate, providing a minimum charge build-up; and signal integration, reducing the effects of random noise and increasing the signal-to-noise ratio" (21).

In the past, one disadvantage of SEM wafer inspection was the necessity to vent and then evacuate the column every time a different wafer was inspected. Also, locating features on a large die with 300,000 transistors wastes valuable inspection time. SEM manufacturers are now offering cassette-to-cassette wafer loading and a user programmable, computer-controlled motorized stage which positions the wafer to pre-programmed cartessian coordinates. Ultimately, the SEM of the future may offer totally automatic wafer inspection by using image analysis computers interfaced to design computers to directly compare photomasks and wafers against design and layout data.

Acknowledgments

The authors gratefully acknowledge Ron Nelson and Carolyn Casoria for their support in gathering data and taking SEM photos for this paper. Also Mark Windland, Connie Krauth and Steve Grams for help in preparing the final manuscript.

Literature Cited

1. D.L. Crosthwait, J.R. Devaney, and E.S. Meicran, "SEM Techniques for Semiconductor Device Studies", SEM/1975 IITRI, p. 716.

2. T.E. Everhart and P.H. Hoff, "Determination of Kilvolt Electron Energy Dissipation versus Pentration Distance in Solid Materials", Journal of Applied Physics, Volume 42, December 1971, p. 5837-5846.

3. Goldstein, Newbury, Joy, Fiori, Lifshin, and Echlin, Practical Scanning Electron Microscopy and X-Ray Microanalysis, Plenum Press, 1981.

4. O.C. Wells, Scanning Electron Microscopy, McGraw-Hill, New York, 1974.

5. Dale E. Newbury, "Fundamentals of Scanning Electron Microscopy for Physicist: Contrast Mechanisms", SEM/1977/I, p. 553-568.

6. Data obtained from Anthony Laudale, JEOL Applications Laboratory, JEOL USA, Inc., Peabody, Mass.

7. T. Mills, "Precision VLSI Cross-Sectioning and Staining", Hewlett Packard Co., CH1846-5/83/000-0324 1983 IEEE/IRPS, p. 324-331.

8. Peter Randklev and A.D. Kostic, "Improved Microsectioning Materials and Their Applications", Medtronic, Inc., ISTFA, 1984.

9. Simon Thomas, "SEM Analysis of Fracture Cross Sections in VLSI Fabrication", Burroughs, April 1982 Semiconductor International, p. 209-222.

10. J. Houskova, K.N. Ho, M.K. Balazs, "Characterization of Components in Plasma Phosphorus Doped Oxides", ACS Symposium on Materials Characterization in Microelectronics Processing (ACS National Meeting), St. Louis, 1984.

11. J.W. Colby, "MAGIC IV, A Computer Program for Quantitive Electron Microprobe Analysis", in Proceedings Sixth National Conference on Electron Probe Analysis, Pittsburgh (Electron Probe Analysis Society of America, 1971).

12. E. Menzel, E. Kubalek, "Fundamentals of Electron Beam Testing of Integrated Circuits", Scanning, Vol. 5 (1983), pp. 103-122.

13. D.R. Hunter, D.H. Paxman, M.R. Burgess, and G.R. Booker, Proc. Scanning Electron Microscopy Conf., Newcastle, (1973).

14. J.D. Schick, "Junction Depth Measurements in a Scanning Electron Microscope", Electron and Ion Beam Science and Technology, 6th Interantional Conference, Electrochemical Society, p. 177 (1974).

15. J.R. Beall and L. Hamiter, Jr., "EBIC-A Valuable Tool for Semiconductor Evaluation and Failure Analysis", IEEE, 15th Annual Proceedings Reliability Physics 1977, p. 61-69.

16. S. Kawado, Y. Hayafuji, and T. Adachi, "Observation of Lattice Defects in Silicon by Scanning Electron Microscopy Utilizing Beam Induced Current Generated in Schottky Barriers", Japan J. Appl. Physics, Vol. 14 (1975), No. 3 pp. 407-408.

17. H. Kressel, "A Review of the Effect of Imperfections on the Electrical Breakdown of P-N Junctions", RCA Review, pp. 175-207, 1967.

18. J.F. Casey, J.W. Meredith, and G.M. Oleszek, "Piped Emitter Delineation Using the Wright Etch", J. Electromem Soc., Solid-State Science and Technology, February 1982, pp. 354-357.

19. W.J. Keery, K.O. Leedy, and K.F. Galloway, "Electron Beam Effects on Microelectronic Devices", SEM/1976 IITRI, p. 508-513.

20. M. Ostrow, E. Menzel, E. Postulka, S. Gorlich, E. Kubalek, "IC-Internal Electron Beam Logic State Analysis", SEM/1982/II, p. 563-572.

21. N. Schneider, "Image Processing in Microscopy", Electronic Imaging, December, 1983, pp. 78-80.

RECEIVED August 30, 1985

Semiconductor Materials Defect Diagnostics for Submicrometer Very Large Scale Integration Technology

G. A. Rozgonyi and D. K. Sadana

North Carolina State University, Raleigh, NC 27695-7916 and Microelectronics Center of North Carolina, Research Triangle Park, NC 27709

> The state-of-the-art analysis methods for the evaluation of structural, chemical and electrical properties of thin layers in processed Si substrates are discussed. The properties of implanted p-n junctions, Si-SiO$_2$ interface, Ge$^+$ implant amorphization of Si and misfit dislocation interface in epitaxial Si are exemplified to illustrate the features and limitations of the techniques.

A variety of information is required to fully characterize thin surface layers incorporated in processed Si in VLSI devices. State of the art designs utilize micron size features and dictate that the location of interfaces and the thickness of thin layers be controlled to tens of nanometers. Detailed information about the crystallographic structure, chemical composition and distribution of dopants and impurities, as well as the electrical properties of micro-regions have been explored by various combinations of experimental techniques (1,2). Some of these techniques are simple and inexpensive while others require elaborate equipment and sophisticated interpretation. In this paper we have attempted to put in perspective the relative merits and capability of interface characterization techniques. The sample preparation techniques presented here have three distinct perspectives, i.e., information is collected from a top surface or plan-view, a variation of a plan-view prepared as a beveled surface, and a cross-sectional edge view (see diagram, Figure 1). The tools to be applied include chemical delineation via preferential etching or staining, angle lapping and surface relief profilometry, spreading resistance, transmission electron microscopy (TEM) and secondary ion mass spectrometry (SIMS).

Optical Microscopy

The most important optical technique for examining semiconductor wafer surfaces is the differential interference contrast microscopy method of Nomarski (N-DIC). First described in 1952, DIC objectives came into general use in the early 1960's at about the same time that vapor phase silicon epitaxy was introduced as a semiconductor materials process. The extremely good depth resolution and high contrast provided by the N-DIC microscope facilitated the study of epi-surface morphology and defects on highly polished silicon wafers, as well as the study of striations introduced during crystal growth. More recently the growth mechanisms operating during liquid phase epitaxy (3) and laser annealing (4,5) have been elucidated with this microscope.

An example of the high contrast, excellent lateral as well as in-depth resolution, and unique three-dimensional character of N-DIC images is shown in Figure 2 (6). The figure shows striations of type I and type II in a longitudinal section of a heavily Sb-doped Czochralski silicon crystal. Type I striations appear as regular horizontal lines. They are deliberately introduced time markers and delineate the instantaneous growth interface of the crystal. The regular pattern of type I striations is intersected by several diagonal type II striations. Striation interactions, or feathering occurs in two varieties: F_{21} feathers denote crossings of type II with type I striations while F_{22} feathers denote regions where type II striations interact with other type II striations. Micrographs of this variety can be used to reconstruct the spatial distribution of impurity incorporation in crystals and leads to a deeper understanding of how liquid-solid interface instabilities occur. It should be evident that a single N-DIC optical micrograph can yield a surprisingly diverse amount of information about how crystals grow and how they respond to thermal treatments. However, in order to obtain and properly interpret micrographs comparable to the one in Figure 2 it is necessary to understand the basic functions and operating principles of the Nomarski DIC microscope.

The important components involved and a simplified schematic of the light path traveled in a Nomarski instrument operating in the reflection or epi-illumination mode are shown in Figure 3, taken from an article by Miller and Rozgonyi (7). The light source is polarized and brought to the optic axis by a semi-reflecting beam-splitting mirror. The light then enters a modified Wollaston or Nomarski double wedge prism, where it is divided into two orthogonally polarized beams. The two beams diverge at the wedge interface with an included angle ε. The Wollaston Prism is inserted between the objective and vertical illuminator in a plane optically equivalent to the back focal plane of the objective. In this way the two beams are collected by the objective and illuminate the sample parallel to one another but with a lateral shear, s, which is a constant for each objective/prism combination. Two important points to note here are that the shear or beam displacement must be less than the resolving power of the objective employed, i.e. a single image is perceived, and the fact that since the two beams are orthogonally polarized when they reach the sample surface, they independently carry phase path information back to the analyzer before interference takes place. Since each beam traverses the same optical components, neither is a reference beam, and the resulting light intensity passed by the analyzer is produced by a Differential Interference Contrast. Therefore, N-DIC images of uniform, flat surfaces such as the object in Figure 3 will have zero contrast. Local changes in intensity for monochromatic illumination, or color changes in white light, only occur when a local change in elevation or variation in optical properties introduces a phase path difference between the two

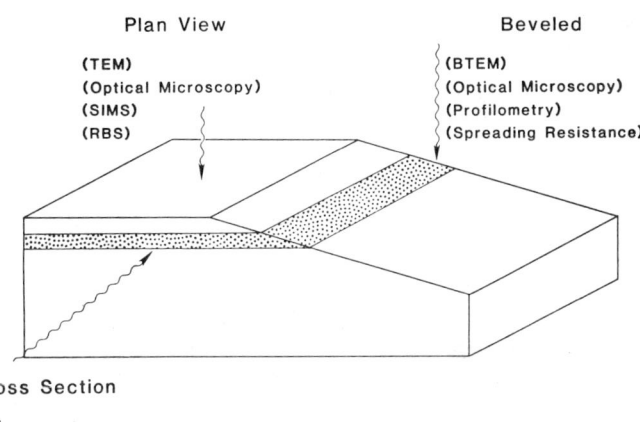

Figure 1. Viewing perspective of various sampling techniques. (Reproduced with permission from Ref. 6. Copyright 1982 *Journal of the Electrochemical Society*.)

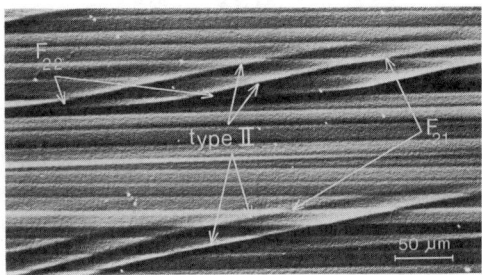

Figure 2. Striations of type I and type II in a longitudinal section of a heavily Sb-doped CZ Si crystal as revealed by N-DIC. (Adapted from Ref. 6.)

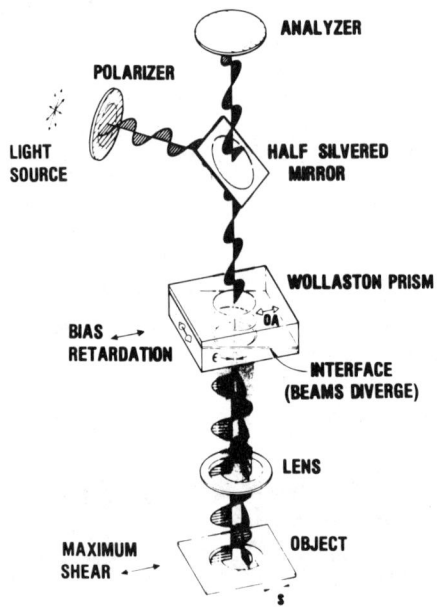

Figure 3. Schematic diagram of essential components in a N-DIC optical microscope. (Adapted from Ref. 7.)

laterally displaced beams. In a given N-DIC micrograph all surfaces having the same slope and optical properties will have identical contrast (or color). This includes all surfaces which are horizontal even if they are at different elevations.

The beautifully vivid colors created in white light N-DIC microscopy by manipulation of the beam splitting prism occur because of its ability to independently control the phase path difference between the two beams. This control function is called bias retardation. With monochromatic illumination bias retardation permits a simple optimization of black and white contrast by enabling the operator to set part of the image at the null or extinction position, e.g., the black line portions of the arrows in Figure 2. An additional enhancement of contrast for small surface elevation changes can often be obtained by evaporation of a thin highly reflecting film of gold or silver.

Preferential Etching

A combination of angle-lapping, chemical delineation and optical microscopic observation has been known for many years as a fast, simple, and inexpensive, although rather crude junction location technique. Significant refinements to this method have been reported recently (1,8). By careful control of the lapping angle, amount and type of staining solution, staining time, intensity and type of light used to illuminate a sample during junction delineation, it has been demonstrated that junction depth measurements with ±20 nm accuracy can be obtained. Furthermore, amorphous or heavily disordered crystalline interfaces in ion implanted Si can also been delineated by Angle Lapping, Etching and Profilometry (ALEP) (9). Figure 4 demonstrates these refinements. Junction/interface delineation by this technique will be correlated with spreading resistance, SIMS and cross-sectional TEM measurements in the subsequent sections of the text.

Electrical Junctions Delineation in a Doubly Implanted n-p-n Structure. High depth resolution (100A) angle lapping is performed by mounting Si chips on a 0° 34' lapping block and polishing directly with Syton on a Plexiglass plate. This angle provides a convenient linear magnification factor of 100 for any distance perpendicular to the sample surface. The usage of a heavy jig assembly prevents undesired rocking of the set during lapping. After demounting, the chips are preferentially etched with the Secco solution (10) for 1 to 15 seconds. Micrographs of the samples are then obtained with an optical microscope equipped with a 40 X Nomarski objective with a numerical aperature of 0.85. These micrographs reveal 10 μm for each mm along the beveled surface when operating at a total system magnification of 1000 X. The delineation process is performed either in the highly concentrated light of a microscope illuminator, in ambient light, or in total darkness.

An n-p-n structure delineated by the procedure described above is shown in Figure 4. The sample was fabricated by first implanting B into (100) Si at 50 keV to a dose of 10^{15} cm^{-2} and then implanting As at 80 keV to a dose of 5×10^{15} cm^{-2}. The sample was subsequently furnace annealed at 1000°C in N_2 for 20 minutes, 900°C in dry O_2 (5 minutes), wet O_2 (5 minutes), dry O_2 (5 minutes) and N_2 for an additional 30

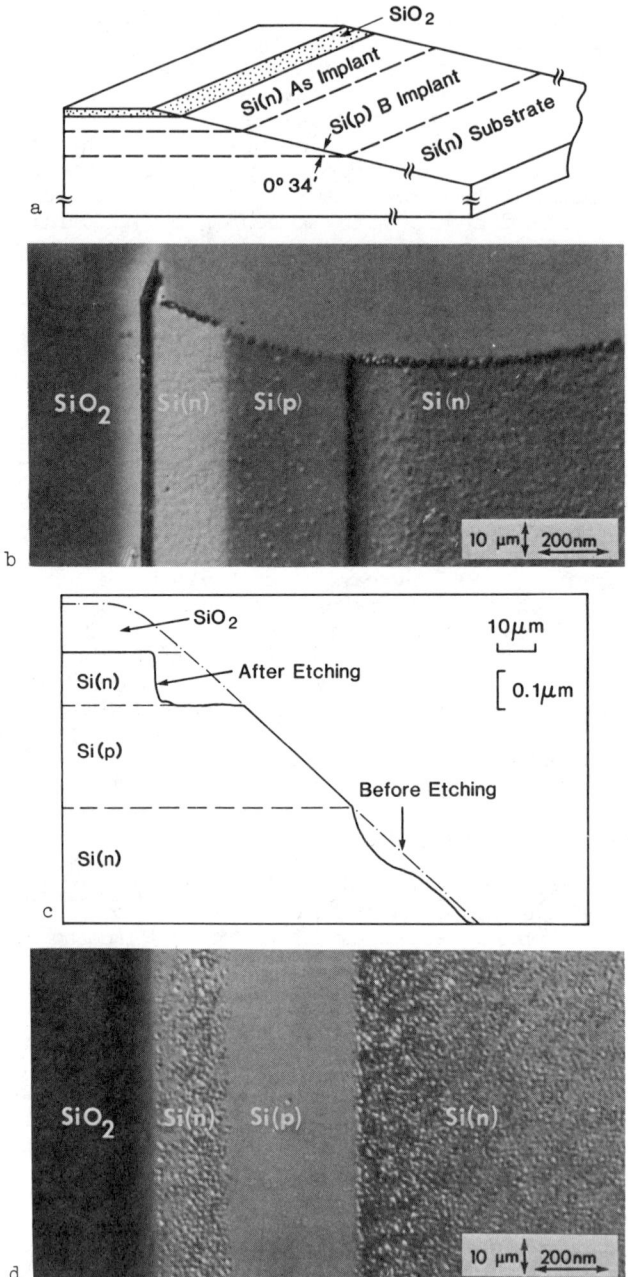

Figure 4. Location of junctions in Si n-p-n structure: a, diagram of the beveled sample; b, optical micrograph of sample after beveling and etching in Secco solution for 4 s; c, surface profile of b measured with Dektak (SiO_2 layer removed); and d, XTEM micrograph of sample after etching with 0.5% HNO_3 in HF for 55 s. (Reproduced with permission from Ref. 1. Copyright 1984 The Electrochemical Society.)

minutes. An SiO_2 layer of 125 nm was created on the surface during the multi-step anneal. The top of the micrograph in Figure 4b shows an area that had been masked with wax prior to etching.

The "goose-bumps" seen on the p-type side of the sample are the result of etch contamination by masking wax. The SiO_2 protective layer is clearly imaged along the left side of the photograph. Clearly visible is an offset of the SiO_2-Si interface demarcation due to the etching of SiO_2 and the n-type layer. The locations of n-p and p-n junctions were found to be about 200 nm and 550 nm below the Si-SiO_2 interface, respectively. Optical evaluation of junction depths at 1000 time magnification allowed us to readily measure delineated interfaces to within ±1 mm, which translates into a ±10 nm accuracy for depth measurements. For the samples with interfaces shallower than 100 nm, this remains a principal source of error in the microscopic method.

Similar to Prussin's study of buried amorphous layers in silicon (9) we have found that a stylus-type profilometer is indeed quite helpful for independent verification of measured depths. As an example, the measurement performed on a sample which had all of the oxide layer removed after Secco etching with concentrated HF is shown in Figure 4c. A profilometer trace measured along the top and beveled surfaces and parallel to the bevel slope is shown in Figure 4c. Exact correspondence of its features to the pattern shown on the microscopic photograph in Figure 4b shows actual displacement of the surface due to selective removal of n-type material and the SiO_2 layer. Location of both junctions is in very good agreement with the values measured from the micrograph. The upper junction was found to be 190 nm below the SiO_2 layer and the lower, p-n junction was located 550 nm below the layer.

Prolonged etching of the sample leads to complete removal of the top n-type layer. Upon reaching the n/p-interface the contour profile levels off, clearly indicating that the etching process is halted there. Accuracy of determination of the step height (i.e., depth of the delineated n/p-interface location) is set by the claimed accuracy of the profilometers, which is ±25Å, or by fluctuations of the interface itself. Determination of the depth of the deeper, p-n junction is less precise due to a shadowing effect of the upper p-type layer upon the lower, n-type substrate, which is being etched. Observation of several profiles measured for the same sample indicate that the variations from profile to profile indicate a more practical value of accuracy of ±10-15 nm.

<u>Junction Delineation by Spreading Resistance Probes.</u> Another relatively inexpensive method of determining the location of electrical junctions utilizes a spreading resistance probe (SRP). This technique used very closely spaced probes to step along the beveled surface of a sample and profiles the electrically active dopant concentration. The results of spreading resistance measurements conducted on the n-p-n structure described above and obtained from two different outside labs are shown in Figure 5. We have noticed that on average the method produced results relatively close to those obtained in the chemical delineation, although substantial differences were observed for the location of the shallower junction obtained from the two labs. The measured depths were about 260 nm and 160 nm for n-p and 520 nm and 470 nm for p-n junctions, respectively, for both labs. Since the measurements

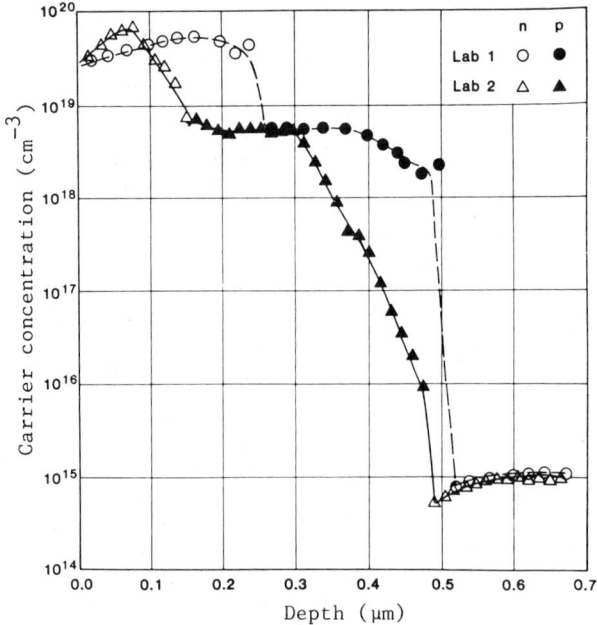

Figure 5. Carrier concentration profile of same Si n-p-n structure as shown in Figure 4. The profile was measured by two different labs using the spreading resistance technique. (Adapted from Ref. 1.)

were taken outside our lab we have not been able to determine whether an improper handling of the data or inherent drawback of the method itself is responsible for the discrepancy. We should point out, however, that the most likely factor contributing to the discrepancy is the quality of the probe tips and amount of load applied. The penetration and stress caused by the probes may significantly alter the data.

External Gettering Via Misfit Dislocations. Removal of unwanted impurities (e.g. metallic) in the active regions of semiconductor devices can be achieved by introducing dislocations away from the electrical junction in a controlled manner (11). Although the precise location and nature of such defects can only be unambiguously obtained by TEM, preferential etching of the same samples (after beveling and polishing) can again provide a quick and reliable estimate of their density and location. In the example shown below, misfit dislocations in a multi-layer Si, (0.1 to 1.0%) Ge alloy grown on a Si substrate are revealed by both XTEM (Figure 6c) and preferential etching (15 sec Secco etch) (Figure 6b). Figure 6b is a N-D/C optical micrograph and illustrates the localization or confinement of the dislocations to the epi/substrate interface. The subsequent epilayer is defect free owing to the fact that most of the lattice strain has been plastically relieved at the interface. After removal of the top Si epilayer, the plan view of preferentially etched surface shows a flat cross-grid network of misfit dislocations at the epi/substrate interface (11). Threading dislocations were found to have a very small density (2×10^3 cm^{-2}) compared to the density of misfit dislocations (10^8 cm^{-2}).

For gettering studies, an Au film of 200Å thickness was vacuum deposited and treated at 900°C for 15 minutes. The sample surface was then cleaned with Aqua-Regia solution. The SIMS profile for Au and electron microprobe profile of Ge given in Figure 6a. The optical and XTEM micrographs along with the Au and Ge profiles are shown in Figure 6. Pronounced gettering of Au was found to occur where misfit dislocations were present (11). The Au concentration stayed at the background level where no dislocations were observed. The amount of gettered Au was proportional to dislocation density i.e., higher the number of dislocations more the gettered Au. This indicated a semiquantitative relationship between gettering efficiency and defect density.

Impurity Analysis in Semiconductor Matrices: Secondary Ion Mass Spectroscopy

Three methods of surface and thin film analysis, viz., Rutherford backscattering (RBS), Auger spectroscopy/microscopy and high performance SIMS, probably form a self-consistent set of techniques for the determination of impurity distributions in electronic materials. The most quantitative of the three is RBS (which is also capable of providing simultaneously structural damage information and lattice location of impurities), but it suffers from poor sensitivity for light elements in heavy matrix, e.g., B or P in Si. In addition, it typically has poor spatial resolution and limited dynamic range of impurity detection. Auger spectroscopy/microscopy has superior surface sensitivity, can readily detect the light elements such as C, O and N, offers the

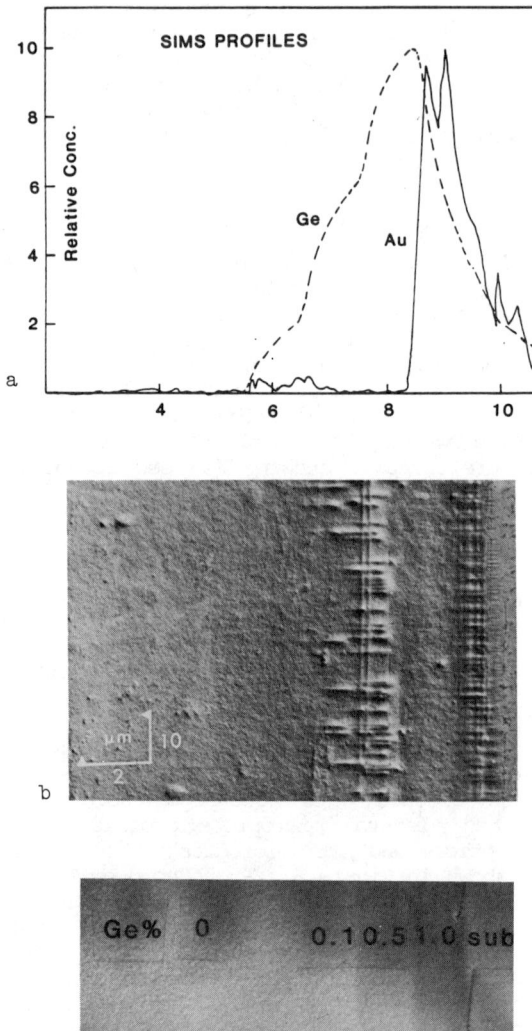

Figure 6. Verification of Au gettering preferentially at misfit dislocations: a, SIMS profile of Au; b, beveled-angled view of dislocations by etching (N-DIC optical micrograph); and c, cross-sectional TEM micrograph showing misfit dislocations. Note the excellent correlation between the three techniques utilized here. (Reproduced with permission from Ref. 11. Copyright 1984 ASTM.)

best spatial resolution (<1000Å) for elemental mapping or microanalysis and is relatively free of matrix artifacts. On the other hand, SIMS is the most artifact-prone and matrix dependent, but is the only technique with trace level sensitivity (< 1 ppm). Furthermore, many of the common matrix effects and/or artifacts disappear when using reactive ion bombardment (such as O_2) and the sample is an electronic grade crystal as is proposed in this study. Considerable work has been reported in the literature on the role and behavior of impurities in Si and GaAs at typical concentrations well below the detection limits of the other two techniques (see for example review by Zinner and references therein) (12). It should be noted, however, that the analytical capabilities of a given technique, particularly in the SIMS case, are strongly dependent upon the instrument used and its exact configuration, e.g., the availability of a Cs primary ion source. Detection limits in the range 10^{12} to 10^{16} ats/cm^3 have been observed (depending on element and matrix) when optimized analytical conditions are employed.

A powerful, however as yet under utilized, method of SIMS analysis for detecting 2 and 3-dimensional elemental distributions in materials is direct ion imaging microscopy. The ion optics of the Cameca IMS-3f maintains the spatial distribution of secondary ions from their origin on the sample surface, through a double focusing mass spectrometer, to a microchannel plate detector. This allows direct imaging of all points within the imaged field simultaneously with a spatial resolution of 0.5 to 1 µm. In generating an ion image, the outer layers of the solid are etched away. Therefore, by acquiring a series of ion images at successively larger depths. 3-dimensional elemental distributions may be characterized. In order to fully utilize the multi-dimensional information generated by the Ion Microscope, the images must be digitized and stored in computer memory. Such a digital imaging system with software specifically designed for ion microscopy is not yet commercially available and is still in developmental stage (13). The prototype digital imaging system is capable of acquiring ion images in real time and processing them at high speeds.

By digitizing ion images, it becomes possible to apply a wide range of computer image processing and reconstruction techniques to the images. Image processing is used to enhance the visual information contained in an image while image reconstruction may be used to correct ion images for artifacts introduced during the acquisition. The image depth profiles of Al and Si acquired through a selected portion of a Si transistor are shown in Figures 7-9 (13). A 4 µA O_2 primary ion beam was rastered over a 250 µm area. The images were acquired from a 150 µm diameter field centered within the rastered area. A 20 second sputtering time was used between each set of Al and Si images. A total sputtering time of 1200 seconds resulted in 40 images of each element.

The schematics of Al metallization pattern shown on the surface of the transistor is shown in Figure 7a along with the circular region which was depth profiled. The diagonal line shows where the cross-section ion images were reconstructed. A diagram of the transistor cross-section across this same line is shown in Figure 7b. Refer to this figure for the interpretation of the cross-section ion images. The ion imaging results are summarized in Figure 8. The secondary ion intensity ranges from 0

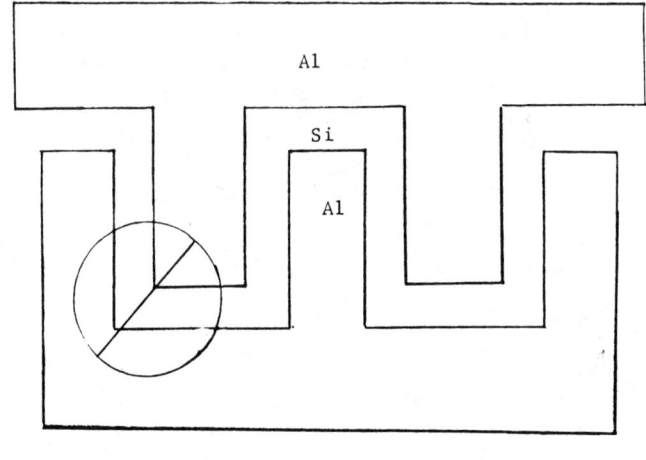

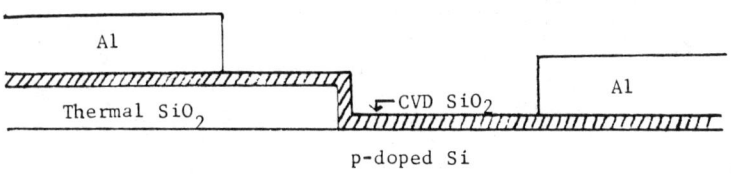

Figure 7. (a) A schematic diagram of metallization pattern on the surface of a Si transistor used for ion probe imaging. The diagonal line shows where the cross-section ion images were reconstructed. (b) Cross-sectional view of (a) across the diagonal line. (Adapted from Ref. 13.)

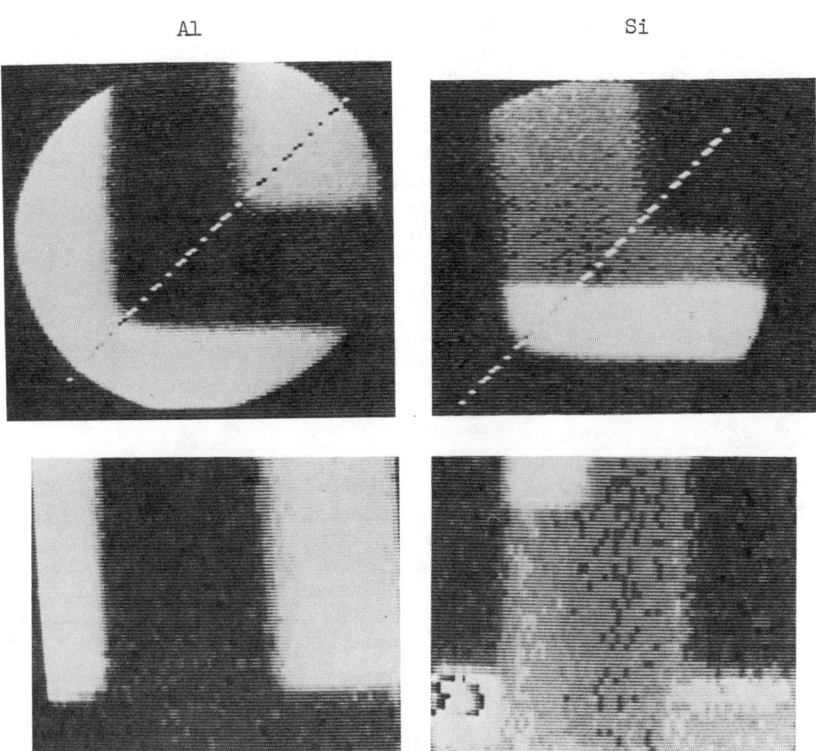

Figure 8. A composite ion image of Al and Si. See the text for details. (Adapted from Ref. 12.)

(black) to 255 (white). The diagonal line through images of Si (upper left) and Si (upper right) shows where a "slice" was taken to generate a cross-section image of the ion intensity across the line versus depth. The respective cross-section images are shown below the surface images. The Si cross-section image shows 3 distinct layers of Si. A thin region of high intensity Si appears across the top of the image, which corresponds to a 830A layer of CVD SiO_2 underneath it. Below the left portion of the CVD layer lies a second high intensity region which is the 3500A thermal SiO_2 layer. The lower intensity region is p-doped Si. The SiO_2 layers are higher in Si intensity due to the oxygen enhancement of the ion yield. Below the left Al layer (which is about 9000A deep) is another high intensity area which is again CVD SiO_2 plus thermal SiO_2.

Figure 9 shows the surface Al (upper left) and Si (upper right) images. The lower right image is the Si distribution after the Al had been sputtered away. The high intensity region near the bottom of the image is the thermal SiO_2 which lies directly below the Al metallization layer (see Figure 7b).

It is clear from above that beside the conventional usage of SIMS for elemental depth profiling in semiconductors, it has a potential of becoming a powerful tool to spatially map two and/or three dimensional distribution of impurity elements. However, spatial resolution at present is rather limited ($\geq 1 \mu m$) especially if the technique has to be extended to sub micron geometries.

Transmission Electron Microscopy

Conventional. In most of the laboratories around the world, TEM specimens are usually prepared in a conventional manner that shows only the "plan" view of the damage distribution, i.e., the plane of the foil being parallel to the implanted or otherwise processed surface. Although much useful information concerning defect structures and their nature can be obtained, the depth distribution of defects is difficult to ascertain even from stereo microscopy. This is especially true in cases where either two or more discrete layers of defects separated by a defect free region are present or where the defects of interest are buried under another more dense band of defects. Therefore, cross-sectional and beveled angle TEM specimens preparation techniques have been developed to circumvent these problems (14,15). Combining the defect depth distribution determined from these specimens with information obtained from the plan view specimens, precise three dimensional damage distributions can be obtained. These defect distributions can be used for correlation with atomic profiling and/or electrical measurements to provide a more complete representation of the material and its characteristics. The power and utility of cross-sectional technique in revealing detailed information on defects lying at different depth levels is shown in Figure 10 (16). We have chosen for demonstration purpose, the cross-sectional TEM micrograph that shows three different defect layers at three different depth levels from the surface. The defects layer labeled I contain a band of small dislocation loops followed by a relatively dense bundle of 'hair-pin' dislocations (II). The surface region (III) contained a high density of clusters, precipitates, etc. The magnified lateral view of each defect layer was revealed in the

5. ROZGONYI AND SADANA *Semiconductor Materials Defect Diagnostics* 89

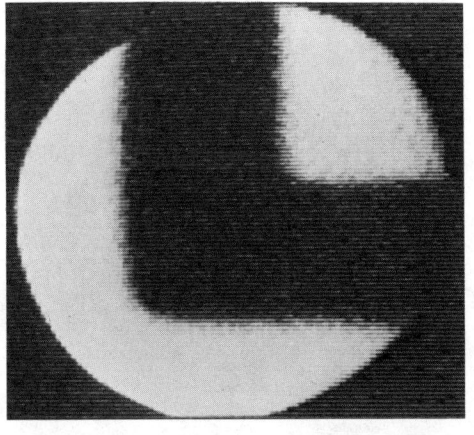

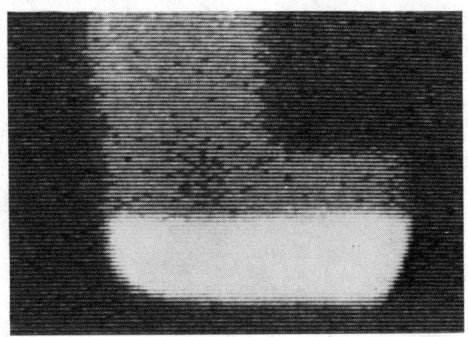

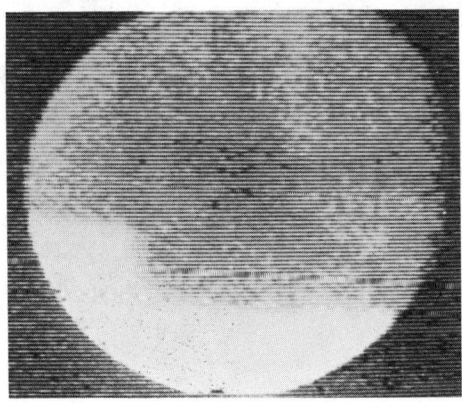

Figure 9. The surface Al (top) and Si (middle) images. The bottom image is the Si distribution after the Al had been sputtered away. (Adapted from Ref. 12.)

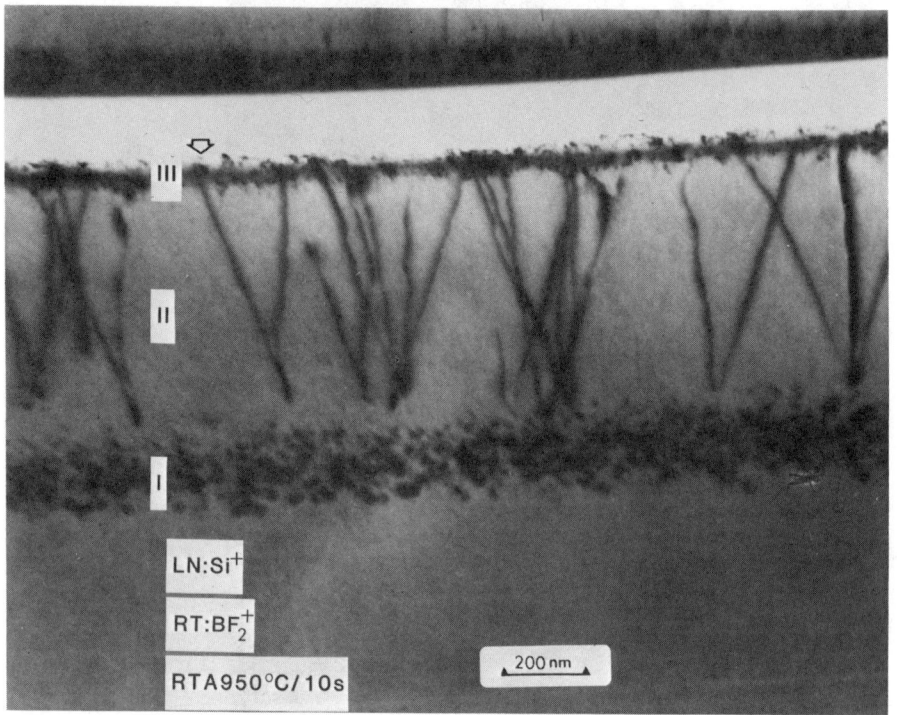

Figure 10. Cross-sectional TEM micrographs from BF_2^+ implanted pre-amorphized (100) Si which was subsequently rapid thermally annealed at 950 °C for 10 s. Three separate layers of defects (types I, II, and III) were observed at three different depths. (Reproduced with permission from Ref. 16. Copyright 1984 Applied Physics Letters.)

B-TEM sample corresponding to the above specimen. From the B-TEM samples, it was possible to conduct detailed TEM analysis which gave the following information on the above defects. Dislocation loops at depth level I were dominantly a/3 <111> type and extrinsic (extra layers) in nature. The tips of the hair-pin dislocations were of the same character as at I, however, the 'arms' of the 'hair-pins' lay along all six <110> directions. The defects in layer III were found to be stacking fault bundles and microtwins by atomic resolution transmission electron microscopy (17).

Atomic Resolution. With the recent advances in superior electron-optical design and high degree of stability in lens current, accelerating voltage, etc., imaging of atomic planes with separation ~3A is becoming a common practice for electron microscopists. The study of interfaces by atomic resolution TEM (AR-TEM) will be key to the understanding of clustering, precipitation, phase transformation, epitaxy, etc., and will dominate many other inherent materials science issues over the next decade. Profound impact of these discoveries on structure-property relationships is anticipated. For example, understanding the gettering phenomenon will obviously benefit a great deal by performing AR-TEM at dislocation cores where impurities have segregated, as well as correlating interfacial atomic microstructure with subsequent growth of epitaxial semiconductor, silicide or dielectric layers.

Two examples of the powerful application of the AR-TEM are shown in Figures 11 and 12. In the first example, structures of interfaces between silicon oxide grown on singular (Figure 11a) and 3° vicinal (Figure 11c) (111) Si is revealed (18). The following atomic details were obtained: (1) The oxide was indeed found to be amorphous (see mottled contrast characteristic of the amorphous material in Figs. 11a and 11c); (2) The SiO_2/Si interface is atomically abrupt. However, there was evidence of steps of the order of one {111} interplanar distance. The width of the terraces between positive and negative steps varies and is dependent upon defocus which indicates that these steps may not extend through the whole TEM specimen thickness. Another interesting feature to be noted is that the last row of crystal image spots is displaced as would be expected if there was a stacking fault parallel to the surface.

The oxide growth on 3° off (111) Si surface (toward [110]) resulted in the interface shown in Figure 11c. Vicinal surfaces in contact with vacuum (vapor) are expected to have steps which connect terraces of minimum surface energy, and the intersection of such steps with the terraces are themselves low energy <110> directions (19). Figure 11b shows that 3° vicinal Si-SiO_2 interface conforms to this model. The interface consists of approximately equally spaced ledges with the width of the terraces 6 nm and their height equal to .314 nm (one interplanar distance). The width of the ledges calculated for the inclination of the surface based on purely geometrical considerations fits well with the observed results. The terraces are atomically flat, however some positive-negative step pairs are also found to be present on some terraces as is shown in the figure. The structure of this interface was found to be independent of the oxidation temperature and time.

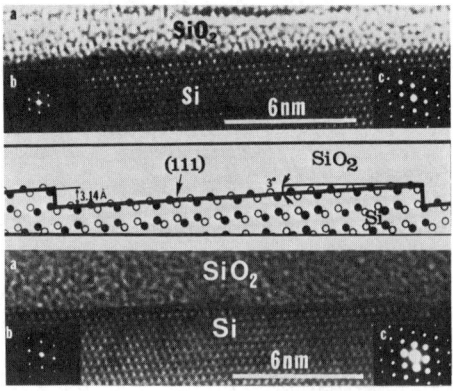

Figure 11. High resolution TEM micrographs of oxide layers on stepped Si substrate: a, native oxide; b, 1000 Å thick oxide thermally grown on Si surface 3° off (111) toward [110] (dry O_2, 1000 °C), structural steps are clearly resolved; and c, model of the stepped Si surface. (Adapted from Ref. 18.)

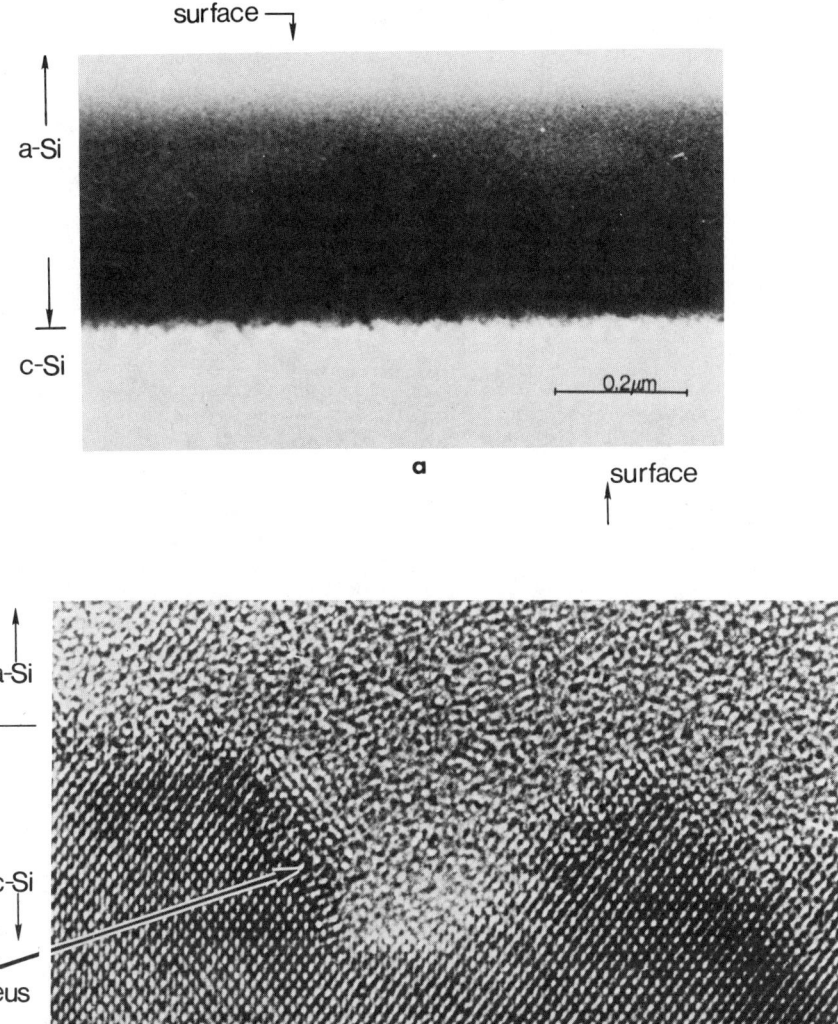

Figure 12. An amorphous-crystalline interface in Ge^+ implanted (100) Si (10^{16} cm^{-2}, 300 keV, RT implant, no anneal). (a) Bright-field cross-sectional TEM micrograph. The dark band represents an amorphous layer. (b) Atomic resolution TEM micrograph showing atomic details of the interface. Note a stacking fault nucleus at the interface. (Adapted from Ref. 22.)

The results shown above demonstrate that oxidation occurs layer by layer very uniformly over large areas of the Si surface. The roughness of the surface corresponds to the height of the individual steps (.314 nm). The oxide growth corresponds to removal of the Si atoms from the substrate surface. The observed structure of the interface suggests that this process occurs at the ledges. In the case of singular (111) surfaces formation of the ledges might be envisioned as occurring by two dimensional nucleation corresponding to the formation of an oxide island in the next layer of silicon atoms. In the case of vicinal surfaces, structural ledges are already present at the interface providing sites for oxidation. However, for too low a density of such ledges two dimensional nucleation still takes place resulting in terraces with additional positive and negative ledges. A similar process is observed for evaporation or dissolution of atoms from a surface into vapor or solution (20,21). Although the Si surface in this case is in contact with solid silica, the interface structure appears to behave very much as it would in contact with a liquid. This is perhaps not surprising because viscous flow of silica occurs above 960°C (14) and oxidations in these experiments were performed at 1100 and 1000°C.

Another example of the application of AR-TEM is shown in Figure 12. Here a 300 keV Ge^+ was implanted into (100) Si to a dose of 10^{16} cm^{-2} at room temperature (22). This produced a continuous amorphous layer of 4200A extending to the surface. Although the boundary between the amorphous and crystalline material appears to be fairly abrupt in the low resolution TEM micrograph (Figure 12a), the precise nature of the amorphous/crystalline interface was not revealed until ARTEM on the interface was performed (Figure 12b). The following interesting observations were made: (1) the interface was indeed abrupt even on atomic scale; (2) the interface showed faceting on {111}; and (3) on some facets, there were stacking faults (see Fig. 12b) presumably due to in-situ annealing. Some small crystallites ($\leq$20A) were also found to be present in the amorphous matrix within 50A from the interface. These findings have enable us to better understand the mechanisms of crystalline to amorphous transformation and the effect of dynamic annealing on such transformation.

Conclusions

Indepth understanding and precise control of shallow layers and interfaces in processed Si is made possible by the state-of-the-art analytical methods and equipment. The techniques, ranging from simple and inexpensive angle-lapping-and-chemical-delineation to sophisticated ion or electron beam probing of semiconductor samples, provide quite precise information about the location, structure, chemical composition and electrical properties of the layers and their boundaries. The top view, the bevel angle top view, and the cross-sectional projections of the sample with respect to the analyzing probe have been utilized in the analyses. Examples of shallow implanted n-p-n junction, $Si-SiO_2$ interface, extrinsic gettering of the Au ions at misfit dislocations, and amorphization of Si surface with Ge^+ implant were used to illustrate the capabilities of the discussed methods.

Acknowledgments

The authors wish to express their gratitude to Witold Maszara and Calvin Carter of North Carolina State University for many useful discussions, for providing a number of figures in this text, and for their active collaboration. We would also like to record our appreciation for Pamella Camp for being patient with typing the manuscript and for doing a fine job.

Literature Cited

1. W. Maszara, C. Carter, G.A. Rozgonyi, VLSI Science Technology, K.E. Bean and G.A. Rozgonyi (eds), The Electrochem. Soc., Inc., 36-47 (1984).
2. C. Carter, W. Maszara, G.A. Rozgonyi, D.K. Sadana, SPIE Conf. 463, D.K. Sadana and C.M. Lampert (eds), 1984.
3. E.A. Bauser, G.A. Rozgonyi, App. Phys. Lett. 37 1001 (1980).
4. G.A. Rozgonyi, H.J. Leamy, T.T. Sheng, G.K. Caller, pp. 492 in Semiconductor Characterization Techniques, P.A. Barnes and G.A. Rozgonyi (eds), The Electrochem. Soc., Inc. (1978).
5. G.A. Rozgonyi, H.J. Leamy, T.T. Sheng, G.K. Celler, pp. 457 in Laser Solid Interactions, S.D. Ferris, H.J. Leamy and J.M. Poate (eds), AIP Proc. 50 457 (1979).
6. E. Bauser, G.A. Rozgonyi, J. Electrochem. Soc. 129 1782 (1982).
7. D.C. Miller, G.A. Rozgonyi, Handbook on Semiconductors 3, S.P. Keller (ed), North Holland Publishing Co., 217-246 (1980).
8. C.P. Wu, E.C. Douglas, C.W. Muellar, R.J. Williams, J. Electrochem. Soc. 126 1982 (1979).
9. S. Prussin, D.I. Margolese, R.N. Tauber, J. App. Phys. 54 2316 (1984).
10. F. Secco D'Argona, J. Electrochem. Soc. 119 948 (1972).
11. A.S.M. Salih, H.J. Kim, R.F. Davis, G.A. Rozgonyi, Semiconductor Processing, ASTM STP 850, D.C. Gupta (ed), American Society for Testing and Materials 1984 (San Jose, CA).
12. E. Zinner, J. Electrochem. Soc. 130 199c (1983).
13. S.R. Bryan, D.P Griffis, W.S. Woodward, R.W. Linton, J. Vac. Sci. & Tech. (to be submitted).
14. H.R. Pettit, G.R. Booker, Electron Microscopy & Analysis, W.C. Nixon (ed), Inst. of Phys. (London) 10 290-293, 1971.
15. T.T. Sheng, R.B. Marcus, J. Electrochem. Soc. 127 737 (1980).
16. C. Carter, W. Maszara, D.K. Sadana, G.A. Rozgonyi, J. Liu, J. Wortman, App. Phys. Lett. 44 459 (1984).
17. T. Sands, J. Washburn, R. Gronsky, W. Maszara, D.K. Sadana, G.A. Rozgonyi, (submitted to App. Phys. Lett).
18. J.H. Mazur, Proc. 41st Ann. Meeting of the Electron Microsc. Soc. of Am. 106 (1983).
19. B.Z. Olshanetsky in Studies in Surface Science & Catalyzers 9, M. Laznicka (ed), Elvisier Sci. Pub. Co. Amsterdam (1982) p.214.
20. D.K. Burton, Phil. Trans. Roy. Soc. 243A 299 (1950).
21. J.P. Hirth in Energetics in Metallurgical Phenomenon 2, W.M. Mueller (ed), Gordon & Beach, NY (1965) p.1).
22. T. Sands and D.K. Sadana (unpublished).

RECEIVED December 1, 1985

6

Applications of Secondary Ion Mass Spectroscopy to Characterization of Microelectronic Materials

Mary Ryan-Hotchkiss

Tektronix, Inc., Beaverton, OR 97077

> In this review the various modes of SIMS and examples of their applications are discussed. SIMS depth profiles are widely used to study dopant profiles and intermetallic diffusion. The extreme surface sensitivity and low concentration detection limits of SIMS make it useful for investigation of substrate and metallization cleaning processes. SIMS elemental imaging is also used in contamination studies. The ability of SIMS to provide isotopic information has allowed elegant mechanistic studies. The identification and determination of the relative abundance of various molecular or elemental species by SIMS is applicable to the development, characterization, and understanding of microelectronic processing. The capability of SIMS in the area of quantitative analyses is also discussed.

The increasing requirements on microelectronic devices to provide larger areas, greater density of circuits, higher production yields, and/or higher speed of device operation necessitates the development of new materials and processes. This in turn requires the application of sophisticated techniques for problem solving, and for materials and process characterization. Secondary Ion Mass Spectrometry (SIMS) is one such technique. This paper provides a description of the three major types of SIMS instruments and the modes in which those instruments can be used. Examples of the types of information that can be obtained using the various modes of SIMS are presented which are drawn from microelectronic technology.

SIMS is a surface analysis technique in which primary ions bombard a solid surface and cause secondary ions to be emitted. The emitted ions are collected, energy filtered, mass analyzed and detected. In this paper surfaces other than solid surfaces, and excitation sources other than ions, e.g. fast atoms or photons, have not been included in the category labeled SIMS. Thorough discussions of the theory and the effect of instrumental and empirical variables in SIMS and diagrams of the various instrument types have been published elsewhere (1-3).

SIMS INSTRUMENTS

Figure 1 shows a block diagram of a generalized SIMS instrument. The essential instrument subunits are the primary ion source, the sample, the mass analyzer and the secondary ion detector, all of which are typically contained in or connected to an ultra high vacuum chamber. Many variations are possible in each of the SIMS instrument subunits. Different primary ion sources may be used to provide different species of primary ions, for example, Ar^+, O_2^+, O^-, N_2^+ and Cs^+. Different ion sources may also be optimal for primary ion beams of a specific energy range, spot size or current density. The primary ion column provides the focussing, rastering and aperturing of the primary ion beam. It is usually necessary to raster the primary ion beam during the analysis to average non-uniformities in the beam over the sputtered area and to sputter a flat bottomed crater. Since some ion sources emit many types of particles the primary ion beam may be electrostatically deflected through a small angle in the primary ion column to separate the primary ions from undesirable uncharged species. Magnetic mass analyzers may also be used in the primary ion column to select the primary ion with the desired mass and charge. Variations in the angle or spacing of the sample from the ion optics may be used to optimize the sputtering yield, the ion yield or the area of the sample to be analyzed. When an inert gas is used as the primary ion, an oxygen jet may be placed near the sample to provide a uniform oxygen coverage of the sample surface. Saturating the surface with oxygen through the use of an oxygen jet or an oxygen primary ion beam enhances the ion yield of many species.

The secondary ion optics portion of the SIMS instrument includes direct imaging optics for the direct imaging instrument, which is discussed later, or extraction lenses and various types of energy filters for other types of SIMS instruments. Magnetic and quadrupole mass spectrometers are both common mass analyzers. The mass analyzed secondary ions may be detected by a Faraday cup, an electron multiplier or a phosphor screen. The phosphor screen may be coupled to film, a TV camera, or a solid state imager. Electrical signals output by the detector may be handled in either analog or pulse counting mode. The signals may be digitized and stored for subsequent signal processing. The data is commonly displayed in the form of images of the sample surface, mass spectra, or plots of ion signal intensity versus time which may be equivalent to depth.

Charge neutralization and secondary electron detection are provided in some instruments. For low conductivity samples that accumulate a positive charge under positive ion bombardment,

charge neutralization may be accomplished by flooding the sample
with electrons from an electron flood gun or by bleeding the
charge off by some other means. Secondary electrons are also
emitted by the sample during ion bombardment. These electrons may
be detected by a channeltron or an electron multiplier. This
secondary electron signal can be used to form a secondary electron
image of the sample. If this signal than be correlated with the
rastering of the primary ion beam the SIMS instrument will
function as a Scanning Ion Microscope (SIM) analogous to a
Scanning Electron Microscope (SEM).

There are three basic types of SIMS instruments: 1) the ion
microprobe, 2) the direct imaging instrument, and 3) the static
SIMS instrument. An early ion microprobe was described by Liebl
(**4,5**). In microprobe instruments a focussed primary ion beam is
rastered over the sample surface to sputter a uniform flat
bottomed crater. An ion image may be obtained by synchronizing
the x-y raster of the primary ion beam on the sample surface with
the x-y raster of a CRT display while the mass analyzer is tuned
to the mass of that ion. The z modulation on the CRT display
represents the intensity of the mass analyzed secondary ion signal
at the corresponding x-y location on the sample. A frame of x-y-z
data could also be digitized and stored directly in a digital
memory and be available for subsequent display or digital signal
processing. Not all ion microprobes are configured to obtain ion
images. In some SIMS instruments the primary ion beam is
asynchronously or semi-randomly rastered over the sample surface.
In these cases, or if the primary ion beam diameter is large in
comparison to the analysis area, it may not be possible or
practical to obtain ion images. Most commercial instruments, and
many laboratory built instruments (**5-8**) have been the ion
microprobe type. Minimum primary ion beam diameter in microprobes
is about 2 μm. Beam diameters over 100 μm are not uncommon for
ion probe instruments.

The first raster scanning ion probe instrument to use a
quadrupole mass spectrometer was reported by Wittmaack in 1976 (**8**)
and now many SIMS instruments incorporate a quadrupole mass
spectrometer (**9**). The quadrupole has the advantage of being
economical and of being able to scan through several hundred mass
units in a period of seconds. Rapid mass scanning is an important
characteristic because the sample surface is being eroded during
SIMS analysis. If the mass spectrum of a thin layer is desired,
the layer may be exhausted before a slow scanning mass
spectrometer has reached the heavier masses. The mass range of
the quadrupole is often less than magnetic mass spectrometers
commonly used on SIMS instruments, and its sensitivity is mass
dependant and variable. An energy prefilter is required before
the quadrupole because good mass resolving power is obtained only
when the ions entering the quadrupole have an energy spread of
only a few eV.

The second type of SIMS instrument, the direct imaging type,
is based on the design of Castaing and Slodzian (**10**). In the
direct imaging instrument an area 1 to 300 μm in diameter is
bombarded by the primary ions. The secondary ions are extracted
by an electrostatic immersion lens which maintains a point to

point location image of the origin of each secondary ion. Thus the lateral resolution is independent of the primary ion beam diameter. A minimum lateral spatial resolution for ion images of about 2 µm is obtained in the commercial instrument. Direct imaging instruments typically use a magnetic mass spectrometer, have very high mass resolution and good sensitivity. A discussion of the relative advantages of the direct imaging and ion microprobe instruments for imaging is presented in the section IV. E.

In static SIMS the primary ion beam energy and current density are sufficiently low that 10^3 or 10^4 seconds are required to remove a monolayer of material (11). Typically a large diameter ion beam is used to bombard a large homogeneous sample surface. Static SIMS is the most surface senstive SIMS mode, and provides the gentlest ionization conditions such that even very large molecules can be ionized essentially intact (12). When a high energy ion beam sputters the sample, surface reactions in the sample or in the gas phase above the sample may result in a variety of molecular species which were not originally present in the sample. The lower energy and low current density (10^{-9} A/cm^2) minimizes these molecular rearrangements. In static SIMS the primary ion beam need not be rastered and depth profiles and ion images are not typically obtained.

CHARACTERISTICS OF SIMS

SIMS is more sensitive than the other common surface or interface analysis techniques of Auger, X-ray Photoelectron Spectroscopy, Rutherford Backscattering Spectroscopy, or Energy Dispersive X-ray Analysis. Detection limits and background signal levels for a large number of semiconductor materials have been reported under typical operating conditions (13). Table I lists the detection limits for a number of dopants used in semiconductors, obtained under optimized conditions.

Other attributes of SIMS include a wide dynamic range, good depth resolution, the ability to detect all elements, the ability to differentiate between different isotopes of the same element, good elemental specificity, and exceptionally good sensitivity for the low atomic number elements. An excellent discussion of the strengths and weaknesses of SIMS has been provided by Magee (15). Comparisons between SIMS and other surface analysis techniques have been published (16).

Because microelectronics problems often require measurement of concentrations that vary over orders of magnitude, efforts have been made to maximize the dynamic range of SIMS measurements. The effects of various instrumental enhancements on the dynamic range of concentrations that can be measured in a depth profile has been illustrated (6). A general discussion of depth profiling and its applications are presented in Section IV.A. Most of the instrumental enhancements allow a greater dynamic range because they allow the instrument to discriminate against crater edge effects. Crater edge effects occur when secondary ions which originate from depths shallower than the maximum crater depth are mistaken for secondary ions from the bottom of the crater. The

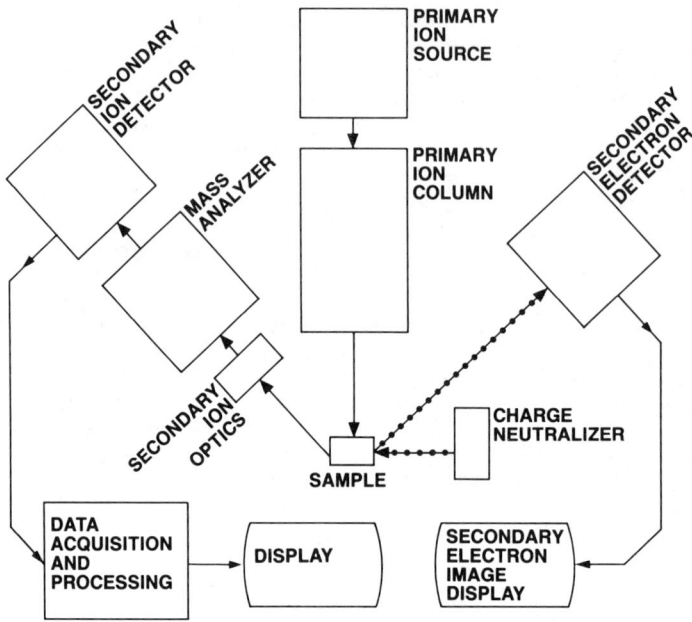

Figure 1. Block Diagram of a SIMS Instrument.

Table I. SIMS Detection Limits for Dopants (3, 14)

Matrix	Element	Primary Ion Beam	Detection Limit (atoms/cm^3)
Si	H	Cs+	10^{17}
	C	Cs+	10^{16}
	N	Cs+	10^{17}
	O	Cs+	3×10^{16}
	B	O_2^+	10^{14}
	P	Cs+	3×10^{15}
	As	Cs+	3×10^{15}
GaAs	O	Cs+	5×10^{17}
	Si	O_2^+	3×10^{15}
	S	Cs+	10^{15}
	Cr	O_2^+	5×10^{13}
	Mn	O_2^+	4×10^{13}
	Fe	O_2^+	5×10^{14}
	Zn	O_2^+	10^{16}
	Se	Cs+	3×10^{13}
	Te	Cs+	2×10^{13}

instrumental improvements include rastering the primary ion beam over an area on the sample larger than the area from which analytical date is taken, use of a secondary ion extraction lens which accepts ions from only the central portion of the flat bottomed crater, and electronically gating the ion counting electronics to be active only when the primary ion beam is rastering the central portion of the crater. These steps decreased the interfering background signal by more than a factor of 650. A boron implant profile was determined from a peak concentration of 10^{20} atoms/cm^3 to the background signal level equivalent to about 10^{16} atoms/cm^3 (6).

A dynamic range of almost 10^6 has been demonstrated for depth profiles of B in Si (17). The dominant process limiting the dynamic range in depth profiles in this study was found to be the presence of energetic neutrals in the primary ion beam which were not focusable and sputtered outside the analysis area. The background signal was related to the number of neutrals in the primary ion beam and was found to be directly proportional to the pressure in the ion gun section. Other factors affecting the dynamic range are the contamination of the extraction lens with sputtered material if it is too close to the sample, and dead time in the counting electronics which was expected to be significant at about 10^7 counts/S (17).

Depth resolution in SIMS is limited to about 1nm (14,15) by the sputtering process and varies with the material being sputtered (14,18). The best depth resolution is obtained for heavier bombarding ions at lower energies (16,18,19). An example of good depth resolution for Ta_2O_5 on Ta is 3.5nm, which was less than 1.1% of the sputtered depth (15).

There are characteristics of SIMS that complicate the interpretation of SIMS data. In comparison to techniques such as Rutherford Back Scattering (RBS) in which the physics are well understood and the scattering cross sections have been very accurately determined, greater care needs to be taken during a SIMS analysis to obtain quantitative or even semi-quantitative results. The concentration sensitivity in SIMS varies widely from element to element and the sensitivity to a single species may vary widely from one matrix to another. Several approaches have been taken to compensate for matrix effects. In some cases saturating the surface of the sample with oxygen during the analysis has been found to minimize the changes in ion yields that sometimes occur at interfaces (20).

The effects of various matrices in the analysis of layered semiconductors was examined by Galuska and Morrison (21). They found that proper calibration of ion yields and sputtering yields made quantitative concentration and depth measurements possible for $Al_xGa_{1-x}As$ matrices.

A variety of approaches to quantative SIMS analysis have been discussed by McHugh (2), Werner (3) and by Wittmaack (22), including the use of sensitivity factors, the calculation of ion yields as a function of a relevant physical parameter, and the use of internal standards. For many sample types and virtually all semiconductors accurate standards closely resembling the analytical sample have been prepared (15). With these standards

absolute accuracies on the order of 5-10% can be achieved
(2,13,15,23). With quadrupole SIMS the relative and absolute
signals for various masses depend on variable instrument
conditions. A metallic glass sample, $Fe_{30}B_{17}W_3$, has been
used to calibrate a quadrupole over a wide mass range (24).
Alkali iodide salts have also been proposed as standards for
quadrupole instruments (25). In summary, to obtain accurate
quantitative results with SIMS, a well prepared similiar standard,
careful instrumental control and characterization, and an
understanding of the sample being analyzed is required.

SIMS is also subject to problems caused by the sputtering
process. The problems associated with all the analytical
techniques which rely on sputter removal have been recently
reviewed and discussed by Zinner (14,26). Some of the ion beam
effects are preferential sputtering, atomic mixing (2), radiation
induced diffusion (2) and charging effects. Reduced primary ion
beam energy can minimize mixing. The charging effects and methods
to compensate for them have been illustrated by Magee (27) and
others (28-30).

MODES OF SIMS ANALYSIS

Six different modes of SIMS analysis have been selected for this
discussion and will be illustrated by applications involving
microelectronics materials and processes. The six modes are:
a) depth profiling, b) mass spectra, c) isotopic studies,
d) static SIMS, e) ion imaging, and the f) positive/negative ion
mode. Some of the modes involve only a variation in the way the
data is obtained or handled but others require hardware
modifications. For example, isotopic studies is only a specific
application of the mass spectral mode, but static SIMS requires a
special low energy, low current density primary ion beam.
Positive/negative ion mode requires the switching of potentials on
the secondary ion optics and changes from the normal electron
multiplier circuitry to collect negative ions instead of the more
commonly detected positive ions. The positive/negative ion mode
may be used at the same time as any of the first five modes.

Depth Profiling. By far the most common mode of SIMS analysis is
depth profiling. Since material must be sputtered from the
surface during SIMS analysis, SIMS is easily applied to situations
where successive layers or interfaces need to be examined.
During depth profiling the primary ion beam is rastered across the
sample to create a flat bottomed crater which is large in
comparison to the primary ion beam. This sputters away successive
layers and allows the ejected secondary ions to be mass analyzed.
To provide optimum depth resolution and dynamic range, only ions
from the central portion of the crater should be allowed to
contribute to the analytical signal. Thorough reviews of depth
profiling by SIMS are available (14,26). The most common
application of SIMS depth profiling to microelectronic materials
is the determination of dopant profiles from diffusion and ion
implantation processes. Empirical SIMS measurements have been

indispensable and continue to be important in the development of models for range statistics in ion implantation. SIMS depth profiles are also used to monitor and develop an understanding of the diffusion of dopants during laser and thermal annealing processes. Metallization and thin films have also been investigated by SIMS. In addition SIMS depth profiles are useful for failure analysis and problem solving.

Dopant Profiles. A compilation of SIMS depth profile studies of dopants in semiconductors and insulators was published by Zinner (26) in 1980. Profiles of very low energy implants by SIMS have been recently reported (31). For the determination of these shallow implants both the sputtering rate and the secondary ion yield must stabilize very near the surface. Both of these conditions could be met by the combined use of a jet of molecular oxygen directed at the sample surface and the use of a 5 KV argon primary ion beam. A 5 KV As implant which peaked at about 40 angstroms depth and a 15 KV As implant which peaked at about 120 angstroms were both successfully characterized (31). In another study of shallow implanted layers SIMS depth profiles showed that lower energy implants experienced greater amounts of channeling (32). The controlling factor in the shape of the implant profile appeared to be the degree of amorphicity of the Si prior to and during implantation. Channeling is less significant at higher implant energies because the Si becomes amorphised during the implantation process. SIMS profiles of B implanted into amorphised Si had a much narrower and more nearly Gaussian distribution than implants into crystalline Si, indicating that channeling and not SIMS ion beam effects were causing the majority of the broadening of the B depth distribution (32). When optimizing the low energy implant process the extent of channeling must be empirically determined because a minor change in process variables such as a slight change in wafer orientation or poor collimation of the implanting beam could cause varying amounts of channeling. Due to the low concentrations and the need for good depth resolution SIMS is the method of choice for this type of study.

A survey of SIMS depth profiles was performed for 25 elements which had been implanted into several semiconductor substrates (33). Good agreement was obtained between theoretical and empirical values for projected ranges and standard deviations of the ranges. These researchers infer that the various ion beam effects, e.g., knock on, that can interfere in depth profiling were not serious and that SIMS could be valuable for studies of matrices where methods for theoretical calculations are not readily available (33) and for the investigation of channeling (34).

SIMS is also predicted to be of benefit with high energy (exceeding 1MeV) ion implantation because the ion ranges are less precisely known at these energies (35). SIMS has been used to verify the calculated dopant profile for a multiple implant of Si into GaAs (36). A nearly flat distribution of Si was obtained to a depth of 1 μm by using 5 successive implants with energies from 40 to 900 KeV. Good agreement was obtained between the calculated and experimental SIMS profile. In another study, range parameters were determined from SIMS depth profiles of S implanted GaAs at energies of 40 to 600 KeV. Significant differences were found from LSS calculations with the greatest differences at the lower ion energies (37). Continued application of SIMS depth profiling will provide a better understanding of these processes.

Annealing. SIMS depth profiles are frequently utilized during the development and investigation of annealing processes. Some of the diffusion processes are too complex and not now well enough understood to be accurately predicted. SIMS depth profiles were used to study the effect of annealing ion implanted phosphorus in Si (38). Profiles from samples annealed at different temperatures show the presence of two distinct components of the P distribution. The first region starts at the surface and extends to the depth originally amorphised by the ion implantation process. The second component extends beyond the amorphous/crystalline (A/C) interface. The presence of these two components of the phosphorus distribution were explained by assuming that during epitaxial regrowth of Si, every atom (P or Si) in the amorphous material is placed on a substitutional site. P atoms in the recrystallized material could diffuse during the subsequent annealing by the diffusion mechanism normally operative at that temperature. However, at the 850 degree C annealing temperature no significant diffusion is expected. On the other hand P atoms below the A/C interface are randomly distributed in interstitial spaces in damaged single crystal Si and migrate rapidly into the Si by a interstitial diffusion mechanism until they occupy substitutional sites. Carrier concentration measurements confirmed the recrystallization and interstitial diffusion model.

Depth profiles of dopants in GaAs have also been used to investigate the effects of annealing. Depth distributions were obtained for S implanted into GaAs as a function of annealing temperature, ion energy, ion fluence and encapsulation (37). It was found that S implants into GaAs did not have predictable depth distributions after annealing and the details of the profiles after annealing varied with the annealing environment, i.e. whether the GaAs had a SiO_2 cap, and with the implant fluence. Empirical measurements were obviously necessary in this study and SIMS is the only surface analysis technique that could have obtained the depth profiles covering a concentration range from 10^{16} to 10^{21} S atoms/cm^3. In a similar study SIMS depth profiles of S in GaAs were obtained before and after annealing (39). As in reference (38) the tail region of the ion implanted dopant profile exhibited faster diffusion than the near surface region. Good agreement between multizone model calculations and the experimental SIMS data was obtained for cases with doses of S in GaAs between $4x10^{13}$ and $4x10^{15}$ atoms cm^{-2} with energies of 120 and 300 KeV.

The effect of encapsulation during annealing was also found to be important in the redistribution of Cr in GaAs (40). SIMS depth profiles showed that Cr was depleted near the surface for both capped and capless annealing. Surprizingly, capped samples exhibited Cr depleted regions about 4 times as deep as those observed in capless annealing. It was proposed that a thermal mismatch between the capping film (SiO_2 or Si_3N_4) and the GaAs caused a stress field that enhanced the Cr out-diffusion. The low detection limit of SIMS for Cr in GaAs (5×10^{14} atoms/cm^3) was useful for these studies. Matrix effects which occurred at interfaces and when going from capped to substrate regions complicated the interpretation of some depth profiles (<u>40</u>).

The out diffusion of dopants from highly doped wafer areas into the epitaxial layer during epitaxial deposition or prebake is called autodoping. SIMS depth profiles during autodoping showed a pile up of As at the Si substrate surface. Spreading resistance measurements showed that the source of the autodoping to be electrically inactive As (<u>41</u>).

The mechanism and rate of diffusion of Te in silicon was investigated by SIMS depth profiles (<u>42</u>). Wafers with Te diffusions that had been annealed for various times and temperatures were examined. SIMS was preferred over electrical measurements for this diffusion study because the electrical techniques yield information on the effect of several dopants rather than just the one of interest. In this study significant levels of Se were contained in the Te source and Se was detected on the sample surface after the Te diffusion. This study produced the first report of the diffusion coefficient of Te in Si. The diffusion coefficient was found to be 2×10^{-15} to 6×10^{-12} cm^2/s which is several orders of magnitude lower than that for the other chalcogens, S and Se, in Si. Some indications were found that the diffusion mechanism for Te in Si is primarily substitutional, whereas interstitial mobility dominates for S and Se. The diffusion of ion implanted Se in GaAs was also measured by SIMS. This work showed that the annealing process could not be modelled as a simple diffusive redistribution. A "dead-time" of 0.25 to 1.5 h, during which the Se diffuses only imperceptibly, was observed at the beginning of the annealing processes (<u>43</u>).

Laser annealing processes are much less well characterized than the thermal annealing processes discussed above. SIMS depth profiles have been used to study these processes (<u>44</u>). The use of excimer lasers for semiconductor processing was recently reported. SIMS depth profiles of B in Si after XeCl laser annealing showed a nearly flat B distribution to a junction depth of about 0.9 μm with an initial implant peak of about 0.3 μm (<u>45</u>). SIMS has also been used to determine the As concentration profile after liquid state diffusion from laser melted spin-on arsenic glass. In this case laser diffusion was desireable because it avoided the thermal stress-induced damage to large thinned wafers that could occur during heating in a furnace, (<u>46</u>).

<u>Thin Films</u>. SIMS depth profiles are also used in the development of new thin film processes. These processes are being investigated to produce higher speed and greater device density in VLSI applications. Molybdenum metal gates are being investigated

because they have the advantage of high intrinsic conductivity compared to polysilicon and can be easily patterned with photoresist and simple metal etchants. However, Mo films need protection from oxidation and mobile ion contamination. Since molybdenum nitride can provide protection against these problems, the formation of molybdenum nitride under various conditions was investigated using SIMS and other techniques (47). SIMS depth profiles showed that the molybdenum nitride film thickness increased with temperature in the RF reactor and was related to the time and concentration of ammonia. Nitride films from 500 to 3000 angstroms thick were formed from the 3000 angstroms of Mo initially deposited (47). Numerous SIMS depth profiles were obtained during the investigation of the effect of phosphorus implantation on molybdenum disilicide films (48). Highly conductive, refractory metal silicides are also under investigation as substitutes for doped polycrystalline silicon. Phosphorus has been reported to help stabilize the silicide-gate threshold voltage and help form reliable contacts between the silicide and silicon substrate. In addition, the silicide gate would be ion implanted if it were used as an ion implantation mask. SIMS analysis determined that argon and carbon, originating from the sputter process which deposited the silicide film, were incorporated in the molybdenum silicide. Depth profiles showed that although the level of phosphorus became fairly uniform within the film, annealing caused it to accumulate at the outer interface. The orignally uniform carbon distribution became very irregular particularly for the implanted samples. In the SIMS analysis a significant mass interference suspected to be SiH affected the measurement of ^{31}P. The plot of the signal at mass 31 versus the P dose was linear but had a y intercept much greater than zero. Since the plot was linear the assumed constant level of interference which caused the positive y intercept could be subtracted out.

In a related study, the deposition of molybdenum silicide from a composite target was compared to that cosputtered from elemental targets (49). SIMS depth profiles showed that the concentration of carbon and argon was lower and the level of oxygen was higher in the cosputtered film. Sheet resistance and C-V measurements for both types of films were also compared and were found to be not grossly different.

SIMS depth profiles have been used in a variety of other microelectronics related applications. Information useful in solving problems can sometimes be provided by SIMS. For example, some lots of nitrided silicon wafers were found to have good etching properties but others had a thin nonconductive film remaining in the etched areas. A depth profile showed that the poor etching lots had an increased Al concentration at the substrate nitride interface. Adding a specific surface cleaning process for Al before the deposition of silicon nitride cured the problem (50). SIMS depth profiles detected H at the interface between a typical MOS gate oxide and the silicon substrate at an areal density estimated to be 2×10^{13} atoms per cm^2 (15). The difficulties involved in H measurement by SIMS have been discussed

(51 - 53) but only SIMS, of all the surface analysis techniques could have performed that analysis (15). SIMS depth profiles have also been used to study the defect/damage density resulting from ion implantation (54), to determine the correlation between surface contaminants and line dislocations in MBE (55), to determine the effect of B concentration on the etch rate of (100) silicon (56), and to measure the Cl concentration and redistribution in SiO_2 (57).

Mass Spectra. Whereas in the depth profiling mode the mass spectrometer is used as a filter to obtain only one or a few masses for detection, in the mass spectral mode the mass spectrometer is used for the identification and comparison of a wide range of masses. A survey mass spectrum is often performed as the first step in any SIMS analysis and determines the presence of various elements and species on or near the surface and the relative intensity of their respective signals. Further work would be required to relate signal intensities to absolute or even relative concentrations. However, the relative intensities observed in isotopic distributions for elements or in cracking patterns for molecules are useful in species identification.

Both organic and inorganic species can be identified from SIMS mass spectra (12). One specialized commercial SIMS instrument was designed to determine the molecular weight and structure of nonvolatile organic compounds (58). Although molecular SIMS has not been widely exploited for semiconductor process investigations, it does have the potential to be valuable for problem solving. SIMS could be used for the identification of specific contaminants or the types of compounds on IC surfaces. For example, the identification of carbon contamination on an IC surface as a hydrocarbon was based on its SIMS mass spectrum showing the $C_nH_m^+$ cracking pattern typical of hydrocarbons (59). Both the type of instrument used and the substrate on which the sample is adsorbed can affect the cracking pattern obtained (12), thus good standards are necessary for accurate compound identification.

Surface cleanliness has been evaluated in a variety of studies using SIMS mass spectra. SIMS has been utilized to define the levels of impurities on critical surfaces at various stages of device manufacture (60). Mass spectra of inorganic contaminants were used in a comparison of four methods to clean Si wafers. The four cleaning methods were FSI-A, aqua regia, fuming nitric acid, and Piranha + HF. The FSI-A cleaning gave the lowest relative values for the most contaminants but Piranha + HF also provided good results. The different cleaning methods removed different elements to various extents (50). In another case anomalous Mo was discovered on the surface of Si wafers by SIMS analysis. A possible source for the Mo, was Mo rods in some of the annealing ovens (61).

SIMS mass spectra also contributed to the development of a molecular beam epitaxy (MBE) system. Initial mass spectra from GaAs from the system indicated contamination from Li, Na, K, Ca, B, Al, Si, Cr, and Fe. By following up on the problem areas which the SIMS spectra suggested and by changing some of the materials used in the system only low levels of Na, AL, K and Ca were finally detected in the GaAs produced in the finished MBE system (31).

Isotopic Studies. In this analysis mode the mass spectrometer is used to determine the relative abundance of different isotopes of the same element. SIMS is the only common surface analysis technique that can distinguish the isotopes of both the light and heavy elements. This mode is advantageously used in mechanistic studies using isotopically labelled samples. By using different isotopes of the same element, experiments can be performed without concern for the differences in chemical effects between the species.

Coleman et al. (62), elegantly utilized isotopically labelled H_2O in a study of the growth mechanism of anodically grown oxides of GaAs. They sought to determine whether oxygen or Ga and As were the mobile species during oxide growth. A GaAs sample was anodically oxidized first in $H_2^{16}O$ then in ^{18}O enriched H_2O. The depth profile for this sample is shown in Figure 2A. Figure 2B is a depth profile for a GaAs sample anodically oxidized in the opposite order, i.e., first in ^{18}O enriched H_2O then in pure $H_2^{16}O$. The figures clearly show that the oxygen isotope near the surface of each anodized specimen is incorporated from the last electrolyte from which the oxide was grown. These depth profiles confirmed that the oxide growth occurred at the oxide-electrolyte interface. This suggests that Ga and As are the mobile species and that mass transport occurs through the interstices of the growing oxide. Similar results were obtained in a study of the oxidation of GaAs in an oxygen plasma (63). SIMS depth profiles of samples generated in $^{16}O_2$ and $^{18}O_2$ plasmas in subsequent steps showed that oxygen was incorporated into the oxide at the oxide-plasma interface.

The ability of SIMS to detect hydrogen and its isotope, deuterium, was used in a study of the diffusion of hydrogen in glow discharge deposited amorphous silicon (15,64). The sample consisted of three layers of amorphous silicon with the center layer deposited from 98% pure SiD_4 and the other two layers deposited from SiH_4. SIMS depth profiles were obtained of the samples before and after heating. From the depth profiles, which covered a dynamic range of over 10^3, both the diffusion constants for H and D in amorphous Si:H as a function of hydrogen concentration, and the activation energy were obtained.

Isotopic information can provide valuable supplemental data in other SIMS studies. For example, for elements with more than one isotope, multiple signals can be monitored. In a study of silicon etching, the relative proportions of ^{10}B and ^{11}B were measured to be 23% and 77%, respectively. That the percentages were close to the naturally occurring abundances of B (20% and 80%) indicated that the calibration factors for the ^{10}B and ^{11}B ions were the same, as would be expected, and that there were negligible interferences at either mass (56). For elements with more than one isotope, a mass interference may be circumvented by monitoring a different isotope. In a study of Te diffusion in Si the ^{130}Te isotope peak was found to have the best signal to background

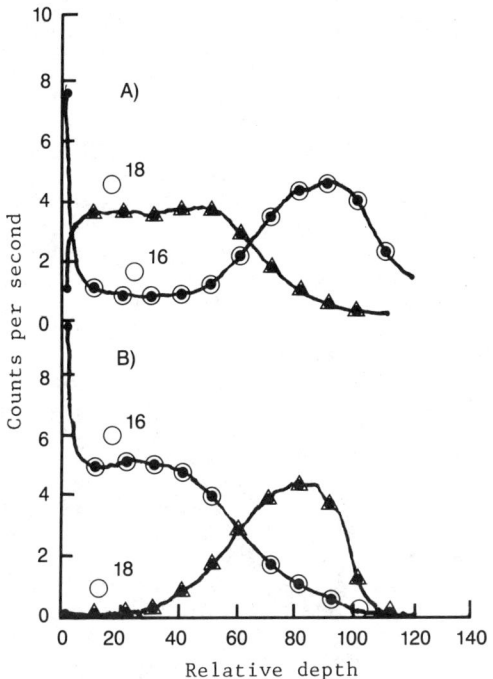

Figure 2. A and B GaAs anodically oxidized in ^{18}O enriched water. (Reproduced with permission from Ref. 62. Copyright 1977 The Electrochemical Society, Inc.)

ratio. In addition the ^{126}Te peak was used to monitor the isotope ratio as an additional check of possible contamination (42). An added piece of information is available in the isotopic fingerprints in a mass spectrum. The isotopic ratios can assist in identifying the species and fragmentation patterns of the substance being analyzed. The ability of SIMS to detect different isotopes of an element adds richness to the data obtained and flexibility to the analysis.

Static SIMS. The static SIMS instrument was briefly described in the instrumental section. Static SIMS has been applied to the study of metal surfaces (65), oxide formation (11), and catalysts (11,66) but static SIMS is not widely used in microelectronics materials characterization. A major limitation of static SIMS for microelectronics applications is the inability to obtain a detectable signal from a very small area on the sample. The low energy, low current density primary ion source used in static SIMS produces a lower count rate per unit area of sample than does the higher energy, higher current density used in dynamic SIMS. Static SIMS should be used in cases requiring utmost surface sensitivity or minimal molecular rearrangements, whereas the higher energy, higher current density ion sources are preferred for most other SIMS applications, especially those involving small areas and requiring good lateral resolution.

Static SIMS has been used to probe in situ the surface of MBE grown Sn-doped film of GaAs (65). The researchers sought to provide experimental evidence that the deviation from the ideal doping profile found in Sn-doped MBE GaAs layers was caused by an enrichment of Sn in the outermost layers during deposition. An erosion rate of only about 50 angstroms per hour was used to detect the Sn in the 10-20 angstrom thick outermost atomic layers. After about 10 minutes the tin peaks in the positive SIMS spectrum had decreased and reached a constant height. By investigating Sn-doped GaAs layers prepared by MBE at different temperatures it was determined that the surface segregation of Sn started at 490°C and increased with increasing substrate temperature. Parallel Auger studies confirmed this result (67) but Auger was not sensitive enough to detect Sn in the substrate.

Static SIMS was also used to evaluate contamination remaining after ion cleaning another III-V substrate, InP. The primary ion beam energy of 0.5 KeV was used to minimize the ion range in the substrate and ensure high surface sensitivity. The primary ion dose was sufficiently small that no change was observed in the spectrum for over 1000s. It was found that contamination levels greater than 100 ppm extended into the substrate and thus ion cleaning the surface even multiple times would leave a high residual contamination level (68).

An investigation of the oxidative behavior of a silicon surface was conducted with static SIMS. Both the increase in SiO_2 intensity during oxidation, and the decrease in SiO_2 intensity during sputter removal of a monomolecular oxide layer were recorded. Negative ion static SIMS spectra were also used to determine the various species present when water is present during the oxidation of the Si surface. (11)

Ion Imaging. A SIMS ion image represents the x-y distribution of a species over the surface of the sample. The usefulness of elemental images has been illustrated for Auger and electron microprobe energy dispersive X-ray (EDAX) maps (69,70). SIMS brings the added usefulness of greater concentration sensitivity. However, the change of secondary ion yields with matrix variations across the sample can make the interpretation of relative intensity differences in SIMS ion images very tenuous.

Two different instrumental approaches to generating an ion image are common (2,71), the ion microprobe and the direct imaging instrument. The lateral resolution and relative advantages of the direct imaging instrument and the ion probe type instruments have been discussed by several authors (2,71,72). For areas less than 20 μm × 20 μm the ion microprobe can have a comparable or shorter time required to obtain an ion image than the direct imaging instrument when both instruments have comparable lateral resolutions and primary ion current densities. For larger areas the direct imaging instrument obtains images more rapidly than the ion microprobe (2). The lateral resolution of the direct imaging instrument is determined by the secondary ion optics. Features in secondary ion images show that the direct imaging instrument, also referred to as an ion microscope, is capable of submicron lateral resolution, ~0.5 μm (71). Refinements in the ion optics have been predicted to yield a resolution limit of about 0.2 μm (5). The lateral resolution of the ion probe instruments is related to the primary ion beam spot size and has been demonstrated to be in the low micron range. Typical ion microprobe primary ion beam diameters are 5-10 μm. Submicron ion beam diameters for field emission sources have been reported (73-76). Whether these sources would be useful for SIMS, including imaging at very high lateral resolution, has not been demonstrated. SIMS images at lower resolution, using a high brightness liquid metal ion source, have been published (77-79). Criegern, et al. (72), presented a graphical comparison between Auger and SIMS detection limits as a function of lateral resolution. SIMS is capable of lower detection limits than Auger with equal lateral resolution down to about 1 μm lateral resolution (72). Image processing of SIMS ion images from the ion microprobe type of instrument has shown that the lateral resolution can be much less than the probing beam diameter (80). Image processing can be used to improve the signal to noise ratio and sharpen the edges of the image. An unusual method of SIMS signal processing has been reported by Griffith and Kowalski (81). Complete mass spectra were collected at each pixel in the image. Subsequent eigen analysis of the data enabled the determination of the number of components present on the surface of the sample. The concentration of each component at each pixel can also be determined and allows the generation of an image of its surface distribution (81).

There are a number of factors that can degrade SIMS images or complicate their interpretation. Wittmaack (82) discussed and presented illustrations on the effect of the time of flight of ions through the mass spectrometer, and of the scan rate on the resolution of the ion image. Detrimental effects can be kept sufficiently small by operating at scan rates of less than 100 lines/s (82).

Variations in secondary ion yield across the sample due to different matrices (2) can make quantitative interpretation of ion images difficult. Rough or nonuniform sample surfaces have the additional problems of orientation contrast, phase contrast and topographic contrast (3). These effects cause the image to be richer in information for well characterized samples, but make straightforward interpretation of images from unique samples questionable. A method of total ion monitoring (TIM) has been demonstrated in a specialized SIMS instruments (83, 84). In TIM the signal from a specific mass analyzed ion is divided by the signal representing the total emitted ions (without mass analysis). This ratioing was demonstrated to lessen the (84) effects of variation in topography, shadowing, matrix, and primary ion beam current or incidence angle.

SIMS ion images can be useful for several types of microelectronics problems. Information on the nature and source of a contaminant can be obtained by examining the lateral distribution in images of the contaminated area. If one detected Na, K, or Li in discrete areas as would be produced by particulate contamination then environmental contamination is indicated. If the suspected ionic contaminants are contained homogeneously within the material one could assume that the contamination was being introduced by the process itself, such as being co-deposited during a deposition process. Since SIMS is exceptionally sensitive to the alkali metals it is ideal for the investigation of ionic contaminants. Secondary ion images were obtained from an integrated circuit with irregular Na and K distributions to illustrate this point (85). SIMS ion images are also useful for failure analysis. In one example IC frames were not always properly wet by solder. EDAX analysis showed the expected distribution for Cu, but the SIMS ion image indicated that some of the conductive paths had little Cu on the surface (86). The greater surface sensitivity of SIMS showed that a contaminant masked part of the Cu surface. In another case SIMS ion images of an IC frame with a contaminated area believed to be a water stain showed nonuniform distributions of Na and Ca which substantiated the identification as a water stain (87).

SIMS ion images could be collected during depth profiling to determine if the analytes were homogeneously laterally distributed (6). A method called Image Depth Profiling (IDP) combines imaging with depth profiling by accumulating successive images at multiple sputter depths (88). IDP was used to analyze an MOS integrated circuit and an implant of $^{115}In^+$ into Si. Successive 3D images were presented in which the z axis on each ion image represents intensity. The method could be considered a 5 dimensional technique; 3 spatial dimensions, elemental identity, and concentration. This technique could be especially useful for ICs since they have structural and compositional changes in 3 spatial dimensions. The potential of IDP was also illustrated by profiling through an Al and SiO_2 layer on the MOS IC. A higher $^{23}Na^+$ signal was observed in the oxide, which could indicate sodium contamination of the oxide. A great deal of information is generated by this technique and the method of data presentation has a strong visual impact. As in all SIMS ion images, changes in ionization efficiency and sputter rate at different regions in the sample require that these ion images receive careful analysis before interpretation. The relative intensities of different species, or from different areas on the sample cannot be simply related to surface concentrations.

<u>Positive/Negative Ion Mode</u>. All of the other SIMS modes can be used while detecting either negative or positive ions, although positive ion detection is more common. Positive and negative ion mass spectra provide complementary information. The complementary nature of the relative positive and negative secondary ion yields has been graphically illustrated by Storms et al. (<u>89</u>). Electronegative elements typically have greater negative ion yields (<u>90</u>). The significance of negative ion formation has been discussed by Cuomo, et al. (<u>91</u>). Many of the elements which control the mechanical and electronic properties of semiconductors are from Groups V to VII in the periodic table. Some of these elements have high ionization potentials and low positive ion yields, but form negative ions quite readily (<u>90</u>). In addition, negative ion spectroscopy often has a lower background signal than positive ion spectroscopy which can mean lower detection limits and less complex mass spectra that are more easily interpreted (<u>92</u>). A Cs^+ ion source can enhance the negative ion yield by two to four orders of magnitude. The use of Cs^+ bombardment and negative ion SIMS allowed the determination of depth profiles of P in Si at sub-ppm levels. $^{11}B^{28}Si^-$ and $^{31}P^-$ were both monitored and thus the electrical junction depths could be determined directly from the SIMS depth profile (<u>93</u>).

Negative ion SIMS has also been used in GaAs studies. Depth profiles of S in GaAs were obtained after implantation and annealing, with typical detection sensitivities down to 2×10^{15} atoms/cm^3. Lower detection limits have been observed for ^{32}S and ^{34}S in Si. The higher detection limit in GaAs could be caused by either native S in the GaAs or the presence of a mass interference of oxygen in the vacuum ambient and on the GaAs surface (<u>37</u>).

Electronegative species such as Cl are seldom observed in positive ion spectra. The presence of Cl in CVD (chemical vapor deposition) $MoSi_2$ was detected by negative ion SIMS spectra obtained on an ion microprobe. The CVD reactants were $MoCl_5$ and SiH_4 and were the source of Cl. The presence of Cl in the deposited film could be advantageous because of the gettering effect of chlorine for mobile ions such as Na^+ (<u>94</u>).

<u>CONCLUSION</u>

The sensitivity and inherent depth profiling capabilities of SIMS make it a valuable tool for characterization and problem solving in microelectronics materials and processes. These and the other capabilities of SIMS which have been described in this review allow SIMS to be applied to a wide range of problems.

An understanding of the capabilities and limitations of SIMS is necessary to obtain optimum results. SIMS is a complex, technique and the instruments and technique are still evolving and being improved. The development of related techniques, such as Laser Microprobe Mass Spectrometry (LAMMA) (95) and Mass and Charge Analysis (MACS)(96), could improve the understanding of the processes involved in SIMS. SIMS is often advantageously combined with other techniques to obtain complementary information (26), e.g. RBS (15,44) Auger (67), X-ray Photoelectron Spectroscopy (97), Nuclear Reaction Analysis (15,98), and Ion Scattering Spectrometry (ISS) (99). A better understanding of SIMS and the further development of standards and methods to compensate for matrix effects will result in even broader application of the SIMS technique.

Literature Cited

1. Wittmaack, K. Surface Science 1979, 89, 668-700.
2. McHugh, J. A. In "Methods of Surface Analysis"; Czanderna, A. W., Ed.; Elsevier Scientific Publishing Co.: New York, 1975; Chap. 6.
3. Werner, H. W. In "Applied Surface Analysis"; Barr, T. L.; Davis, L.E., Eds.; ASTM STP 699, American Society for Testing Materials: Washington, D. C., 1980; pp. 81-110.
4. Liebl, H. J. App. Phys. 1967, 38, 5277.
5. Liebl, H. Nucl. Instr. and Meth. 1981, 187, 143-151.
6. Magee, C. W.; Harrington, W. L.; Honig, R. E. Rev. Sci. Instrum. 1978, 49 (4), 477-485.
7. Orloff, J., private communication, Oregon Graduate Center, Beaverton, OR., 1982.
8. Wittmaack, K. Rev. Sci. Instrum. 1976, 47, 157-158.
9. Rudenauder, F. G.; Steiger, W.; Kraus, U. Mikrochimica Acta [Wien], 1979, Suppl. 8, 51-58.
10. Castaing, R.; Slodzian, G. J. Microsc. 1962, 1, 395.
11. Benninghoven, A. Surface Science 1973, 35, 427-457.
12. Schiefers, S. M.; Hollar, R. C.; Busch, K. L.; Cooks, R. G.; Am. Lab. 1982, 14, 19-33.
13. Leta, D. P.; Morrison, G. H.; Anal. Chem. 1980, 52, 514-519.
14. Zinner, E. J. Electrochem. Soc. 1983, 130, 199C-22C.
15. Magee, C. W. Nucl. Instr. and Meth. 1981, 191, 297-307.
16. Ryan, M. A.; McGuire, G. E. In "Industrial Applications of Surface Analysis"; Casper, L. A.; Powell, C. J., Eds; ACS SYMPOSIUM SERIES No. 199, American Chemical Society: Washington, D. C., 1982; pp 229-249.
17. Wittmaack, K.; Clegg, J. B. Appl. Phys. Lett. 1980, 37 (3), 285-287.
18. Clegg, J. B.; O'Connor, D. J. Appl. Phys. Lett. 1981, 39 (12), 997-999.
19. Helms, C. R. J. Vac. Sci. Technol. 1982, 20 (4), 948-952.
20. Slusser, G. J.; Slattery, J. S. J. Vac. Sci. Technol. 1981, 18 (2), 301-304.
21. Galuska, A. A.; Morrison, G. H. Anal. Chem. 1983, 55, 2051-2055.

22. Wittmaack, K. <u>Nucl. Instrum. and Meth.</u> 1980, 168, 343-56.
23. Leta, D. P.; Morrison, G. H. <u>Anal. Chem.</u> 1980, 52, 277-280.
24. Riedel, M.; Gnaser, H.; Rudenauer, F. G. <u>Anal. Chem.</u> 1982, 54, 290-294.
25. Rawls, R. <u>Chem. Eng. News</u> 1981, 59 (41), 24-25.
26. Zinner, E. <u>Scanning</u> 1980, 3 (2), 57-70.
27. Magee, C. W.; Harrington, W. L. <u>Appl. Phys. Lett.</u> 1978, 33(2), 193-196.
28. Hunt, C. P.; Stoddart, C. T. H.; Seah, M. P. <u>Surface and Interface Anal.</u> 1981, 3(4), 157-160.
29. Reuter, W.; Yu, M. L.; Frisch, M. A.; Small, M. B. <u>J. Appl. Phys.</u> 1980, 51, 850-855.
30. Conlin, B. D.; Kulkarni, M. V.; Smith, T. L.; Wey, S. S. <u>IBM Tech. Disc. Bull.</u> 1977, 20, 615-616.
31. Phillips, B. F. <u>J. Vac. Sci. Technol.</u> 1982, 20(3), 793-6.
32. Sheperd, F. R.; Robinson, W. H.; Brown, J.D.; Phillips, B. F. <u>J. Vac. Sci. Technol.</u> 1983, A1(2), 991-5.
33. Leta, D. P.; Morrison, G. H.; Harris, G. L.; Lee, C. A. <u>Int. J. Mass Spectrom. Ion Phys.</u> 1980, 34, 147-57.
34. Hofker, W. K.; Politiek, J. <u>Philips Technical Review</u> 1980, 39, 1-14.
35. Rosen, A.; Caulton, M; Stabile, P.; Gombar, A. M.; Janton, W. M.; Wu, C. P.; Corboy, J. F.; Magee, C. W. <u>IEEE Trans. Microwave Theory Tech.</u> 1982, Vol. MTT-30(1), 47-53.
36. Liu, S. G.; Douglas, E. C.; Wu, C. P.; Magee, C. W.; Narayan, S. Y.; Jolly, S. T.; Kolondra, F.; Jain, S. <u>RCA Review</u> 1980, 41, 227-262.
37. Wilson, R. G.; Jamba, D. M.; Deline, V. R.; Evans, C. A., Jr.; Park, Y. S. <u>J. Appl. Phys.</u> 1983, 54, 3849-54.
38. Sadana, D. K.; Washburn, J.; Magee, C. W. <u>J. Appl. Phys.</u> 1983, 54, 3479-84.
39. Kwor, R.; Paz de Araujo, C. <u>J. Electrochem. Soc.</u> 1983, 130, 1580-86.
40. Ackerman, G. K.; Ebert, E. <u>J. Electrochem. Soc.</u> 1983, 130, 1910-15
41. Oshima, M.; Watanabe, K.; Miyazawa, S. <u>J. Electrochem. Soc.</u> 1984, 131, 130-6.
42. Janzen, E.; Grimmeiss, H. G.; Lodding, A.; Deline, C.; <u>J. Appl. Phys.</u> 1982, 53, 7367-71.
43. Lidow, A.; Gibbons, J. F.; Deline, V. R.; Evans, C. A., Jr. <u>Appl. Phys. Lett.</u> 1978, 32(1), 15-17.
44. White, C. W.; Christie, W. H. <u>Solid State Technol.</u> 1980, 23, 109-116.
45. Young, R. T.; Narayan, J.; Christie, W. H.; Van der Leeden, G. A.; Levatter, J. I.; Cheng, L. J. <u>Solid State Technol.</u> 1983, 26, 183-189.
46. Wu, S.-y. <u>J. Electrochem. Soc.</u> 1983, 130, 199-202.
47. Kim, M. J.; Brown, D. M.; Katz, W. <u>J. Electrochem. Soc.</u> 1983, 130, 1196-1200.
48. Chow, T. P.; Grant, C. S.; Katz, W.; Gildenblat, G.; Reihl, R. F. <u>J. Electrochem. Soc.</u> 1983, 130, 933-8.
49. Chow, T. P.; Bower, D. H.; Van Art, R. L.; Katz, W. <u>J. Electrochem. Soc.</u> 1983, 130, 952-6.

50. Phillips, B. F.; Burkman, D. C.; Schmidt, W. R.; Peterson, C. A. J. Vac. Sci. Technol. 1983, A1, 646-9.
51. Magee, C. W.; Botnick, E. M. J. Vac. Sci. Technol. 1981, 19(1), 47-52.
52. Magee, C. W.; Wu, C. P. Nucl. Inst. and Meth. 1978, 149, 529-33.
53. Satake, T.; Tsukakoshi, O.; Shimizu, A.; Komiya, S.; Jap. J. Appl. Phys. 1981, 20, 1541-52.
54. Wilson, R. G.; Sadana, D. K.; Sigmon, T. W.; Evans, C. A., Jr. Appl. Phys. Lett. 1983, 43, 549-61.
55. McFee, J. H.; Swartz, R. G.; Archer, V. D.; Finegan, S. N.; Feldman, L. C. J. Electrochem. Soc. 1983, 130, 214-216.
56. Raley, N. F.; Sugiyama, Y.; Van Duzer, T. J. Electrochem. Soc. 1984, 131, 161-171.
57. Beard, B. C.; T. Titcomb, S.; Butler, S. R. J. Electrochem. Soc. 1983, 130, 1959-61.
58. Indust. Res. Dev. 1983, 25, 84.
59. Fontana, P. V.; Decosterd, J. P.; Wegmann, L. J. Electrochem. Soc. 1974, 121, 146-150.
60. Lowry, R. K.; Masters, R. G. Reliabilitiy Physics 19th Annual Proceedings 1981, 19, 157-62.
61. Sparrow, G. R. 4th International Symposium on Contamination Control Wash. D. C. Sept. 10-13, 1978.
62. Coleman, D. J.; jr.; Shaw, D. W.; Dobrott, R. D. J. Electrochem. Soc. 1977, 124(2) 239-241.
63. Koshiga, F.; Sugano, T. Thin Solid Films 1979, 56, 39-49.
64. Carlson, D. E.; Magee, C. W. Appl. Phys. Lett. 1978, 33, 81-83.
65. Hewitt, R. W.; Shepard, A. T.; Baitinger, W. E.; Winograd, N.; OTT, G. L.; Delgass, W. N. Anal. Chem. 1978, 50, 1286-1290.
66. Buhl, R.; Preisinger, A. Surface Science 1975, 47, 344-357.
67. Ploog, K.; Fischer, A. J. Vac. Sci. Technol. 1978, 15 (2), 255-259.
68. Dowsett, M. G.; King, R. M.; Parker, E. H. C. Appl. Phys. Lett. 1977, 31, 529-531.
69. Buono, J. A.; Wisniewski, A. W.; Andrus W. S. Solid State Technol. 1982, 25, 95-101.
70. Shappirio, J. R.; Cook, C. F., Jr. Solid State Technol. 1979, 22, 89-94.
71. Evans, C. A., Jr. Anal. Chem. 1972, 44, 67A-80A.
72. Criegern, R. v.; Hillmer, Th.; Weitzel, I. Fresnius Z. Anal. Chem. 1983, 314, 293-299.
73. Seliger, R. L.; Ward, J. W.; Wang, v.; Kubena, R. L. Appl. Phys. Lett. 1979, 34, 310-2.
74. Orloff, J.; Swanson, L. W. Proc. 16th Electron Ion Photon Beam Technol. Symp. 1981.
75. Ishitani, T.; Tamura, H.; Todokoro, H. J. Vac. Sci. Technol. 1982, 20, 80-83.
76. Levi-Setti, R. Ind. Res. Dev. 1982, 24, 124-30.
77. Prewett, P. D.; Jeffries, D. K. Inst. Phys. Conf. Ser. No. 54, The Institute of Physics, 1980, Chap. 7.
78. Bayly, A. R.; Wells, M. G.; Smith, A. I. C. Pittsburgh conference on Analytical Chemistry and Applied Spectroscopy, 1981, paper No. 665.

79. VG Scientific pamphlet, 1981, VG Instruments, Inc., Stamford, CT.
80. Vandeginste, B. G. M.; Kowalski, B. R. Anal. Chem. 1983, 55, 557-564.
81. Griffith, M.; Kowalski, B. R. Pittsburg Conference on Analytical Chemistry and Applied Spectroscopy, 1983, paper No. 315.
82. Wittmaack, K. Scanning 1980, 3, 133-8.
83. Rudenauer, F. G.; Steiger, W. Mikrochimica Acta [Wien] 1981, II, 375-389.
84. Kobayashi, H.; Suzuki, K. K.; Yukawa, K.; Tamura, H.; Ishitani, T. Rev. Sci. Instrum. 1977, 48, 1298-1302.
85. Blattner, R. J.; Evans, C. A., Jr. Scanning Electron Microscopy, 1980, IV, 55-68.
86. Leys, J.; Ruscica, R. Indust. Res. Dev. 1978, 20.
87. Leys, J. A.; McKinney, J. T. Scanning Electron Microscopy 1976, I, 231-238, 762.
88. Patkin, A. J.; Morrison, G. H. Anal. Chem. 1982, 54, 2-5.
89. Storms, H. A.; Brown, K. F.; Stein, J. D. Anal. Chem. 1977, 49, 2023-30.
90. Blattner, R. J.; Evans, C. A., Jr.; Thin Solid Films 1978, 53, 39-40.
91. Cuomo, J.J.; Gambino, R.J.; Harper, J.M.E.; Kuptis, J. D.; Webber, J. C. J. Vac. Sci. Technol. 1978, 15, 281-8.
92. Purser, K. H.; Litherland, A. E.; Rucklidge, J. C. IEEE Trans. Nuc. Sci. 1979, NS-26, 1338-46.
93. Magee, C. W. Springer Ser. Chem. Phys. 1979, 9 (Second. Ion Mass Spectrom., SIMS 2), 88-90; Chem. Abstr. 1981, 95, 34824y.
94. Inoue, S.; Toyokura, N.; Nakamura, T.; Maeda, M.; Takagi, M. J. Electrochem. Soc. 1983, 130, 1603-7.
95. Denoyer, E.; Van Grieken, R.; Adams, F.; Natusch, D. F. S. Anal. Chem. 1982, 54, 26A-41A.
96. Purser, K. H. Indust. Res.Dev. 1979, 21, 126-130.
97. Frisch, M. A.; Reuter, W.; Wittmaack, K. Rev. Sci. Instrum. 1980, 51, 695-704.
98. Hubler, G. K.; Comas, J.; Plew, L. Nucl. Inst. and Meth. 1978, 149, 635-8.
99. Bubert, H.; Klockenkamper, R.; Spectrochim. Acta 1981, 36B, 61-70; Chem. Abstr. 1981, 95, 34737X.

RECEIVED October 25, 1985

7

Applications of Auger Electron Spectroscopy in Microelectronics

Paul A. Lindfors[1], Ronald W. Kee[2], and Douglas L. Jones[3]

[1] Physical Electronics Division, Perkin-Elmer Corporation, Eden Prairie, MN 55344
[2] Northwest Integrated Circuit Division, Hewlett-Packard, Corvallis, OR 97330
[3] Tektronix, Inc., Beaverton, OR 97077

Auger electron spectroscopy (AES) has a broad range of applications in the microelectronics industry because it is able to detect light elements, and its analysis volume is compatible with the dimensions of circuit features in integrated circuits (IC's). In this chapter the authors approach the issue of applications of AES from the viewpoint of an analytical laboratory with discussions of failure analysis and process development. The entire scope of applications is too extensive to be covered in a single chapter. Other recent reviews have approached applications of AES to IC's from other viewpoints such as materials flow in fabrication of IC's (1 - 4).
    The chapter is divided into four sections: introduction, failure analysis, process development and future uses. The introductory section discusses fundamentals of AES. Applications are included under two headings: failure analysis and process development. The authors have placed applications under "failure analysis" when a process line was operational and failures have occurred. The applications are included under "process development" when the analysis was used to improve or characterize an existing or future process line. Of course, there is considerable bridging between the categories. Process development may reduce the need for failure analysis, and failure analysis may indicate the need for process development. AES has also been used for fundamental studies of materials used in microelectronics (5 - 9).
    It has been assumed by the authors that the reader is already familiar with the materials and processes used in the fabrication of microelectronic devices. Persons not familiar with this information will find it available in the literature (10 - 12).

0097-6156/86/0295-0118$07.25/0
© 1986 American Chemical Society

Detection of Elements

Auger electrons result from a three electron process in atoms of a specimen being analyzed (13). Elements can be identified by measuring the energies of emitted Auger electrons, and handbooks are available to aid in proper identification of elements (14). All elements except H and He can be detected directly using AES. The presence of H can sometimes be inferred as the result of changes in the spectral features of an Auger peak for another element with which it is chemically combined (15). Peak overlap is generally not a problem in AES.

Analysis Volume

The analysis volume for AES is compatible with the dimensions of features in microelectronic circuits. The depth of the analysis volume depends upon the escape depth of the Auger electrons of interest, and the escape depth is determined by the kinetic energy of the electron and electron scattering of the material being analyzed (16 - 17). The Auger electrons of typical interest have energies from 30 to 3000 electron volts (eV). The dependency of the escape depth of electrons upon energy for elements and compounds is shown in Figure 1. Note that the range of escape depths extends from 0.4 to 7 nanometers (nm). These escape depths are less than the thickness of most layers in integrated circuits.

The lateral dimension of the analysis volume is determined mainly by the diameter of the incident electron beam (18). Although an energetic electron beam will be scattered laterally once it enters a solid, most of the lateral spreading will occur beneath the surface at a depth that is greater than the escape depth of the Auger electrons. This is shown schematically in Figure 2. When submicron geometries are probed, scattering of the incident electron beam because of surface topography is often the most important factor limiting chemical spatial resolution. For specimens that are essentially atomically flat across a chemical interface, state-of-the-art scanning Auger microprobe (SAM) systems now allow lateral definition of a chemical interface of approximately 50 nm. The spatial resolution of AES has kept pace with the reduction in the dimensions of features on integrated circuits in current production and for the next generation as shown in Figure 3.

The analysis volume for AES is substantially smaller than the analysis volume for systems that detect emitted x-rays. This is because the x-rays can generally escape from any point in the excitation volume shown in Figure 2. However, Auger electrons initated beneath the top layers of atoms either cannot escape from the solid or they lose energy such that they cannot contribute to the Auger electron signal.

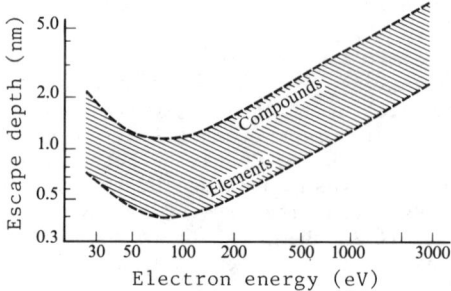

Figure 1. Electron escape depths vs. energy. (Adapted from Ref. 16.)

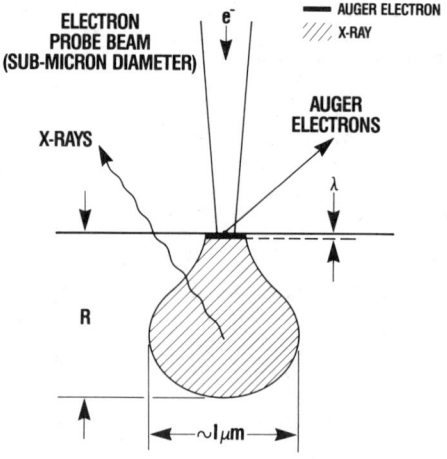

Figure 2. Analysis volume vs. excitation volume for AES. (Reproduced with permission. Copyright Perkin-Elmer Corporation.)

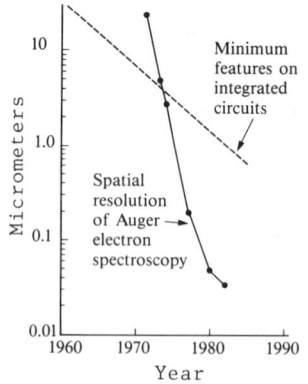

Figure 3. Spatial resolution of AES compared to minimum dimensions of features on microelectronic devices.

Qualitative Aspects of AES

Survey spectra are generally utilized to determine what elements are present within the analysis volume. They are displayed as either the electron energy distribution, $N(E)$, or its derivative, $dN(E)/dE$.

Sputter depth profiles combine analysis with sputtering (ion etching). They are utilized to determine the composition of specimens as a function of depth. The units along the abscissa are normally sputter time. Sputter time can be converted to depth if the sputter rate of the material is known. The units along the ordinate are normally Auger signal intensities or calculated atomic concentrations.

Auger electron maps are displays of signal strength of a particular Auger peak as a function of position of the incident electron beam. The brightest areas on the displays represent locations of greatest signal strength.

The spatial resolution of scanning Auger microprobe systems is now equivalent to most scanning electron microscopes. High magnification secondary electron images or absorbed current images are a routine part of the data from such systems.

Quantitative Aspects of AES

AES can be quantitative, if proper calibration is done using standards. The accuracy of quantitative analysis can be excellent on an absolute basis (19 - 21). However, many quantitative questions are answered by comparison of relative quantities without use of absolute values. Calculations of concentrations are usually based on the peak-to-peak amplitudes in the $dN(E)/dE$ displays. Calculations of concentations based upon $N(E)$ data appear in the literature (22), but widespread use of $N(E)$ data has been hampered by lack of agreement regarding ackground substraction.

Detection Limits

The minimum concentration that can be detected is nominally 0.1 atomic percent of the top atomic layer, but this can vary depending upon the material being analyzed and the conditions for analysis. The ability of AES to detect an elemental species varies across the periodic table. This variation in detectability is illustrated in Figure 4 which shows relative sensitivity factors (14). Points higher on this graph indicate higher sensitivity to detection. The incident electron beam current and the time allowed for analysis can also influence the detection limits. Questions regarding detection limits can occur, especially when one is probing small features. Incident electron beams with the smallest diameters generally have the least current, and signal-to-noise can be an important limitation. The graph shown in Figure 5 can be used to estimate the detection limits depending upon the analysis conditions (23). The diagonal lines are lines of constant "H"

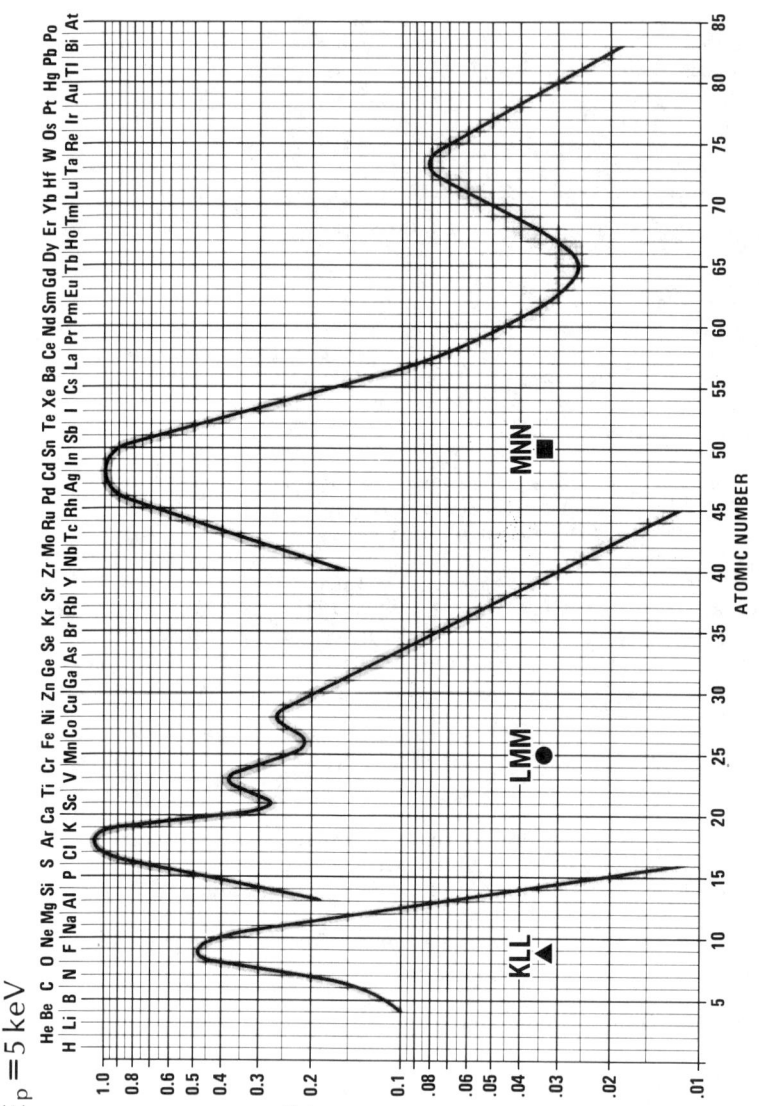

Figure 4. Relative sensitivity factors for detection of elements using AES. (Reproduced with permission from Ref. 14. Copyright 1976 Perkin Elmer Corporation.)

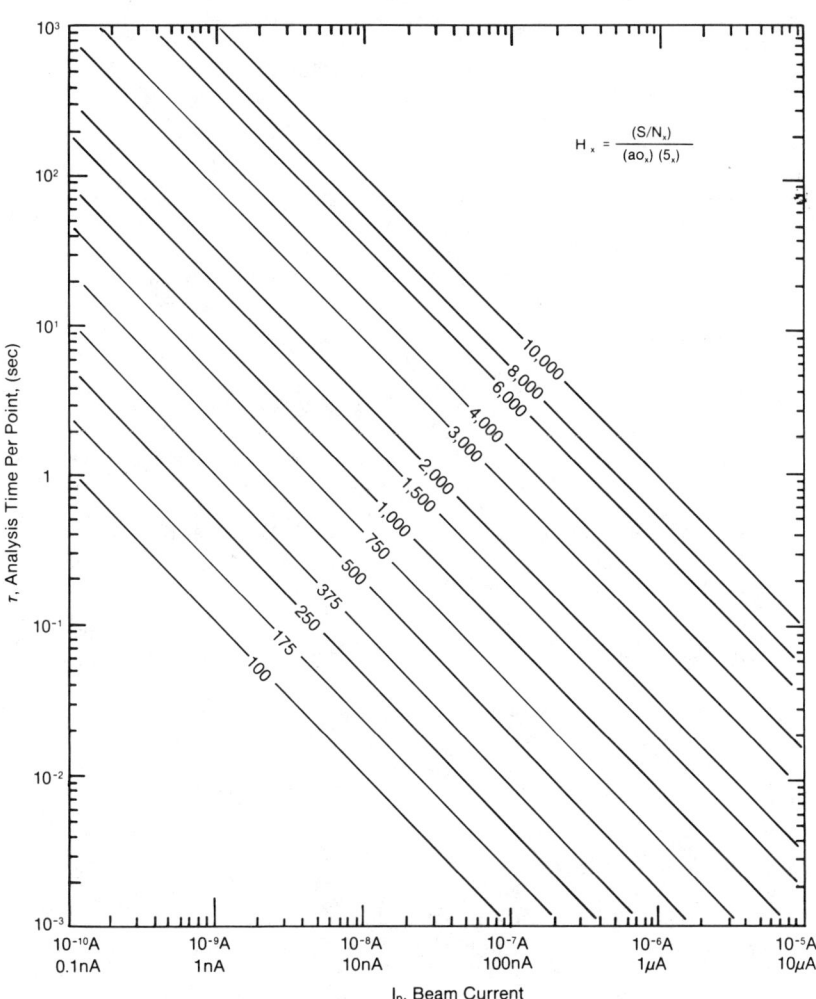

Figure 5. Relationship of detection limits, analysis time, and electron beam current. (Reproduced with permission Copyright Perkin-Elmer Corporation.)

factor. The "H" factor combines the signal-to-noise ratio desired (S/N), the atomic fraction of the element to be detected (A), and the relative sensitivity factor (SR) taken from a chart as shown in Figure 4. The graph shown in Figure 4 can be useful to estimate the time required for the information sought.

Effects of the Incident Electron and Ion Beams

Although the results of bombardment of the electron and ion beams are generally straightforward, understanding possible artifacts sometimes is important in the interpretation of AES data. The incident electron beam used to stimulate emission of Auger electrons can also alter the specimen being analyzed. Some compounds can be modified by the incident beam if the electron dosage is too great. Insulators can be difficult to analyze due to charging. Poor thermal conductors can suffer damage due to localized heating. The possible effects of incident electron beams are discussed more thoroughly in other sources (24 - 26).

The ion beam used to sputter clean a specimen or to obtain a sputter depth profile alters the top atom layers on the surface of a specimen in addition to removing material. The depth of the altered layer will depend upon the specimen being sputtered, the species of the bombarding ion, its energy, and its angle of incidence on the surface of the specimen. In most cases the depth of the mixed layer will approximate the depth of analysis of AES. Artifacts due to the ion beam could be preferential sputtering, formation of cones and other topographical features, and chemical changes. The effects of ion beam bombardment are also discussed elsewhere (27 - 33).

FAILURE ANALYSIS

AES can normally provide answers to many commonly asked questions in failure analysis regarding 1) identification of physical defects such as particles, 2) reasons for high contact resistance, 3) reasons for problems in lead bonding, and 4) reasons for delamination in structures.

Preparation of Specimens

Not all failure analysis concerns devices that have been completed, but this is often the case. At the very least, tests which indicate a failure are often several steps downstream from what would be the optimum point for AES analysis. This means that preparation of specimens is often crucial to efficient and unambiguous analysis. Specimen handling for AES usually requires more care than is required for other analytical techniques (34).

Removal of encapsulation and/or passivation layers may be necessary to expose the surfaces/interfaces of interest. Most IC's are packaged in either ceramic/metal packages or in an epoxy resin. Access to circuits in ceramic/metal packages is normally quite convenient, requiring only the simple removal of the lid. Circuits in epoxy packages are more difficult to access, but procedures have been developed (35 - 36). When etching is done using acids, there is a tendancy to remove contaminants of interest and complicate the analysis by depositing trace materials from the epoxy (e.g. Sb, Br) onto the IC surface. In spite of these difficulties, "wet" decapsulation has been used successfully.

The probing electron beam can result in specimen charging. Therefore care must be taken to ground appropriate pins and to mask those sections of the specimen that are composed of insulating materials.

Characterization of Physical Defects Such as Particles

Particles often consist of materials that are also found in the sample background, so differentiation of a particle from its background is important. High spatial resolution is becoming more important as smaller defects become troublesome in reduced geometries.

Following is an example of identification of a particle. Electrical measurements indicated a short between an aluminum metallization run and an adjacent polysilicon line. Secondary electron images indicated a particle was present at a location that could have accounted for the short if the particle was of a conductive material. Both the passivating layer and the metallization were removed by wet chemistry. Secondary electron images of the particle before and after the chemical treatments are shown in Figure 6. An AES survey spectrum was obtained from the particle and another survey spectrum from the surrounding field oxide. These spectra are displayed in Figure 7. In the spectrum from the particle the Si peak at approximately 80-90 eV has a doublet structure indicating both elemental and oxidized forms of silicon were present in the particle. Care was taken during the measurement to insure that the peak indicative of elemental silicon was not the result of reduction of silicon oxide by the icident electron beam. In the spectrum from the field oxide the same Si peak has only one feature, and it is at an energy that is indicative of oxidized silicon. These data indicate the short was caused by a particle of elemental silicon. The chemical state information available in AES data is discussed in more detail later.

In some instances physical defects covering an area are detected. An example of an area defect would be voids in a Au film plated on Pd. An absorbed current image of such a void is shown in Figure 8. An AES survey spectrum was obtained from a void area and showed peaks for P, Ti, O, and Pd. An AES sputter

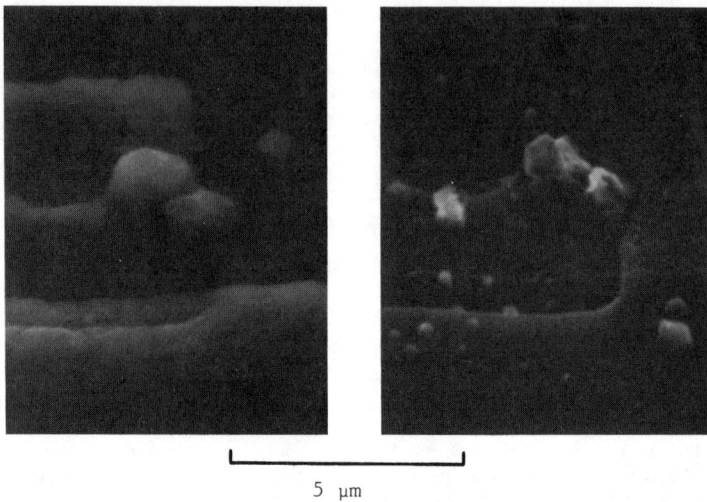

5 μm

Figure 6. Secondary electron image of a defect particle before (left) and after (right) removal of passivation and metallization by wet chemical means.

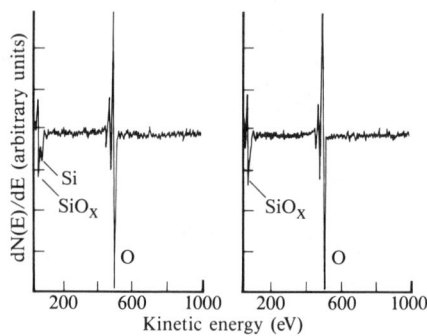

Figure 7. AES survey spectra from defect particle (left) and adjacent field oxide (right).

depth profile was obtained following Ti, O, P and Pd and is shown in Figure 9. The profile indicates a layer rich in Ti and O and approximately 20 nm thick was present on the surface of the Pd. Survey spectra and sputter depth profiles obtained at locations adjacent to the voids did not show peaks for P, Ti or O. Prior to plating of the Au, a Ti/W layer had been removed from the Pd using an etch containing phosphoric acid. The AES data indicate that some Ti, probably in the form of TiO_x, reacted with the acid to form a product that was insoluble in the etch and would not accept the subsequent gold plating.

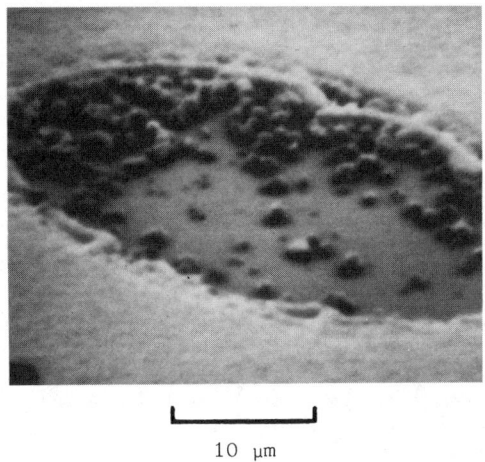

Figure 8. Absorbed current image of a void in a Au film plated on Pd.

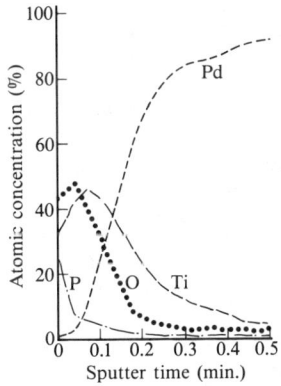

Figure 9. AES sputter depth profile from a void in a Au film plated on Pd. Sputter rate = 600A/minute.

High Contact Resistance

High resistivity of contacts is a problem that is often investigated using AES. The contaminants causing the high resistance are usually oxygen and/or carbon. These contaminants may be found on the surface, within a thin film layer, or at a subsurface interface.

At an IC fabrication facility high contact resistance resulted in the expensive scrapping of completed IC's. After chemical etching of the metal conductor, AES survey spectra were obtained from both "good" and "bad" examples. These spectra are displayed in Figure 10. Note in the spectra the much greater concentrations of oxygen and carbon from the "bad" example. Once the contaminants were identified it was possible to trace their source back to sporatic extinguishing of the plasma in a sputter deposition system. This problem with the sputtering systems had been considered minor by production personnel because the plasma could be restarted.

Lead Bonding

Interconnections between IC die and package leadframes or printed circuit boards are usually accomplished by wire bonding with Al or Au wire. The most widely used techniques are ultrasonic/thermosonic and thermocompression ball bonding. Ball bonding is much more sensitive to contamination at the bonding interface than is ultrasonic bonding. However, impurities at the bond interface will cause both types of bonds to fail. Bonding failures have been investigated using AES (37 - 38). Shown in Table I are the most commonly found contaminants in cases of problem bonding.

Normally, the reason for bonding failure is apparent on the surface of the bonding pads. An AES survey spectrum obtained from an Al bonding pad demonstrating poor bonding is shown in Figure 11. This spectrum contains large peaks for C, O and F. Only after an estimated 70 nm of material was sputtered away did another spectrum indicate a clean Al surface. The large concentrations of C, F and O indicated a plasma etching step using fluorocarbons was not carried out under proper conditions. If conditions are not correct it is possible to deposit material rather than remove it in a plasma "etching" process (39).

For a limited number of problems the reasons for bonding failure may lie just beneath the surface. For example, bonding failure occurred to pads on a printed circuit board using a Au/Ni/Cu structure. Comparison of AES survey spectra from good and bad bonding pads indicated the presence of Cu contamination on both. Surface concentrations of Cu were equivalent in both good and bad cases. However, sputter depth profiles indicated the penetration of Cu was more extensive on pads with poor bonding (40 nm) than those pads with good bonding (5nm). The sputter depth profiles are displayed in Figure 12.

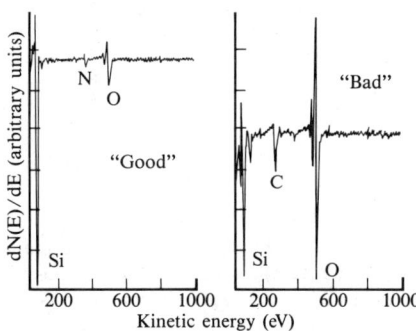

Figure 10. AES survey spectra from contacts with "good" and "bad" resistivity.

Table I. Common Contaminants in Lead Bonding Failures

Location	Contaminant Source and Species
IC bond pads Si_3N_4	Residue from passivation: SiO_2, Residue from photoresist: C, N, Cl, S Residue from plasma etch: C, F, O, S, Cl Residue from chemical etches: Cl, F, S, N, P Metallization oxides: O, AlO_x, SiO_x, CuO_x
Leadframe or PC Board Cr	Organic residues: C, F, Cl, N Impurities in Au plating: Cu, Ni, Tl, Fe, Co, Oxides: Metal oxides

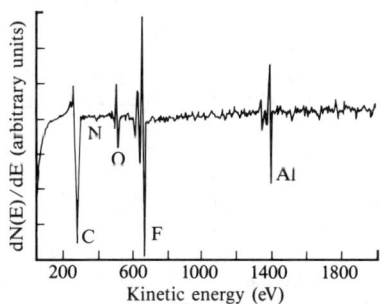

Figure 11. AES survey spectrum from Al pad with bonding problems.

Bond failures that occur following encapsulation of IC's have been successfully investigated using AES. For example Au ball bonds failed after thermal cycling and thallium was detected on the failed surfaces (40). The thallium contamination was traced to transfer from lead frames during processing.

Delamination

Because of its surface sensitivity, AES has been a valuable tool for identification of contaminants at interfaces where delamination has occurred. Sometimes no contaminants are detected at interfaces where delamination has occurred. Lack of evidence for contaminants can indicate mechanical reasons for the delamination. Specimens can be prepared by using adhesive tape to separate films or using a sharp point to peel back blister like features. Whenever possible both sides of the failed interface should be analyzed.

For example, AES was used to analyze delamination in a multilayer structure composed of several different metals. Following completion of the devices a tape pull test resulted in delamination of the metal stack. Both sides of the locus of failure were analyzed. The AES survey spectra obtained are displayed in Figure 13, and they include peaks for C, N, Ti, O, F and Al. These spectra indicate the delamination occurred at the Ti/Al interface in the multilayer structure. Furthermore, the peaks for C, N and F suggest that contamination at that interface was the reason for the delamination. Sputtering and additional analysis indicated the layer rich in C and F was approximately 5 nm thick, and the Al was covered with an oxide approximately 10 nm thick. In the processing there was a dry etch step using CF_4 to expose the Al prior to deposition of the Ti. The source of the contamination and subsequent delamination was improper termination of the plasma etch.

PROCESS DEVELOPMENT

Proper use of AES during process development may eliminate much of the need for subsequent failure analysis. In addition, any characterization of surfaces at crucial process steps during the development of those steps can aid in quickly locating a problem if a fabrication line does fail at some future date. In process development the analyst generally has greater opportunity to define the specimens that will be characterized.

Characterization of Thin Films

Thin films are used for a variety of applications in the microelectronics industry. As the technology has advanced, the control of contaminants, thicknesses, and interfaces of thin films has become more critical. Following deposition of multicomponent structures, further processing or thermal stress after packaging may result in a redistribution of the constituents.

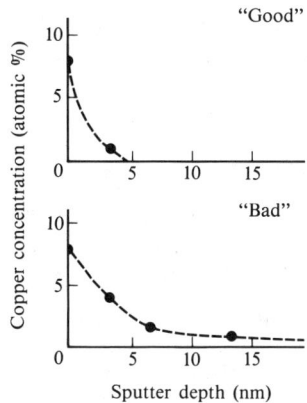

Figure 12. AES sputter depth profiles of printed circuit boards showing the distribution of Cu in Au/Ni/Cu bonding pads with "good" and "bad" bonding.

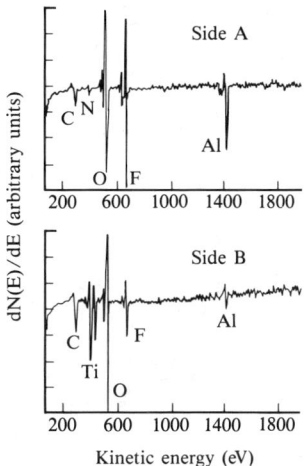

Figure 13. AES survey spectra from both sides of a delaminated multilayer structure.

Studies of the Formation of Silicides

Interdiffusion would be desired for the formation of metal silicides using deposition of a suitable metal layer on silicon and subsequent annealing to form the silicide. Several metal silicide systems have been studied using AES (41 - 45).
Formation of Ti silicide requires annealing be done in an oxygen free atmosphere or formation of Ti oxide will dominate formation of the silicide. Displayed in Figure 14 is a sputter depth profile of Ti/Si that was annealed in forming gas, 80% H_2 and 20% N_2. In the profile there is no evidence of diffusion of Si into the Ti. There was enough oxygen in the furnace tube to prevent any formation of a silicide. Following annealing in a vacuum environment a silicide was formed as shown by the sputter depth profile in Figure 14. A result similar to that of the forming gas case would occur if annealing was done under poor vacuum conditions.
Pt silicide will not form if there is a native oxide on the surface of the Si prior to deposition of Pt. The lack of any diffusion of Pt into the Si is illustrated in the sputter depth profile shown in Figure 15. Note the presence of the oxygen at the Pt/Si interface. Pt is not able to diffuse through the oxide layer on the surface of the Si.

Study of Si-Cr Thin Film Resistors

Redistribution of components can also occur under thermal stresses following packaging. In this example Si-Cr thin film resistor elements were tested under high current load and they subsequently failed electrical tests. Visual examination under a light microscope indicated some modification of the resistors had occurred under test conditions as shown in Figure 16. During testing, the resistors had been covered by SiO_x and SiN_x passivation layers, and initially it was theorized that the visual and electrical changes had occurred as the result of chemical reactions involving carbon, nitrogen or oxygen. Later it was found that resistors stressed with high current prior to passivation showed the same electrical failure and visual changes. AES survey spectra were obtained from Areas A and B identified in Figure 16. These spectra indicated the Si/Cr ratio in Area B was greater than the Si/Cr ratio in Area A. Auger electron maps for Cr and Si indicated that the smallest Si/Cr ratio occurred at the pointed end of Area A. These Auger electron maps are shown in Figure 17. AES survey spectra obtained from Area B were equivalent to spectra obtained from unstressed resistors. Sputter depth profiles were obtained from both areas and they indicated that the resistor film in Area A was only 70% as thick as the film in Area B. Sputter depth profiles obtained from an unstressed resistor were equivalent to those from Area B. The conclusion reached was that Si was diffusing away from a portion of the resistor when high currents were passed through the film.

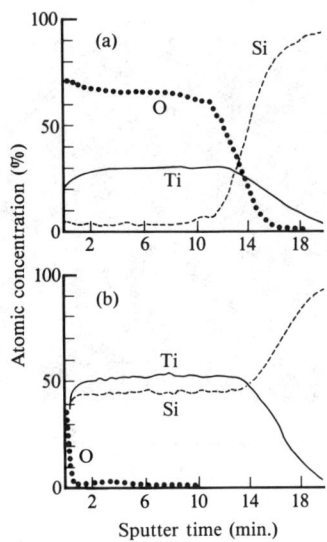

Figure 14. AES sputter depth profiles for Ti/Si annealed in forming gas (a) and vacuum (b).

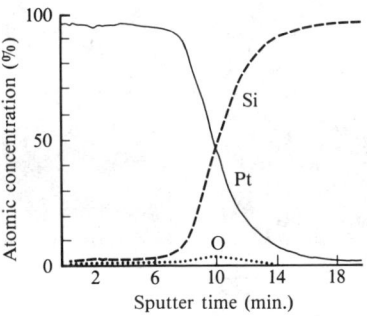

Figure 15. AES sputter depth profile for annealed Pt/Si with a native oxide on the surface of the Si.

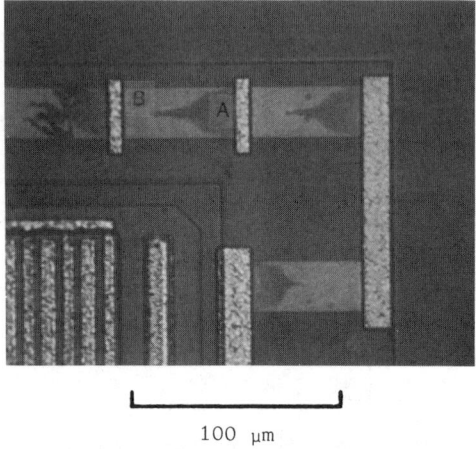

Figure 16. Light optical photograph of Si-Cr thin film resistors following stress at high currents.

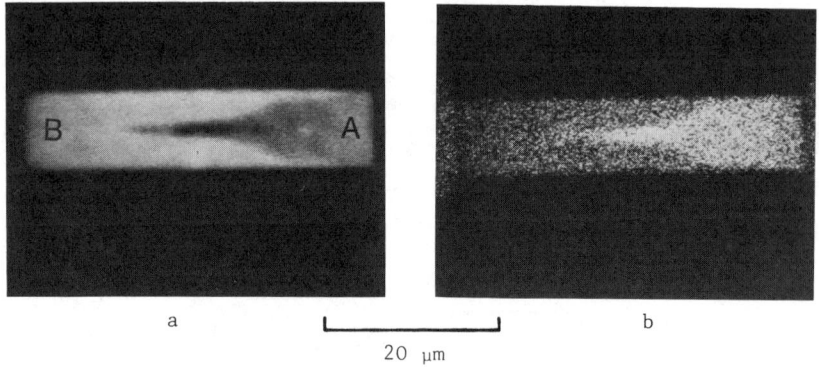

Figure 17. Auger electron maps for Si (a) and Cr (b) in a stressed Si-Cr resistor.

Studies of Diffusion Barriers

To limit diffusion of components in IC structures barrier films are used. An example structure is the Au/Ni/Cu system that is often used for metallization on printed circuit boards and ceramic substrates. Without a Ni barrier there was diffusion of Cu through the Au layer. Copper oxides could then form on the outer surface when this structure was heated. Introduction of a Ni barrier reduced problems with diffusion of Cu, but the improved structure still demonstrated uneven performance. To understand the variations in performance this structure has been studied using AES (46). From a survey spectrum obtained while the incident electron beam was rastered over an area it was learned that Au, Ni, O and C were all present on the surface. Auger electron maps obtained for Ni, O and Au indicated that the Ni and O had the same distribution, and their distribution was complementary to the distribuion of Au. The Auger electron maps are shown in Figure 18. A sputter depth profile obtained from a location initially showing high concentrations of Ni/O indicated the Ni/O patches were approximately 5 nm thick. In this case the diffusion barrier material itself became the diffusing element. The sputter depth profile from this example shows high concentrations of Ni on both sides of the Au layer but a low concentration within the Au layer. Such a profile is the signature of grain boundry diffusion. Diffusion along grain boundaries is normally the dominant diffusion mechanism in thin film structures.

Study of Defect Stringers Following Plasma Etching

As the dimensions of features on IC's have been reduced, the use of anisotropic plasma etching has increased in order to achieve more vertical sidewalls on circuit features. The chemistry and physics of plasma etching is complex and many variables need to be better understood so that the process can be controlled. For Al metallizations it was noted that stringers remained at the base of steps following plasma etching. This problem was studied using AES (47).

The structure used in this investigation is shown in Figure 19. The jagged top edge of stringers can be seen at the bottom of the steps. Examination at high magnification indicated that the stringers were not tucked into the base of the step, but were removed by a distance that corresponded to the thickness of the Al film that had been etched away. This is shown schematically in Figure 20. Near the beginning of the etch cycle survey spectra were obtained from a midpoint on a sidewall as well as from locations that were normal to the direction of the plasma flux. All spectra obtained from surfaces normal to the ion flux were similar with large peaks for Al, O and C and lesser peaks for F and Cl.

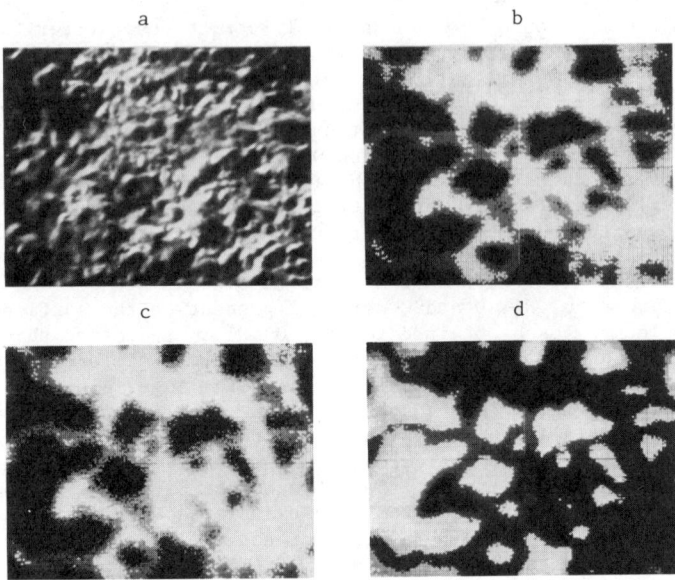

Figure 18. Secondary electron image (a) and Auger electron maps for Nickel (b), oxygen (c), and gold (d) from Au/Ni/Cu metallization following heating. (Reproduced with permission. Copyright Perkin-Elmer Corporation.)

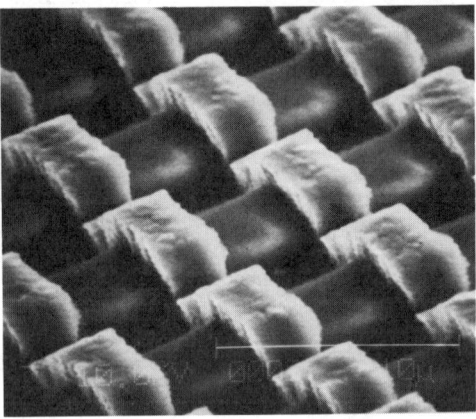

Figure 19. Secondary electron image of aluminum metallization pattern used for study of formation of stringers. (Reproduced with permission. Copyright Perkin-Elmer Corporation.)

A representative spectrum is shown in Figure 21. In contrast, the survey spectrum obtained from a sidewall had peaks for only Al, O and C. This spectrum is similar to those obtained from surfaces that had not been exposed to the plasma etch and is shown in Figure 21. A sputter depth profile was obtained from a point on the sidewall, and it indicates that a layer rich in O and C and approximately 5 nm thick was present on the sidewall. The sputter rate of the material in the sidewall was less than that for Al metal, and the anisotropic etching is directed normal to the plane of the circuit. This means that in the location of the sidewall the etching was proceeding at a slower rate with stringers being the result. Re-examination of Figure 20 may help to visualize the process.

Chemical State Information in AES Data

As discussed above it is possible to distinguish between elemental and oxidized Si. For many elements AES spectral data can provide chemical state information. Two elements that are important in microelectronics, Si and Al, have dramatically different spectral features that allow relatively easy identification of elemental vs. oxidized or nitrided states. A spectrum displayed in the N(E) form and showing peaks for both elemental and oxidized Si is displayed in Figure 22.

The shifts for Al and Si are so large that one can treat the separate chemical states like separate elements and map for the distributions of the chemical states. Auger electron maps for elemental and oxidized Si showing their distributions in a 64 RAM are shown in Figure 23.

Many other elements also demonstrate changes in spectral features that can be useful to the analyst. Not all of the changes are as dramatic as those discussed for Al and Si so greater effort may be required on the part of the analyst to obtain and interpret the information. Reviews of chemical state information are available in the literature (15 , 48).

Changes in the details of AES spectral features have been used to study the stoichiometry of silicide films (49). This approach is called principle factor analysis and it is applicable to several elements (50).

Future Uses of AES in Microelectronics

The use of AES in microelectronics will expand in the future. The chemical spatial resolution of state-of-the-art scanning AES systems coupled with the surface sensitivity of the technique make them uniquely capable for characterization of modern microelectronic devices. This will mean greater dependence on AES for both failure analysis and process development. The demands of greater production

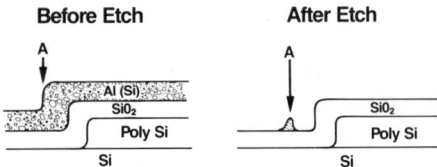

Figure 20. Schematic drawing showing geometry of Al metallization and location of stringers. (Reproduced with permission. Copyright Perkin-Elmer Corporation.)

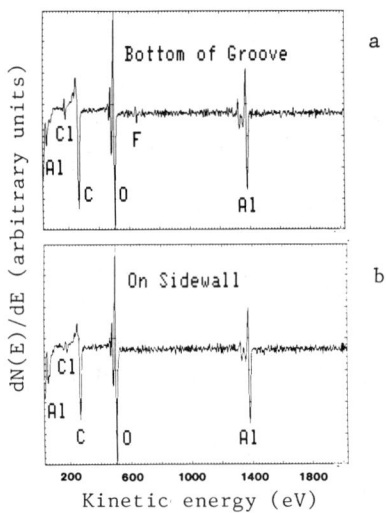

Figure 21. AES survey spectra from Al surfaces exposed to plasma etching (a) and not exposed to plasma etching (b). (Reproduced with permission. Copyright Perkin-Elmer Corporation.)

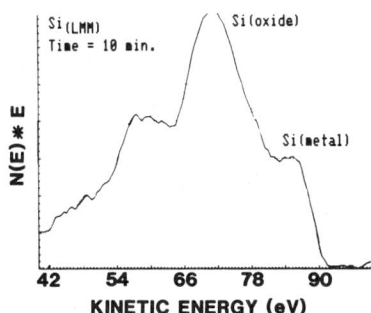

Figure 22. AES spectrum showing peaks for both elemental and oxidized silicon. (Reproduced with permission. Copyright Perkin-Elmer Corporation.)

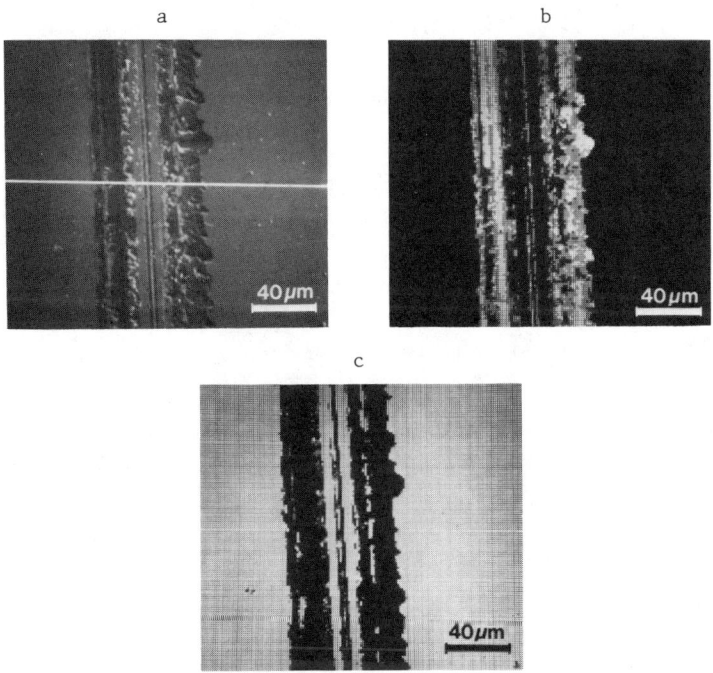

Figure 23. Secondary electron images (a) from a 64K RAM plus Auger electron maps for elemental Si (b) and oxidized Si (c).
(Reproduced with permission. Copyright Perkin-Elmer Corporation.)

control will require the use of AES to monitor the results of critical process steps such as cleaning, integrity of thin films, etc. AES systems will become a part of the process line. Computer controlled analytical systems will provide highly automated analyses. These analyses will not be "in-line" for several years but improved specimen introduction hardware will soon be capable of handling wafers. The wafers themselves will be patterned with test areas designed for AES analysis. These test areas will probably be of a larger geometry than the minimum features on the circuits. As such they will allow more efficient analyses. There will be more efficient manipulation of data with improved software routines and expanded use of the chemical state information available in AES data.

Literature Cited

1. Olson, R. R.; Palmberg, P. W.; Hovland, C. T.; Brady, T. E.; in Practical Surface Analysis ; Briggs, D.; Seah, M. P., Ed.; John Wiley and Sons: New York, 1983; pp 217-246.

2. Ryan, M. A.; McGuire, G. E.; in Industrial Applications of Surface Analysis ; American Chemical Society Symposium Series 199: Washington, D. C., 1982; pp 229-249.

3. McGuire, G. E.; Church, L. B.; Jones, D. L.; Smith, K. K.; Tuenge, D. T.; Journal of Vacuum Science and Technology A 1983, 1, pp 732-738.

4. Holloway, P. H.; Applications of Surface Science 1980, 4, pp 410-444.

5. Kazmerski, L.L.; Jamjoum, O. J.; Ireland, P. J.; Whitney, R. L.; Journal of Vacuum Science and Technology 1981, 18, pp 960-964.

6. Helms. C. R.; Johnson, N. M.; Schwarz, S. A.; Spicer, W. E.; Journal of Aplied Physics 1979, 50, pp 7007-7014.

7. Wildman, H. S.; Bartholomew, R. F.; Pliskin, W. A.; Revitz, M.; Journal of Vacuum Science and Technology 1981, 18, pp 955-959.

8. Shiota, I.; Motoya, K.; Ohmi, T.; Miyamota, N.; Nishizama, J.; Journal of the Electrochemical Society 1977, 124, 155-157.

9. Cheng, K. Y.; Cho, A. Y.; Wagner, W. R.; Bonner, W. A.; Journal of Applied Physics 1981, 52, pp 1015-1021.

10. Colelaser, R. A.; Microelectronics; Processing and Device Design ; John Wiley and Sons; New York, 1980.

11. Oldhams, W. G.; Scientific American 1977, 237, pp 111-128.

12. Gise, P. G.; Blanchard, R.; Semiconductor and Integrated Circuit Fabrication Techniques ; Reston Publishing Co.: Reston, VA, 1979.

13. Joshi, A.; Davis, L. E.; Palmberg, P. W.; in Methods of Surface Analysis ; Czanderna, A. W. Ed.; Elsevier: New York, 1975; pp 159-222.

14. Davis, L. E.; MacDonald, N. C.; Palmberg, P. W.; Riach, G. E.; Weber, R. E.: Handbook of Auger Electron Spectroscopy 1976, Perkin-Elmer/Physical Electronics Division, Eden Prairie, MN .

15. Madden, H. H.; Journal of Vacuum Science and Technology 1981, 18, pp 677-689.

16. Seah, M. P.; Dench, W. A.; Surface and Interface Analysis 1979, 1, pp 2-11.

17. Massopust, T. P., Ireland, P. J., and Kazmerski, L. L.; Journal of Vacuum Science and Technology A 1984, 2, pp 1123-1128.

18. El Gomatis, M.M.; Janssen, A. P.; Prutton, M.; Venables, J. A.; Surface Science 1979, 85, pp 309-316.

19. Burrell, M. C., Kaller, R. S.; and Armstrong, N. R.; Analytical Chemistry 1982, 54, pp 2511-2517.

20. Holloway, P. H.; Scanning Electron Microscopy 1978, I, pp 361-374.

21. Mathiew, H. J.; Landolt, D.; Surface and Interface Analysis 1981, 3, pp 153-156.

22. Farber, W.; Journal of Vacuum Science and Technology 1978, 15, pp 1139-1142.

23. Clough, S. P.; Application Note No. 8401 , Perkin-Elmer/ Physical Electronics Division: Eden Prairie, MN, 1984.

24. Pantano, C. G.; Madey, T. E.; Applications of Surface Science 1981, 7, pp 115-141.

25. Ahn, J.; Perleberg, C. R.; Wilcox, D. L.; Cobern, J. W.; Journal of Applied Physics 1975, 46, pp 4581-4583.

26. Carriere, B.; Lang, B.; Surface Science 1977, 64, pp 209-223.

27. Dawson, P. T.; Heavens, O. S.; Pollard, A. M.; Journal of Physics C 1978, 11, pp 2183-2193.

28. Holloway, P. H.; Madey, T. E.; Campbell, C. T.; Rye, R. R.; Houston, J. E.; Surface Science 1979, 88, pp 212-220.

29. Ohuchi, F.; Ogino, M.; Holloway, P. H.; Pantano, C. G.; Surface and Interface Analysis 1980, 2, pp 85-90.

30. Wehner, G. K.; in Methods of Surface Analysis ; Czanderna, A. W. Ed.; Elsevier: New York, 1975, pp. 5-37.

31. Hofman, S.; Surface and Interface Analysis 1980, 2, pp 148-160.

32. Lea, C.; Metal Science 1983, 17, pp 357-367.

33. Sputtering by Particle Bombardment ; R. Bershich, Ed., Springer-Verlag, New York, 1981.

34. Lindfors, P. A.; in Methods of Surface Characterization Vol 3 Specimen Handling and Depth Profiling; A. W. Czanderna, et. al.,Ed.; Elsevier, New York, to be published.

35. Byrne, W. J.; Proceedings Reliability Physics 1980, pp. 107-109.

36. Doyle, E., Jr.; Morris, B.; Microelectronics Failure Analysis Techniques, A Procedural Guide ; Rome Air Development Center, Air Force Systems Command, Griffiss Air Force Base, NY 13441, Section III-E.

37. McGuire, G. E.; Jones, J. V.; and Dowell, H. J.; Thin Solid Films 1977, 45, pp. 59-68.

38. Lowry, R. K.; and Hogrefe, A. W.; Solid State Technology January 1980, pp. 71-75.

39. Coburn, J. W.; and Winter, H. F.; Journal of Vacuum Science and Technology 1979, pp. 391-403.

40. James, H. K.; IEEE Transactions Vol. CHMT-3, No.3, Sept. 1980, pp. 370-374.

41. Thomas, S.;and Terry, L. E.; Applied Physics Letters 1975, pp 433-437.

42. Schmid, P. E.; Ho, P. S.; Foll, H. and Rubloff, G. W.; Journal of Vacuum Science and Technology 1981, pp. 937-943.

43. Roth, J. A.; and Crowell, C. R.; Journal of Vacuum Science and Technology 1978, pp. 1317-1324.

44. Ho, P. S.; Tan, T. Y.; Lewis, J. E.; and Rubloff, G. W.; Journal of Vacuum Science and Technology 1979, pp. 1120-1124.

45. Roth, J. A.; and Crowell, J. E.; Journal of Vacuum Science and Technology 1978, pp. 1317-1324.

46. Brady, T. E.; Hovland, C. T.; Journal of Vacuum Science and Technology 1981, 18, pp 339-342.

47. Ward, C. M.; Olson, R. R.; Brewer, R.; Johnson, S. G.; Microbeam Analysis 1983, Cooley, R. Ed.; San Francisco Press: San Francisco, CA, 1983, pp 5-10.

48. Turner, N. H., Murday, J. S., Ramaker, D. E.; Analytical Chemistry 1980, 52, pp 84-92.

49. Atzrodt, V.; Lange, H.; Physics and Statistics of Solids 1983, 79, pp 489-496.

50. Gaarrenstroom, S. W.; Journal of Vacuum Science and Technology 1982, 20, pp 458-496.

RECEIVED September 16, 1985

8

X-ray Photoelectron Spectroscopy Applied to Microelectronic Materials

William F. Stickle and Kenneth D. Bomben

Physical Electronics Division, Perkin-Elmer Corporation, Mountain View, CA 94043

> This review describes some of the recent studies that have used X-ray Photoelectron Spectroscopy to investigate materials commonly used in the microelectronics industry. It is divided into two sections: first, an introduction that is intended to give a general overview of the technique and second, a review, divided into five categories, of some of the work done over the last five years characterizing material systems of interest to the semiconductor industry.

Photoelectron spectroscopy has its origins in a physical principle that has been understood for nearly eighty years -- a principle that was partially responsible for a revolution in chemistry and physics. The principle is the photoelectric effect; the revolution arose because of the concept of quanta of energy.

Einstein's account (1) of the photoelectric effect can be expressed as:

$$E = h\nu - w \qquad [1]$$

where E is the kinetic energy of the photoelectron, $h\nu$ is the incident photon energy and w is the minimum energy required to remove an electron from the sample and is characteristic of the material being studied. Although the early work on the photoelectric effect dealt mainly with the conduction band electrons of metals, the above equation can be applied to the description of the photoejection of core electrons.

Electron spectroscopy can be divided into several categories. These would include X-ray Photoelectron Spectroscopy (XPS), also known as Electron Spectroscopy for Chemical Analysis (ESCA), Ultraviolet Photoelectron Spectroscopy (UPS) and Auger Electron Spectroscopy (AES). Other electron spectroscopies include Penning ionization and ion neutralization. XPS uses soft X rays as the ionizing source of radiation for the ejection of core electrons. UPS makes use of ultraviolet light and is generally used to study the valence electrons of atoms and molecules. AES is a secondary electron process where electron emission occurs because of the coulombic rearrangement induced by a core hole created by a photon or a high-energy electron beam.

Historically, electron spectroscopy has matured in two separate but related areas. One has been the use of electron spectroscopy as applied to analytical problems, especially those that relate to surfaces, such as failure analysis, corrosion, catalysis, or tribology. In such studies, the technique is often used in conjunction with other techniques such as low energy electron diffraction (LEED), secondary ion mass spectrometry (SIMS), or ion scattering spectroscopy (ISS). Another related area is the use of electron spectroscopy to examine the electronic structure of materials or chemical species.

In the past, the study of matter and its interaction with radiation was largely confined to the measurement of electromagnetic radiation. The excitation and de-excitation of atoms and molecules by photons emitted or absorbed when electrons made transitions between different discrete states has been well studied. Scanning electron microscopy (SEM), for example, makes routine use of energy dispersive X-ray analysis (EDX). In this case the sample emits X rays as a by-product of the technique.

High interest in the applications of photoelectron spectroscopy was stimulated by Kai Siegbahn and his group at Uppsala (2-4) in the 1950's. Their major initial contribution was to increase the kinetic energy resolution of their spectrometer over that of existing instruments. The photoelectron spectra were characterized by distinctly narrow features which could be energetically located with great accuracy as illustrated in Figure 1. Assuming an orbital description of the electronic structure of matter, each feature corresponded to photoionization of successive orbitals or structure associated with the photoionization process. In this example, the 2s, 2p, 3s and 3p levels of titanium are ionized by Mg X rays and appear in the binding energy spectrum. Additional transitions, attributable to relaxation processes and discussed below, also occur.

Over the past several years electron spectroscopy has established itself as a powerful technique in the electronic studies of matter and has gained wide acceptance as an unparalleled tool for surface analysis. The escape depth of the electrons is material and energy dependent but, in general, the information depth is about 50 A thus making electron spectroscopy a uniquely surface-sensitive analytical technique. This is because electrons that travel through a material for distances greater than this have a relatively high probability of suffering energy losses due to inelastic collisions with bound electrons. Hence, electrons emitted from deep within the sample (caused by X rays which can penetrate several micrometers) loose energy before leaving the surface and thereby contribute only to the background.

The purpose of this chapter is to discuss the principles of photoelectron spectroscopy and its applications in the semiconductor and microelectronics industries. Other recent reviews (5-10) have dealt with the use of XPS, AES, SIMS and ISS for failure analysis and materials characterization for these and related industries.

X-Ray Photoelectron Spectroscopy

XPS offers a direct method for measuring the binding energies of core orbitals. The ejected electron leaves the atom with a well-defined kinetic energy unique for that orbital and that atom. Since the dipole selection rules allow for electron emission from any orbital, photoejection occurs as long as the photon energy is greater than the binding energy of a particular level. Thus XPS allows the study of all accessible electronic levels regardless of symmetry (11).

XPS is a surface sensitive technique but its greatest strength lies in its ability to provide information on the surface chemistry of a sample. The binding energy can be measured to high resolution with today's spectrometers and is sensitive to changes in the chemical environment. Thus, in many cases the position of the XPS peak, after correction for static charging, can assign the chemical environment.

A large fraction of the XPS studies to date have involved the precise measurement of core-electron binding energies and the measurement of the chemical shifts of these energies. The term "chemical shift" refers to the fact that core-orbital ionizations often vary measurably due to changes in the chemi- cal environment. For example, the presence of a native oxide on silicon can be easily demonstrated by XPS as is shown in Figure 2. The silicon 2p photoelectron spectrum shows two distinct peaks, corresponding to the different environments of the silicon atoms, one, at higher binding energy, oxidized and the other, at lower binding energy, elemental. The asymmetry of the lower binding energy peak arises because of spin-orbit splitting in the 2p orbital. Final-state effects, discussed below, obscure the spin-orbit effect in the oxide peak.

XPS is well suited for these types of studies because the common excitation sources (Mg Kα = 1253.6 eV and Al Kα = 1486.6 eV) can easily cause photoejection. The more common ultraviolet sources are limited to the valence region and weakly-bound core levels. X-ray monochrometers can be used to narrow the excitation spectrum and remove satellite and bremsstrahlung contributions, however, they suffer a concomitant loss in intensity that increases the data-gathering time. Synchrotron radiation can be used to get around these limitations, although such light sources are not common and are not in widespread use among chemists. Studies that use tunable light sources often use the term Soft X-ray Photoelectron Spectroscopy (SXPS).

Core levels may be considered as representative of the filled subshells of an atom and are often found by XPS to be relatively sharp in energy. The width of the core photoelectron line depends on several factors both inherent and instrumental. Inherent sources include the lifetime of the subshell core hole created by photoionization and the various possible values of the final-state energy. The final-state energy will be influenced by multiplet splittings, multielectron effects or vibrational broadening. Another source of broadening is the presence of unresolvable chemically-shifted peaks. Instrumental sources of broadening include the width of the X-ray beam, the finite resolving power of the spectrometer, and non-uniform sample charging.

Multi-electron effects are often termed "satellite structure". Due to the sudden change in the potential of an atom, as experienced

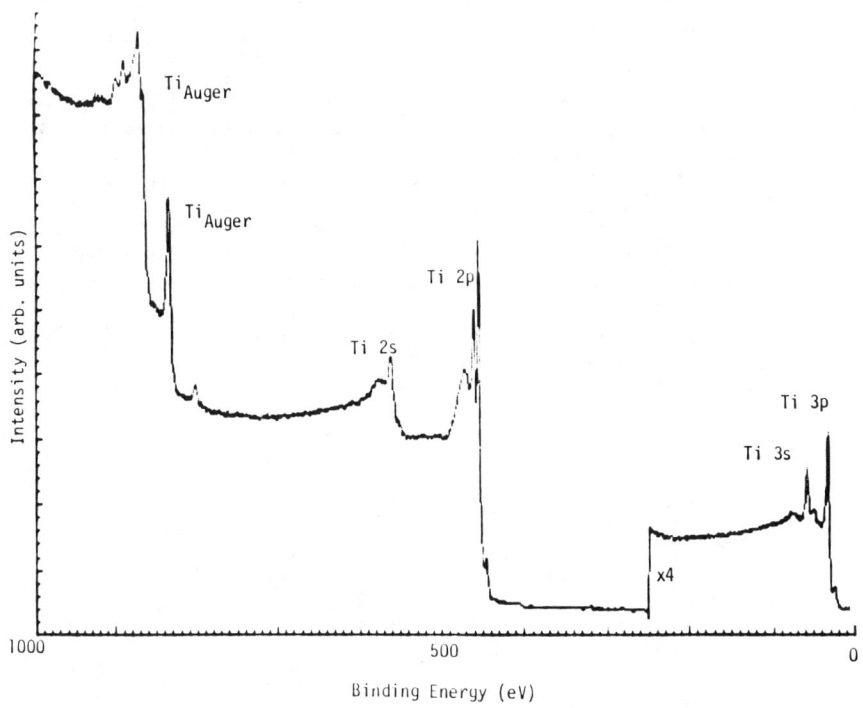

Figure 1. Photoelectron spectrum of clean Ti metal.

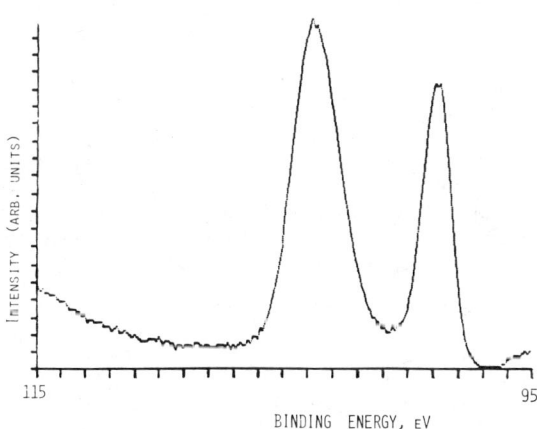

Figure 2. Silicon 2p photoelectron region showing oxide (left) and metal (right).

in photoionization, an electron in a valence orbital may go into an unoccupied bound or continuum state. These phenomena have been termed "electron shake-up" and "electron shake-off" and appear as satellites on the high binding energy side of the primary ionization. Shake-up structure appears as discrete satellite lines in the photoelectron spectrum while shake-off has no structure and appears as a continously rising background. Both types of structure occur at lower kinetic energy than the main photoelectron peak.

The final-state energy of a photoionized species will also be influenced by multiplet splitting. Multiplet splitting will occur if there are one or more unpaired electrons in the valence shell with unpaired spins. Photoionization in another shell can lead to more than one final state depending on how the unfilled shells couple.

In XPS, bound electrons are ejected to free states outside the atoms. The kinetic energy of these photoelectrons is well defined and is a measure of the electron's binding energy. From conservation of energy:

$$E_{h\nu} = E^f - E^i + E'_{kin} \qquad [2]$$

where E^f and E^i are the total energies of the final and initial states and E'_{kin} is the kinetic energy of the photoejected electron. $E^f - E^i$ can be defined as the binding energy, E_b, of the photoelectron. For gaseous samples this relationship is unambiguous because it is referenced to the vacuum level, but for solids this is true only when referenced to the vacuum level of the spectrometer. Consider, for example, photoejection from the K shell in solids. From equation 2 it is seen that:

$$E_{h\nu} = w' + E'_{kin} \qquad [3]$$

where w' is the sum of the work function and the binding energy of the sample. Entering the spectrometer slits, the kinetic energy of the electron is slightly altered due to an electrostatic field between the sample and spectrometer. This field is from the differences in the work function of the sample and the material from which the spectrometer is constructed. Common grounding merely requires the Fermi levels be equal and so any difference in work function will result in such a field.

Upon entering the spectrometer the electron acquires a new kinetic energy, and it is this energy which is measured and consequently related to the binding energy. Choosing the Fermi level as a reference, conservation of energy requires:

$$E_b^F = E_{h\nu} - E_{kin} - \phi_{sp} \qquad [4]$$

where E_b^F is the binding energy referenced to the Fermi level and ϕ_{sp} is the work function of the spectrometer and is thus independent of the excitation source. Therefore, the same work function may be applied to all measurements made on the same instrument.

<u>Auger Electron Emission.</u> In addition to the primary photo-process, secondary processes can occur. Following core excitation, the atom can relax by emitting either a photon or an electron. The electron-

ejection phenomenon is called the Auger process and the electron is an Auger electron. This process was discovered by Pierre Auger (12) while using a Wilson cloud chamber. The presence of a double track along the X-ray beam was indicative of a two-electron process. The first track was the electron due to photoionization, while the second track was due to what is now called the Auger electron.

Energetically, the core hole is filled by an electron cascading from a higher energy orbital along with the simultaneous ejection of yet another electron. The process is a simultaneous two-electron coulombic readjustment by the remaining electrons to the core hole. As such, it competes directly with X-ray fluorescence (XRF) but it is not limited by the dipole operator selection rules. All energetically allowed transitions are observed in an Auger electron spectrum. In addition, the electron escape depth is also a few tens of angstroms, unlike XRF where typical escape depths are on the order of tens of thousands of angstroms.

Different approximations can be used to estimate the energy of the Auger electrons (13,14). In general, the energy region for a particular transition can be easily calculated. The energy of a KL_IL_{II} transition, for example, is given by:

$$E_A = E^f - E^i = E(K) - E(L_I) - E(L_{II}) \quad [5]$$

plus a (negative) term of magnitude 10-20 eV which accounts for the interactions of the final holes. E_A is the kinetic energy of the Auger electron and $E(K)$, $E(L_I)$ and $E(L_{II})$ are the binding energies of the K, L_I and L_{II} shell electrons, respectively.

By precisely measuring the kinetic energy of the Auger electrons, chemical information may be deduced. In fact, the chemical shift of Auger electrons is, in some cases, larger than that of the associated photoelectron lines (13,15-17).

Auger Parameter. The energies of the X-ray induced Auger electrons, in conjunction with photoelectron energies, have been used by Wagner to develop the concept of the Auger parameter (18), which is defined as the kinetic energy of the Auger electron minus that of the photoelectron. This parameter, measurable without the necessity of static charge correction, is a quantity characteristic of a molecular or solid state. Chemical shifts in this quantity represent, approximately, differences in the polarization energy of the final state (19).

When an ion is created with a core vacancy, relaxation can occur by the emission of an X-ray photon, by the emission of an Auger electron, or by the emission of an electron in a Coster-Kronig process (19). The probability of each of these processes depends on several factors and the resulting particle emission may be detected by well-known methods.

The initial state for each of these processes contains a core hole on an atom. In AES, the vacancy is usually created either by a high-energy electron (5-10 keV) or by a photon. Commonly used photon sources include Mg and Al, as noted above, Zr (2042 eV), Au (2123 eV) and Ag (2984 eV). The higher energy sources are useful for creating deeper lying core holes than are conveniently accessible with Mg or Al, but are not too useful for routine photoelectron studies because of the large linewidth (20). Once the hole is created, an electron kinetic energy spectrum can be measured which can yield chemical, analytical, and electronic structure information.

In fact, it is impossible to avoid the appearance of spectral peaks due to Auger electrons during the course of a photoelectron experiment as seen in Figure 1. Some workers in XPS have viewed these lines as a nuisance, but the presence of the Auger lines increases the amount of information available in a photon-induced electron kinetic energy spectrum. Several workers have made use of the bremsstrahlung radiation from Al and Mg sources to look at Auger electron emission due to core holes of energy higher than the characteristic line of the X-ray source (13,20-23).

Wagner has also developed a two-dimensional chemical state plot (24-25) that makes use of the Auger parameter to distinguish between compounds of an element that have the same, or nearly the same, binding energy. In cases where the XPS chemical shift alone is not enough to determine the chemical state, these plots or the Auger parameter by itself can be used to help identify the type of compounds present in the surface.

Depth Profiles. By ion milling a surface with inert gas ions, the chemical composition can be followed as a function of depth, yielding a "Chemical Depth Profile". For example, the change from oxide to metal near an interface can be charted unambiguously by examining the changes in peak shape and position for each element at a series of depths. Care must be taken, however, in the interpertation of such data because ion-induced chemical changes, such as reduction, may occur. The chemical interactions at interfaces govern the width of the interface, the rate of diffusion across it and influence the electrical and physical properties of interfaces.

Figure 3 shows a depth profile of tantalum silicide on silicon. The sputter rate is approximately 200A/min. In this example, the interface can be easily identified at a sputter time of ten minutes. Figures 4 is the Ta 4f photoelectron region prior to the start of sputtering (spectrum a) and after five minutes (spectrum b). Spectrum a shows that Ta is predominately in the form of a silicide with only a samll amount of oxide evident. In spectrum b, only the silicide is present indicating that Ta is only oxidized at the sample surface. The Si 2p region (Figure 5) shows the Si to be heavily oxidized prior to sputtering (spectrum a) as can be seen in the height of the oxide peak. However, after five minutes of sputtering, spectrum b shows only the silicide.

Instrumentation. The basic instrumentation needed to do photoelectron spectroscopy is shown in Figure 6. The components include an X-ray source, sample or source chamber, electron energy analyzer, electron detector or multiplier, counting electronics, the all-encompassing vacuum system and a data output device. Samples are introduced through a load lock mechanism into the ultra-high vacuum region of the spectrometer. Some commercial spectrometers are now equipped with X-ray sources with multiple anode materials, allowing simple switching of excitation lines to "move" interference from Auger transitions or to induce more efficient ionization of deep-lying core levels to observe Auger lines for Auger parameter experiments.

Recently, equipment manufacturers have begun producing "small

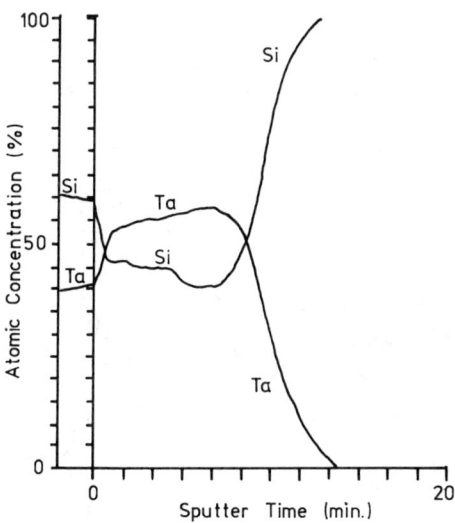

Figure 3. Depth profile of tantalum silicide on silicon.

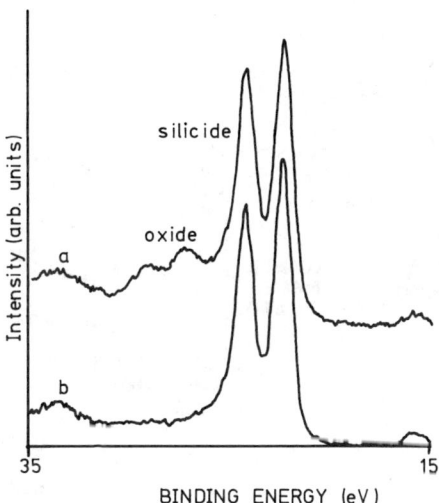

Figure 4. Ta 4f photoelectron region at time=0 (a) and time=5 (b) minutes.

152   MICROELECTRONICS PROCESSING: INORGANIC MATERIALS CHARACTERIZATION

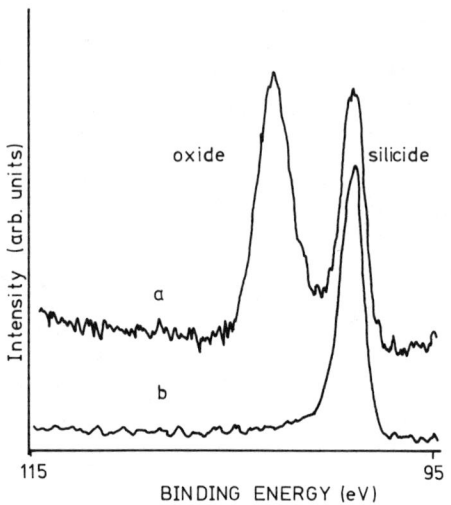

Figure 5. Si 2p photoelectron region at time=0 (a) and time=5 (b) minutes.

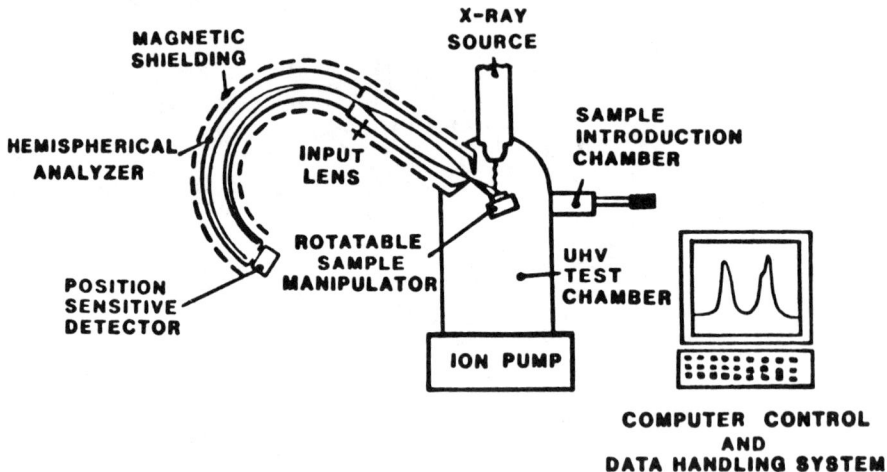

Figure 6. Schematic of a typical X-ray photoelectron spectrometer.

spot" XPS systems with the ability to examine, roughly, a 150 micrometer diameter spot on a sample surface (26-30). These instruments have two advantages; first they bring the major strength of XPS, the ability to determine chemical information, to bear on smaller areas and, second, it means that XPS depth profiles can be done at sputter rates that approach those commonly used in AES, as shown in Figure 3, above. Some recent instruments also allow the opportunity to do continuous sputter profiles which is a significant time saver.

<u>Soft X-Ray Photoelectron Spectroscopy.</u> Typically, SXPS uses synchrotron radiation to tune the X-ray energy for maximum absorption by one element. The chemical shift information available via this technique is exactly analogous to the chemical shifts discussed above. Examples using SXPS are included in the examples below and are meant to indicate further how XPS can be used to solve problems in the microelectronics industry.

Because of the ability to tune the excitation source to a particular element, data can be gathered rapidly. SXPS is often used to investigate the interactions of materials being deposited at an interface. The one drawback to the technique is that it requires time on a synchrotron.

<u>Applications</u>

The application of XPS to microelectronic materials typically focuses on two areas. First, as a surface investagative technique, XPS can be used to establish the chemical interactions between two materials or between a material and the ambient atmosphere. Second, in conjunction with depth profiling, the changes in chemical composition with depth can be followed.

The following sections give examples of the kinds of information that can be gained by using XPS to investigate a sample surface. The first three sections cover microelectronic materials and describe recent work in the characterization of these materials. The next section covers interactions at interfaces and the last section is a potpourri of topics including packaging and cleaning. While an attempt was made to keep these categories from overlapping, some of the cited work contains material that is applicable to more than one section.

<u>Silicon and its Oxides.</u> In the recent past silicon and silicon dioxide have become very important because of their widespread use in the semiconductor industry. Most devices currently being manufactured use silicon as a primary material, making silicon and its oxides the subject of many studies that characterize these materials and their interfaces.

Hofmann and Thomas (31) have discussed the effect of ion bombardment on thermally grown silicon dioxide. XPS showed that small changes in the surface chemistry were observable. The Si 2p and O 1s peaks were seen to broaden due to ion damage which presumably increased bond-angle disorder resulting in charge redistribution.

Photoelectron spectroscopy has also been applied to the study of thin SiO_2 layers on Si (32,33). Angular resolved XPS (34-36) can be used to produce non-destructive depth profiles and thereby avoid the ion-induced damage described by Thomas (5). Moulder and Hammond (37) found that by changing the angle between the sample surface and the entrance slit of the analyzer (Figure 7) an effective depth profile could be generated. Multiple oxidation states of silicon were identified in the interfacial region (Figure 8) and followed as a function of escape depth (Figure 9).

Bertrand and Fleischer (38) have studied the chemical deposition of silicon dioxide on indium phosphide. They found that on unoxidized, etched surfaces, oxide coverage was "always patchy". If the InP had a monolayer of chemisorbed oxygen, an approximately 60A-thick oxide film could be formed at room temperature through the formation of Si-O-P bonds at the interface.

Interdiffusion of metals in silicon has also been studied by SXPS (39) where it was revealed that Au and Al interact in very different ways with silicon. The Au-Si interface exhibits a strong chemical interaction while the Al-Si interface shows much weaker interactions. A fully reacted Si-Cr interface was found to be an effective barrier for Au-Si interdiffusion (40).

Thin films on Si or SiO_2 can also be investigated by XPS. The Cu-Si interface (41) was found to be similar to the diffuse Au-Si interface study done by SXPS. Comparison of Cu-Si with Cu-InP shows strong similarities between the two systems (42). Nefedov et al. (43) examined the $FeNi/SiO_2$ interface and found that the uppermost layer of the film was enriched with iron oxides while the next layer was depleted of iron atoms. Torrisi et al. (44) used XPS to study the adhesion of various metals to polysiloxane in an attempt to understand adhesion of films to common semiconductor materials.

Silicides. The most commonly used gate material in the technology of MOS integrated circuits is doped polycrystalline silicon. As device dimensions become vanishingly small the high sheet resistance of polysilicon becomes a limiting factor in device performance. The well-established technology of polysilicon gates has led to the incor- poration of metal silicides into the devices. The silicides of tantalum, molybdenum, titanium and tungsten (45) are often used. Using XPS to examine silicides allows for the examination of the chemistry of the surface either after treatment or upon reaction. Tungsten silicide, for example, has been investigated by XPS (46). The experimental results suggest bonding similar to that found in metal carbonyls in terms of σ donation and π backbonding.

Dubois and Nuzzo (47) have observed the reaction of silane with clean Ni, Rh, Pt, Mo, Ta, W, Co and Au. Surface compound formation is strongly suggested by the observation of shifts to higher binding energy of the silicon and metal core levels. These shifts are indicative of silicide formation (48,49). In another report (50), ruthenium, rhodium and palladium silicides were studied by XPS. Binding energies for the various silicides are tabulated, showing these differences in binding energy between the silicides and the metals are very small. Therefore, for some metals, the transition from metal to silicide can be followed by XPS but for other metals it is not so straightforward.

Rare earth silicides have been studied for several reasons. Of fundamental interest is the fact that the rare earth silicides will form solid solutions showing both bivalent and trivalent rare earth atoms as well as compounds showing mixed valence behavior (51-53). More practical reasons for such studies include the fact that rare

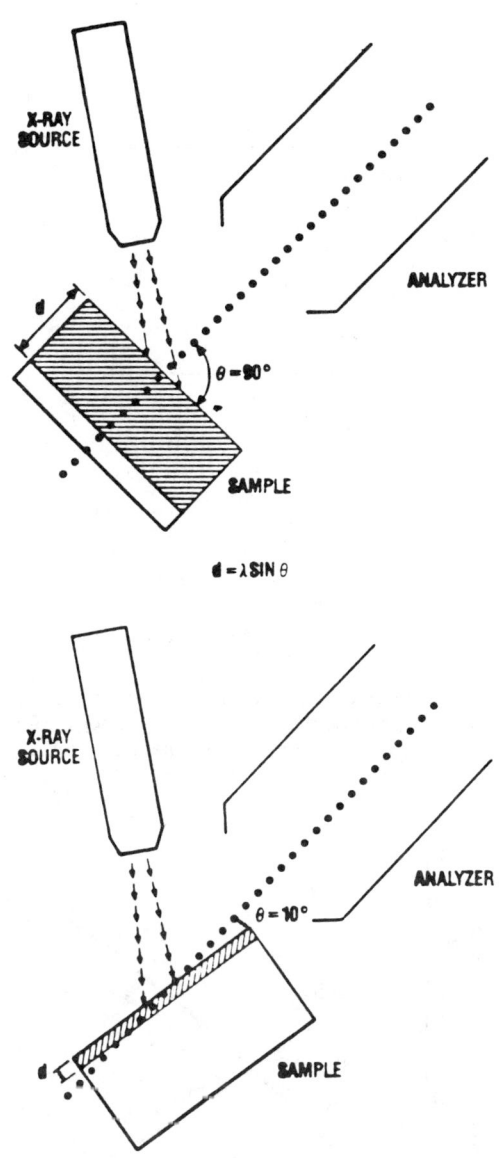

Figure 7. Experimental principles of Angular Resolved X-ray Photoelectron Spectroscopy (ARXPS). Reproduced with permission from Ref. 37. Copyright 1985 Research and Development.

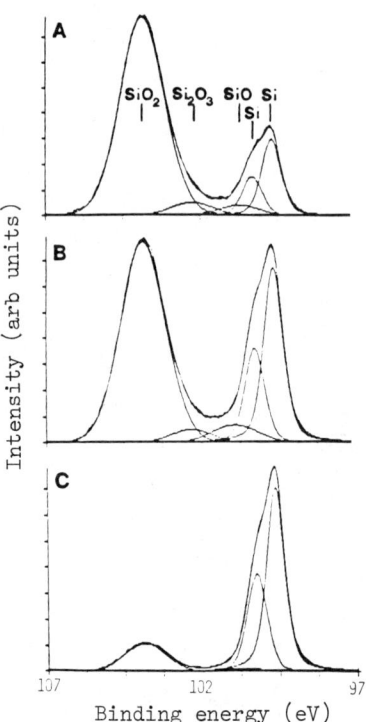

Figure 8. ARXPS spectra of silicon 2p region of silicon with a native oxide showing multiple oxidation states. A) θ = 10; B) θ = 20; C) θ = 90. Reproduced with permission from Ref. 37. Copyright 1985 Research and Development.

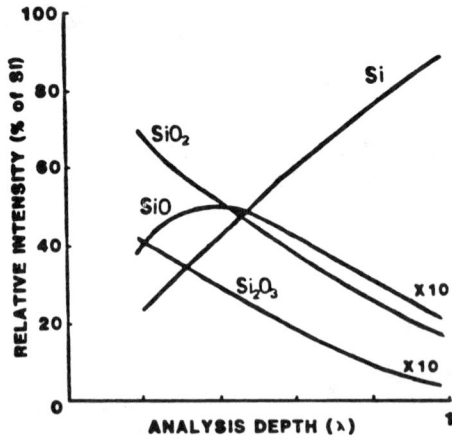

Figure 9. Relative intensities of the silicon species versus analysis depth. Reproduced with permission from Ref. 37. Copyright 1985 Research and Development.

earth contacts exhibit a small Schottky barrier. One such system studied was the Yb-Si interface (54). Mixed valence compounds were found where two stable and uniform compositions were observed.

Pirri et al. (55) studied cobalt disilicide epitaxial growth on the silicon (111) surface. At room temperature, a strong reaction forms CoSi during the deposition of up to four monolayers of Co. However, more than four monolayers of Co results in metallic cobalt being deposited. Annealing this surface results in the formation of the disilicide. XPS gives confirmation of the sp^3 covalent nature of the Co-Si bond in $CoSi_2$.

Non-Silicon Semiconductor Materials. III-V compounds are finding wider uses in the semiconductor industry. One area of primary interest is the concern over the cleanliness of substrates used in Molecular Beam Epitaxy (MBE) studies. Substrate preparation is important because the substrate may play a role in failure that occur after additional processing steps. The surface sensitivity of XPS makes it ideally suited for such investigations.

Various cleaning procedures have been studied by several groups using XPS. It was observed (56) that low energy ion etching combined with simultaneous annealing was an efficient cleaning procedure. Vasquez et al. (57) examined different cleaning procedures for GaAs (100) substrates. The GaAs oxide decomposition was observed to proceed by the thermal reduction of As_2O_5 which led to the formation of Ga_2O_3 and As metal. Minimum carbon contamination was observed by combining the generally accepted oxide passivation growth sequence with an HCl/EtOH strip. Stoichiometric GaAs could be obtained at temperatures as low as 350 C.

Woodall et al. (58) have also described a wet chemical technique for passivating air-exposed GaAs surfaces. XPS showed that there was little or no Ga metal or Ga and As compounds in the As film created during passivation or at the As/GaAs interface.

Interdiffusion of metals into GaAs has also been studied with metals such as Pd (59) and Au (60). The effect of oxygen on the intermixing of the GaAs/Au interface has also been studied (61) where oxygen was found to inhibit interdiffusion.

Kendelwicz et al. (62) have found evidence for the formation of palladium phosphide at the Pd/InP (110) interface. They found that, between 2.5 and 15 monolayers of Pd coverage on InP, deposition leads to the formation of a stable Pd_3P compound. They add that "for the first time for a non-elemental semiconductor interface the detailed nature of the chemical bond has been studied by following the evolution of the....core lines" and they note similar results for Ni on InP. This leads them to deduce that transition metal phosphide formation may be a general phenomenon on III-V materials.

XPS has also been used in an attempt to investigate Schottky barrier heights (SBH) and the effects of heating on semiconductor materials. Petro et al. (63) found that the SBH increases during the deposition of up to 25 monolayers of Au on GaAs and attribute this to intermixing of the Au with the GaAs. XPS was used to demonstrate that the Au is not chemically bound or alloyed with the GaAs. It was also found that heating inhibits the SBH increase by removing defect states from the surface region.

Pan et al. (64) used XPS to demonstrate the preferential surface segregation of As during the intermixing of noble metals on the GaAs (110) surface. Petro et al. (64) used SXPS to find significant Au intermixing in the Au-GaAs (110) interface. Furthermore, As moves to the surface at the highest Au coverages and heating the sample causes the Ga to become metallic. In an SXPS examination of Au on InP (110), Babalola et al. (66) found that, between 1 and 37 monolayers, InP + Au -> Au + In and P. At less than one monolayer of Au, they saw no measurable chemical changes. Skeath at al. (67) were able to use SXPS to examine Sb on GaAs (110) while Kendelewicz et al. (68) examined room temperature exchange reactions at the Al-InP (110) interface by SXPS.

Interfaces. Interfaces play an important role in determining the characteristics and responses of semiconductors. Typically, a layer or layers of a material is laid down on the substrate as a metal film, as an intermetallic or as a compound in an attempt to impart particular electrical properties to the device. In doing so, close attention must be paid to the interface, because the surface of a material is not like the bulk. Fortunately, films on semiconductor surfaces produce strong atomic and charge rearrangements at the microscopic interface that changes can be characterized by XPS. As Brillson (69) notes, XPS reveals that the "magnitude and stoichiometry of (substrate dissociation and diffusion of cation and anion into the metal) depends systematically on the strength and nature of the interface chemical bonding and that metal indiffusion also takes place."

As an example of the information that can be acquired by using XPS to study interfaces, consider the result of some recent work by Hirokawa et al. (70). In looking at Cu or Fe on SiO_2 they found that Cu, as a metal, diffused slightly into SiO_2 at 500-800 C. Upon further heating, CuO formed on SiO_2 and changed to Cu_2O or Cu_2O plus Cu. Fe began reacting with SiO_2 at about 600 C and migrated into the SiO_2 as Fe^{+2}. Franciosi et al. (71) demonstrated that Si-Cr interface formation at room temperature results in reacted phases that differ from both bulk chromium silicide and a Si-rich chromium silicide.

Miscellaneous Applications. The packages that incorporate semiconductor devices are often the source of failures. Alumina is commonly used in packaging substrates using Cr/Au or Cr/Cu metallizations. As smaller geometries are employed the substrate roughness becomes a considerable factor in terms of reliability. Changes in surface topography can cause reduction in adhesion, but, in an XPS study by Orent and Wagner (72), they found that sputtered Cr/Au and Cr/Cu thin films show chemical bonding to the substrate which may contribute to adhesion. They also found silicon nitride to be an acceptable replacement for alumina. Further, titanium metallizations on silicon nitride were studied and they observed that titanium nitride had been formed during the titanium deposition.

It is also possible to use XPS to analyze the effects of mechanical handling and cleaning on surfaces. Wasche et al. (73) investigated Al samples and films in an attempt to determine if mechanical treatment of Al in vacuo resulted in clean metal surfaces. They were able to observe the disappearance of the oxide peak and concluded that mechanical pretreatment of Al results in surfaces that compare favorably with surfaces prepared by other techniques.

Sputter-deposited or sputter-etched surfaces can also be examined. Reactively sputter-ion plated TiN films showed the surface to be oxidized to TiO_2 (74). Surface films deposited during plasma etching of SiO_2 on Si by CHF_3 were studied by a combination of XPS and AES (75). It was found that two type of films were deposited. A "non-persistent" film could be removed by an oxygen plasma and consisted of a fluorocarbon polymer. The "persistent" film formed oxygen and fluorine compounds of silicon and could not be removed by an oxygen plasma.

The effects of ion bombardment were investigated by Christie et al. (76) on a range of Groups II and IV compounds. They were able to observe stoichiometric changes induced by the ion bombardment of surfaces. This has important consequences for inert gas ion bombardment and plasma etching processes in the fabrication of semiconductor devices.

Summary

This review has attempted to demonstrate how the chemical information that is inherent in XPS can be used to investigate problems of interest in the semiconductor and microelectronics fields. A number of areas have been discussed where XPS was found to play an important role in attempting to understand the chemical and physical processes that occur in semiconductor materials and devices, however, this review is not all-encompassing. Furthermore, the areas that can be investigated by XPS are not limited to these examples.

XPS offers unique advantages that complement or surpass other analytical techniques. The ability to analyze insulators as well as conductors makes this technique particularily useful. In addition, there is usually little damage to the sample. With the spatial resolutions that are now available on small spot instruments, the technique is no longer limited to large area analysis.

Literature Cited

1. Einstein, A. Ann. Physik 1905, 31, 983.
2. Siegbahn, K.; Nordling, C.; Fahlman, A.; Nordberg, R.; Hamrin, K.; Hedman, J.; Johansson, G.; Bergmark, T.; Karlsson, S. E.; Lindgren,I.; Lindberg, B. "ESCA : Atomic, Molecular and Solid State Structure by Means of Electron Spectroscopy"; Nova Acta Regiae Soc. Sci. Upsalienius, Ser IV, Vol. 20 (Almqvist and Wiksells, Stockholm, 1967).
3. Hagstrom, S.; Nordling, C.; Siegbahn, K. Z. Physik 1964, 178, 439-444.
4. Siegbahn, K.; Nordling, C.; Johansson, G.; Hedman, J.; Heden, P.-F.; Hamrin, K.; Gelius, U.; Bergmark, T.; Werme, L. O.; Manne, R.; Baer, Y. "ESCA Applied to Free Molecules"; (North Holland, Amsterdam, 1969).
5. Thomas III, J. H. Chem. Anal. 1982, 63, 37-59.
6. Fuchs, E. Microelectronic Engin. 1983, 1, 143-159.

7. Torrisi, A.; Pignataro, S. Appl. Surf. Sci. 1982, 13, 389-401.
8. Drummond, I. W.; Hutton, D. R.; Thompson S. P.; Carrick, A. Int. Sympos. Test. Fail. Anal. (ISTFA), Proceed., 1984, 40-42.
9. Bowling, R. A.; Shaffner T. J.; Larrabee, G. B. Anal. Chem. 1985, 57, 130R-151R.
10. Anderson D. G.; Vandeberg, J. T. Anal. Chem. 1985, 57, 15R-29R.
11. See for example Appendices 1 and 3 of "Handbook of X-ray and Ultraviolet Photoelectron Spectroscopy"; D. Briggs (ed.), Heyden & Son Ltd., London, 1977.
12. Auger, M. P. Compt. Rend., 1925, 180, 65-68; J. de Phys. Radium 1925, 6, 205-208; Compt. Rend. 1926, 182, 773-775; Compt. Rend. 1926, 182, 1215-1216.
13. Carlson, T. A. "Photoelectron and Auger Spectroscopy"; Plenum, New York, 1975, p. 280.
14. Coghlan, W. A.; Clausing, R. Surf. Sci. 1972, 33, 411-413.
15. Fahlman, A.; Hagstrom, S.; Hamrin, K.; Nordberg, R.; Nordlng, C.; Siegbahn, K. Phys. Lett. 1966, 20, 159-160.
16. Wagner, C.; Biloen, P. Surf. Sci. 1973, 35, 82-85.
17. Shirley, D. A. Phys. Rev. A 1973, 7, 1520-1528.
18. Wagner, C. Faraday Disc. Chem. Soc. 1975, 60, 291-300.
19. Wagner, C. D. in "Handbook of X-ray and Ultraviolet Photoelectron Spectroscopy"; D. Briggs (ed.), Heyden & Son Ltd., London, 1977, p. 249.
20. Bearden, J. Rev. Mod. Phys. 1967, 39, 78-124.
21. Wagner, C. D.; Passoja, D. E.; Hillery, M.; Kinisky, T.; Six, H.; Jansen, W.; Taylor, J. J. Vac. Sci. Technol. 1982, 21, 933-944.
22. Castle, J.; West, H. J. Elec. Spec. Rel. Phen. 1979, 16, 195-197.
23. Castle, J.; West, H. J. Elec. Spec. Rel. Phen. 1980, 18, 355-358.
24. Wagner, C. D.; Gale,L. H.; Raymond, R. H. Anal. Chem. 1979, 51, 466-482.
25. Wagner, C. D.; Zatko, D. A.; Raymond, R. H. Anal. Chem. 1980, 52, 1445-1451.
26. Yates, K.; West, R. H. Surf. Interface Anal. 1983, 5, 217-221.
27. Keast, D. J.; Downing, K. S. Surf. Interface Anal. 1981, 3, 99-101.
28. Kelley, M. A.; Scharpen, L. H.; Cormia, R. D. Int. Sympos. Test. Fail. Anal. (ISTFA), Proceed., 1984, 35-39.
29. Wagner, C. D.; Joshi, A. Surf. Interface Anal. 1984, 6, 215-220.
30. Stickle, W. F.; Bomben, K. D. unpublished results.
31. Hofmann, S.; Thomas III, J. H. J. Vac. Sci. Technol. 1983, B1, 43-47.
32. Grunthaner, F. J.; Grunthaner, P. J.; Vasquez, R. P.; Lewis, B. F.; Maserjian, J.; Madhukar, A. J. Vac. Sci. Technol. 1979, 16, 1443-1453.
33. Grunthaner, F. J.; Grunthaner, P. J.; Vasquez, R. P.; Lewis, B. F.;Maserjian, J.; Madhukar, A. Phys. Rev. Lett. 1979, 43, 1683-1686.
34. Fadley, C. S.; Baird, R. J.; Siekhaus, W.; Novakov, T.; Berstrom, S. A. L. J. Elec. Spec. Rel. Phen. 1974, 4, 93.
35. Paynter, R. W. Surf. Interface Anal. 1981, 3, 186-187.
36. Ebel, M. F. Surf. Interface Anal. 1981, 3, 149-152.

37. Moulder, J. F.; Hammond, J. S. Res. and Develop. 1985, 27, 144-147.
38. Bertrand, P. A.; Fleischer, P. D. J. Vac. Sci. Technol. 1983, B1, 832-836.
39. Brillson, L. J.; Katnani, A. D.; Kelly, M.; Margaritondo, G. J. Vac. Sci. Technol. 1984, A2, 551-555.
40. Fraciosi, A.; O'Neill, D. G.; Weaver, J. H. J. Vac. Sci. Technol. 1983, B1, 524-529.
41. Ringeisen, F.; Derrien, J. J. Vac. Sci. Technol. 1983, B1, 546-552.
42. Kendelewicz, T.; Rossi, G.; Petro, W. G.; Balboa, I. A.; Landau, I.; Spicer, W. E. J. Vac. Sci. Technol. 1983, B1, 564-569.
43. Nefedov, V. I.; Pozdeyev, P. P.; Dorfman, V. F.; Pypkin, B. N. Surf. Interface Anal. 1980, 2, 26-30.
44. Torrisi, A.; Marletta, G.; Puglisi, O.; Pignataro, S. Surf. Interface Anal. 1983, 5, 161-166.
45. Mattausch, H. J.; Hasler, B.; Beinvogl, W. J. Vac. Sci. Technol. 1983, B1, 15-22.
46. Akimoto, K. Appl. Phys. Lett. 1982, 41, 49-51.
47. Dubois, L. H.; Nuzzo, R. G. J. Vac. Sci. Technol. 1984, A2, 441-445.
48. Grunthaner, P. J.; Grunthaner, F. J.; Madhukar, A J. Vac. Sci. Technol. 1982, 20, 680-687.
49. Abbati, I.; Rossi, G.; Galliari, L.; Braicovich, L.; Landau, I.; Spicer, W. E. J. Vac. Sci. Technol. 1982, 21, 409-412.
50. Didyk, V. V.; Zakharov, A. I.; Krivitski, V. P.; Narmonev, A. G.; Senkevich, A. I.; Yupko, L. M. Izv. Akad. Nauk., SSSR, Ser. Fiz., 1982, 46, 802-806.
51. Iandelli, A.; Palenzona, A.; Olcese, G. L. J. Less Common Metals 1979, 64, 213-220.
52. Baglin, J. E. E.; d'Heurle, F. M.; Petersson, C. S. J. Appl. Phys. 1981, 52, 2841-2846.
53. Thompson, R. D.; Tu, K. N. Thin Solid Films 1982, 93, 265-274.
54. Rossi, G.; Nogami, J.; Yeh, J. J.; Landau, I. J. Vac. Sci. Technol. 1983, B1, 530-532.
55. Pirri, C.; Peruchetti, J. C.; Gewinner, G.; Derrien, J. Phys. Rev. B 1984, 29, 3391-3397.
56. Oelhafen, P.; Freeouf, J. L.; Pettit, G. D.; Woodall, J. M. J. Vac. Sci. Technol. 1983, B1, 787-790.
57. Vasquez, R. P.; Lewis, B. F.; Grunthaner, F. J. J. Vac. Sci. Technol. 1983, B1, 791-794.
58. Woodall, J. M.; Oelhafen, P.; Jackson, T. N.; Freouf, J. L.; Pettit, G. D. J. Vac. Sci. Technol. 1983, B1, 795-798.
59. Oelhafen, P.; Freouf, J. L.; Kuan, T. S.; Jackson, T. N.; Batson, P. E. J. Vac. Sci. Technol. 1983, B1, 588-592.
60. Narusawa, T.; Watanabe, N.; Kobayoohi, K. L. I.; Nakashima, H. J. Vac. Sci. Technol. 1984, A2, 538-541.
61. Lu, Z. M.; Petro, W. G.; Mahowald, P. H.; Oshima, M.; Landau, I.; Spicer, W. E. J. Vac. Sci. Technol. 1983, B1, 598-601.
62. Kendelwicz, T.; Petro, W. G.; Landau, I.; Spicer, W.E. Phys. Rev. B 1983, 28, 3618-3621.
63. Petro, I.; Babalola, A.; Skeath, P.; Su, C. Y.; Hino, I.; Lindau, I.; Spicer, W. E. J. Vac. Sci. Technol. 1982, 21, 585-589.
64. Pan, S. H.; Mo, D.; Petro, W. G.; Lindau, I.; Spicer, W. E. J. Vac. Sci. Technol. 1983, B1, 593-597.

65. Petro, W. G.; Babalola, I. A.; Kendelewicz, T.; Lindau, I.; Spicer, W. E. J. Vac. Sci. Technol. 1983, A1, 1181-1184.
66. Babalola, I. A.; Petro, W. G.; Kendelewicz, T.; Lindau, I.; Spicer, W. E. J. Vac. Sci. Technol. 1983, A1, 762-765.
67. Skeath, P.; Su, C. Y.; Harrison, W. A.; Lindau, I.; Spicer, W. E. Phys. Rev. B 1983, 27, 6246-6262.
68. Kendelewicz, T.; Petro, W. G.; Babalola, I. A.; Silberman, J. A.; Lindau, I.; Spicer, W. E. J. Vac. Sci. Technol. 1983, B1, 623-627.
69. Brillson, L. J. Appl. Surf. Sci. 1982, 11/12, 249-267.
70. Hirokawa, K.; Yokokawa, Y.; Oku, M. Surf. Interface Anal. 1981, 3, 81-85.
71. Franciosi, A.; Weaver, J. H.; O'Neill, D. G.; Bisi, O.; Calandra, C. Phys. Rev. B 1984, 28, 7000-7008.
72. Orent, T. W.; Wagner, R. A. J. Vac. Sci. Technol. 1983, B1, 844-849.
73. Wasche, M.; Linke, E.; Richter, K. Surf. Interface Anal. 1982, 4, 178-179.
74. Robinson, K. S.; Sherwood, P. M. A. Surf. Interface Anal. 1984, 6, 261-266.
75. Tuppen, C. G.; Heckingbottom, R.; Gill, M.; Helsop, C.; Davies, G. J. Surf. Interface Anal., 1984, 6, 267-273.
76. Christie, A. B.; Lee, J.; Sutherland, I.; Walls, J. M. Appl. Surf. Sci. 1983, 15, 224-237.

RECEIVED July 11, 1985

Application of Neutron Depth Profiling to Microelectronic Materials Processing

R. G. Downing, J. T. Maki, and R. F. Fleming

Center for Analytical Chemistry, National Bureau of Standards, Gaithersburg, MD 20899

> Thermal neutron depth profiling (NDP) provides an isotope specific, nondestructive technique for the measurement of concentration versus depth distributions in the near-surface region of solids. The profiles are generated in real-time, analyzing depths of up to tens of micrometers. The method is currently applicable to the investigation of He, Li, Be, B, Na and Bi profiles. Demonstrative applications are presented for the technique, including: ion implantation-anneal sequence profiling; diffusion studies in a number of microelectronic materials; and homogeneity analysis of thin glass film overcoats. Comparisons are made for NDP and other profiling techniques such as secondary ion mass spectrometry (SIMS), Rutherford backscattering (RBS) and spreading resistance profiling (SRP).

Electronic and optical properties of a material are sensitive to subtle differences in composition and spatial distribution of many chemical species, whether they have been purposely introduced into the matrix or are present as contaminants. The capabilities of analytical techniques are challenged to quantify these differences adequately in the near-surface region and across interfacial boundaries. It is clear that collaboration among scientists with different techniques of analysis is required to correlate the properties of a material with specific compositional structures. In particular, this is true for evaluating the significant effect that light elements have upon a material. Only a few methods exist that can accurately measure and map, with respect to depth, these elements in sufficient detail. The methods commonly used include secondary ion mass spectrometry (SIMS), Rutherford backscattering (RBS) and for electrically active impurities, spreading resistance profiling (SRP).

In 1972, Ziegler, et al. (1) first reported the development of a near-surface technique complimentary to those mentioned above. The technique uses neutron reactions to measure absolute concentration versus depth profiles of a number of the light elements. Neutron

This chapter not subject to U.S. copyright.
Published 1986, American Chemical Society

depth profiling (NDP) allows the first few micrometers of nearly any material to be probed nondestructively. Biersack and co-workers (2,3) at the ILL facility in Grenoble have advanced the technique to its present capabilities.

Since its introduction, over fifty articles have been published (see references list) describing the use of NDP to investigate materials and effects directly relating to semiconductor research and device processing. The widespread application of NDP has been limited by the number of intense neutron sources available - nuclear reactors. At present, the U.S. has only two NDP facilities; one at the University of Michigan Ford Nuclear Reactor (4-6), and a facility at the National Bureau of Standards Research Reactor (NBSR) in Gaithersburg, Maryland (7). Nevertheless, the literature indicates NDP has significant potential in the semiconductor industry.

Foundations of NDP

Physics. Lithium, beryllium, boron, sodium, and a number of other elements, each have an isotope that, upon capturing a thermal neutron, undergoes an exoergic reaction. These reactions produce energetic charged particles, either a proton or an alpha particle depending upon the isotope, and a recoil particle. Each particle emitted has a specific kinetic energy defined by the Q-value of the reaction which in turn serves to identify the element. For the case of lithium,

$$^{6}Li + n \rightarrow {}^{4}He(2055 \text{ keV}) + {}^{3}H(2727 \text{ keV}) \tag{1}$$

while for the boron case 94 percent of the reactions are

$$^{10}B + n \rightarrow {}^{4}He(1472 \text{ keV}) + {}^{7}Li(840 \text{ keV}) + \text{gamma-ray } (478 \text{ keV}) \tag{2}$$

and 6 percent of the reactions (8) proceed as

$$^{10}B + n \rightarrow {}^{4}He(1776 \text{ keV}) + {}^{7}Li(1013 \text{ keV}) \tag{3}$$

Four elements, Li, Be, B, and Na, are particularly well suited for the NDP technique since their neutron cross sections are quite large, relative to other particle-producing reactions. In principal, there are essentially no interferences and profiling is permissible for all host materials. In practice, however, there are background contributions from energetic electrons and photons when analyzing materials that contain elements with large (n,photon) cross sections.

To obtain a depth profile, a well-collimated beam of thermal neutrons is used to uniformly illuminate a sample volume. While most of the neutrons pass through the sample without interacting, those sites containing reactive atoms will capture an occasional neutron and act as an isotropic source of mono-energetic charged particles. The particles travel outward in essentially straight paths and lose energy through numerous interactions with the electrons of the matrix. The difference between the well-known initial energy of the particle and its residual energy upon emerging from the surface of the sample is a direct measure relating the depth of origin for the particles (i.e., the site of the parent atom). The target chamber is kept under vacuum so that no additional energy is lost from the particle traveling between the sample surface and the detector. The low-energy neutron beam, about 10^{-2} eV per neutron, carries very little momentum, consequently; the reaction center of mass is concentric with the site of the parent atom and matrix damage is minimized during the analysis. In a worst case, where an entire 10^8 n/s beam is stopped by boron reactions in a sample, the temperature increase would only be a few °C/hr, even assuming there is no heat removal. Such a example would contain the equivalent of a few millimeters thickness of pure ^{10}B. The amount of boron consumed during a typical analysis is only a few tens-of-thousand atoms.

The depth corresponding to the measured energy for the emitted particle is determined by using the characteristic stopping power of the material, as compiled by Ziegler (9) and others (10), or by estimating the stopping power for compounds using Bragg's law (11) for the addition of the stopping powers of individual elemental constituents. Mathematically, the relation between depth and residual energy can be expressed as

$$x = \int_{E(x)}^{E_0} dE/S(E) \qquad (4)$$

where

x is the path length traveled by the particle through the matrix material, E_0 is the initial energy of the particle, $E(x)$ is the energy of the emerging particle, and $S(E)$ represents the stopping power of the material.

Spectra. The energy spectrum is collected from the particles emitted from all depths simultaneously using a silicon surface barrier detector, electronic amplifiers, an analog-to-digital converter and a multichannel analyzer. A reference pulse is fed into the electronics to monitor the stability of the system thus allowing corrections to be made should electronic drift occur during the course of the measurement. Specific systems are described in the references (1-4,6,7,12-17). By using computer-based data acquisition systems, the depth profile can be displayed at the time of analysis.

Examples of the detected energy spectra from three boron containing structures are shown in Figure 1. Figure 1(a) is the energy spectrum of a 2 nanometer thick, surface deposit of boron on a nickel substrate. Figure 1(b) shows the energy distribution of particles from a 660 nanometer thick borosilicate glass (BSG) film on a silicon wafer substrate. Both figures show the four-fold redundancy (see Equations (2) and (3)) of depth profiles for a boron containing material. The 1472 keV alpha particle or its 840 keV ^{7}Li recoil particle are typically used for the profile determinations because of their higher intensity, however the remaining two peaks can serve to confirm the results. Figure 1(c) shows only the energy spectrum of the 1472 keV alpha for a borophosphosilicate (BPSG) film with a periodic concentration variation from the surface down to the glass-

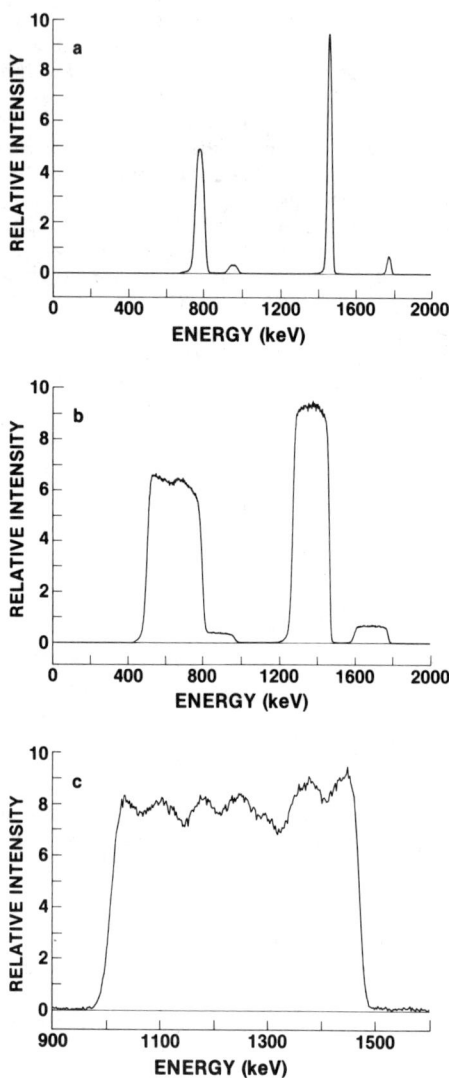

Figure 1. Energy profiles of particles emitted by the ^{10}B reaction for (a) a 2 nm thick surface deposition, (b) a 660 nm thick borosilicate glass film on Si, and (c) the excited state alpha particles from a borophosphosilicate glass film, 1200 nm thick.

silicon interface. The total thickness of this film is about 1.2 micrometers.

Resolution. The broadening of the peaks in Figure 1(a) is primarily due to the energy resolution of the detector and associated electronics. In addition to the detector resolution, other factors that contribute to the depth resolution include: i) small angle scattering of particles ii) energy straggling of the particles and iii) the non-zero acceptance angle of the detector giving a spread in pathlengths for particles from the same depth. These contributions to the resolution are treated by Biersack, et al. (18).

Each material has a characteristic stopping power and therefore the resolution and the depth of profiling will vary in different materials. The lithium particle from the boron reaction is heavier than its alpha counterpart and usually loses energy more rapidly allowing greater depth resolution; however, the alpha particles have the greater range and consequently allow deeper profiles to be obtained.

The full width at half maximum (FWHM) resolution in the depth profile obtained from the 1472 keV alpha of a boron reaction in silicon is typically a few tens of nanometers. On the other hand, protons from the $^{22}Na(n,p)^{22}Ne$ reaction give a resolution on the order of a few hundred nanometers, but can be used to profile 30 to 40 micrometers in depth. However, since particle emission is isotropic, the detector can be placed at an angle with respect to the normal of the sample surface to view the longer particle pathlengths. The depth resolution is improved in this fashion (18,19). Small concentration variations in the first nanometer of a sample surface can be identified by comparing the spectrum of a homogeneous sample, differentiated with respect to concentration, to the differentiated spectrum of an unknown sample. Deconvolution algorithms used to unfold the system response function from collected energy spectra (1,12,20-23) have provided improvement in depth resolution by greatly reducing system resolution broadening.

A noteworthy aspect of the NDP technique is that the chemical or electrical state of the target atoms has an inconsequential effect on the measured profile. Only the concentration of the major elements in the material need to be known to establish the depth scale through their stopping powers.

Elemental Sensitivities. The number of counts collected in a data channel, of energy width ΔE, is directly proportional to the concentration of target atoms located within that corresponding depth interval. Upon calibrating a facility for a given isotope, concentrations can be measured for that isotope in subsequent samples, independent of the matrix, the concentration level, or location within the depth that induced particles can escape the sample surface.

Table I lists several properties for target atoms and the detection limits using the NDP facility at the NBS 20 MW reactor. Isotopes with charged particle cross sections of about a barn or greater are given. The detection limits listed in Table I were calculated assuming 0.1 counts per second detected and an acceptance solid angle of 0.1 percent.

Table I. Properties and Detection Limits of Target Atoms Determined by Using the NDP Facility at the NBS 20 MW Reactor

Element	Reaction	% Abundance or (atoms/mCi)*	Energy of Emitted Particles (keV)	Cross Section (barns)	Detection Limit (atoms/cm^2)+
He	^{3}He(n,p)^{3}H	0.00014	572	5533	3.1×10^{13}
Li	^{6}Li(n,α)^{3}H	7.5	2055	940	1.8×10^{14}
Be*	^{7}Be(n,p)^{7}Li	(2.5×10^{14})	143	48000	3.5×10^{12}
B	^{10}B(n,α)^{7}Li	19.9	1472	3837	4.3×10^{13}
N	^{14}N(n,p)^{14}C	99.6	584	1.83	9.1×10^{16}
O	^{17}O(n,α)^{14}C	0.038	1413	0.24	7.1×10^{17}
Na*	^{22}Na(n,p)^{22}Ne	(4.4×10^{15})	2247	31000	4.7×10^{12}
S	^{33}S(n,α)^{30}Si	0.75	3081	0.19	1.2×10^{18}
Cl	^{35}Cl(n,p)^{35}S	75.8	598	0.49	3.4×10^{17}
K	^{40}K(n,p)^{40}Ar	0.012	2231	4.4	3.8×10^{16}
Ni*	^{59}Ni(n,α)^{56}Fe	(1.3×10^{20})	4757	12.3	1.4×10^{16}

* Radioactive species.
+ Detection limit based on 0.1 counts per second, 0.1% detector solid angle, and a neutron intensity of 6×10^8 s^{-1}.

Figure 2 depicts the depth versus residual energy curves obtained by evaluating equation (4) for particles emitted by 2(a) beryllium in gallium arsenide, 2(b) boron in silicon, and 2(c) sodium in silicon dioxide. Assuming a practical profiling depth of two micrometers for the case of boron in silicon and using the detection limit value from Table I, boron concentrations down to the 4 ppm level can be accurately measured. The time required for an analysis is a function of the element and the desired accuracy. A boron implant dose of 1×10^{15} atoms per cm^2 typically takes about 8 hours to obtain 1 percent precision at most points along the curve of the profile. Since the background signal is almost negligible, a sample could be counted for tens of hours to obtain the required definition in the profile shape.

Application

The development of the neutron depth profiling technique has been motivated by the importance of boron in both optical and microelectronic materials. Boron is widely used as a p-type dopant in semiconductor device fabrication and in the insulating oxide barriers applied as an organometallic or in vapor phase deposition glasses. NDP has both good sensitivity for boron and an adequate spatial resolution to the depth of a few micrometers. It is used both as a stand-alone technique and in a complimentary role with a variety of other analytical methods. A brief survey is made using applications to demonstrate the capabilities of NDP.

Implantation. Ziegler and co-workers (1,14,24,25) introduced NDP by determining the range and shape of boron implantation distributions in intrinsic and doped silicon wafers. With the resultant profiles, they were able to calculate diffusion coefficients for boron in crystalline, amorphous, and arsenic-doped silicon. Since little experimental data existed for the case of boron to judge the validity of the current range theories, the shape of the boron profiles from NDP were of great interest. NDP and other techniques have since been able to show that a Pearson IV model rather than a gaussian profile is required to describe accurately the implant distribution (21,26-28).

In subsequent experiments, Biersack, et al. (29) used the boron (n,alpha) reaction to show the effect of pre- and post-irradiation damage on boron implantation profiles. By post-irradiating a boron implant in silicon with 200 keV H_2^+, a migration of the boron to the induced damage sites was observed. In the same paper, diffusion and trapping of lithium ions in niobium were reported. Using the lithium (n,alpha) reaction, they mapped irradiation induced crystal defects through a depth of several micrometers with respect to several sample treatment conditions.

An advantage of NDP is brought out in both of the above applications. The thermal neutron probe induces negligible damage to the material either through sputtering of the sample surface, as observed with SIMS, or by alteration of the matrix. The neutrons carry an insignificant amount of momentum into the material and the induced reactions are of such low intensity that radiation damage is also negligible. This allows samples to be subjected to different processing conditions and to be studied at each stage. The sample

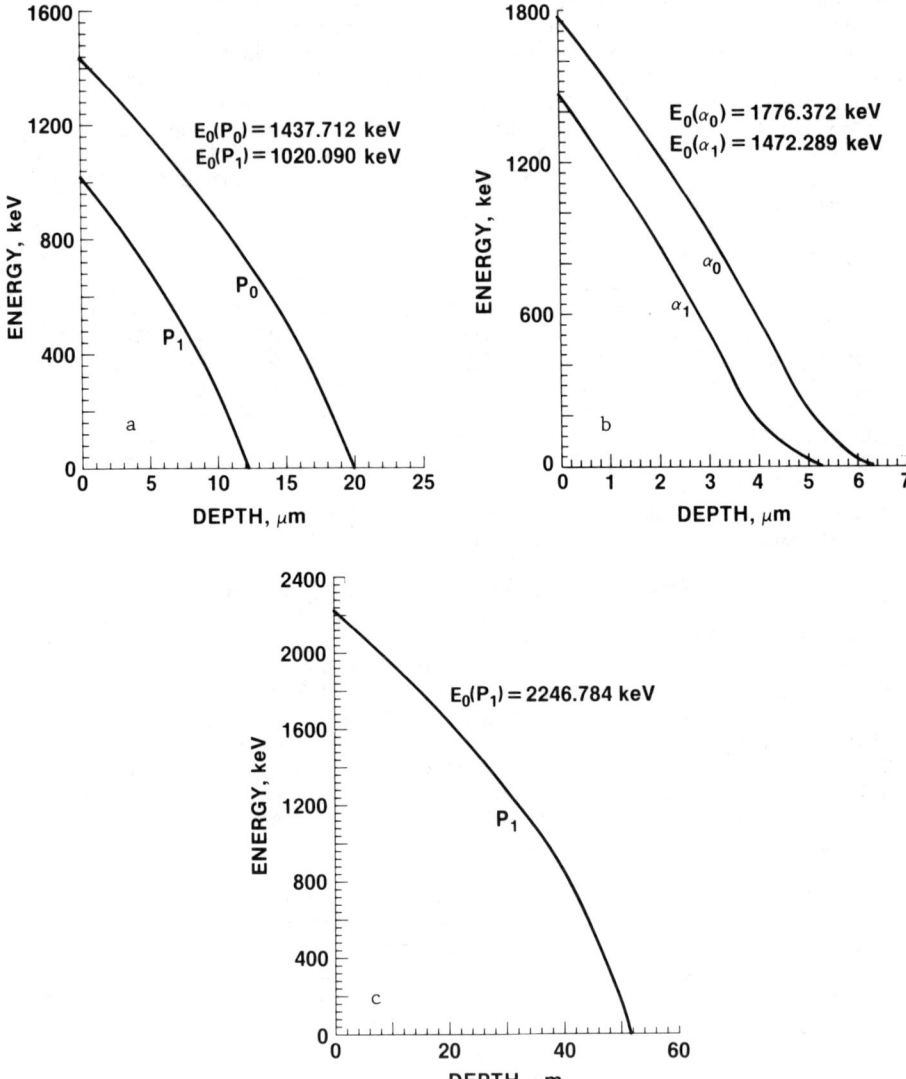

Figure 2. Plot of the residual particle energy versus depth for (a) ^{7}Be in GaAs, (b) ^{10}B in silicon, and (c) ^{22}Na in SiO$_2$.

may alternatively be passed to another analytical method to obtain complimentary data on the material. Analysis of the same sample by different methods allows extensive experimental testing of possible variability between samples or even across a single sample. As a result, NDP has been used as a referee technique for other methods of analysis (30). If radioactivity from activatable elements is formed, the sample should not be placed in a sputtering-type instrument thus avoiding possible contamination of sensitive detectors. This is certainly not the case for silicon wafers and most other electronic materials due to their small neutron activation cross sections.

Some of the features observed for the 'as-measured' NDP profile are illustrated in Figure 3. Curve 3(a) is an NDP profile for a 70 keV ^{10}B in silicon at a total dose of 4 x 10^{15} atoms per cm^2. To prevent channelling, the implant (30) was made nominally 7° off normal in silicon cut perpendicular to the <111> surface. Curve 3(b) is the same wafer after being annealed at 1000 °C for 30 minutes. The diffusion broadening bounded by the surface is clearly apparent. The apparent boron concentration above the surface is an artifact of the detector resolultion. Of particular interest is the small peak near the surface. A small unintentional air leak into the nitrogen backfilled annealing furnace allowed a thin film of SiO_2 to grow on the silicon wafer surface. The segregation coefficient of boron between Si and SiO_2 favors movement in the SiO_2 direction. Boron, as a consequence, was extracted from the bulk Si wafer, giving rise to the boron peak in the SiO_2.

Boron profiles by NDP in cadmium mercury telluride, an important infrared detector material, have been measured by Ryssel, et al. (31) and Vodopyanov, et al. (32). Cervena, et al. (23) used NDP to study the implantation profiles of ^{10}B in several photoresists used in masking operations and to determine range values for implants in several types of grown or deposited SiO_2 films.

Interfacial Profiling. Neutron depth profiling is well suited for measurements across interfacial boundaries. Kvitek et al. (20) and others (16,17,21,30) have studied profiles of boron implanted and diffused across the interfacial region of Si/SiO_2. Other NDP experiments (33,34) have been described for interfaces of silicon, silicon dioxide or metal on metal, where diffusion distributions and segregation coefficients were studied.

Knowledge of stopping powers for the major elemental constituents is the primary requirement to establish the depth scale. Figure 4 depicts an NDP profile of boron across an SiO_2-Si interface. Boron was implanted to a dose of 1 x 10^{16} atoms per cm^2 at 70 keV into a silicon wafer that had 0.2 micrometers of thermally grown SiO_2 covering the surface. The ^7Li particle energy spectrum from the $^{10}B(n,alpha)^7Li$ reaction was used for this profile to increase the depth resolution. Notice the smooth transition of the as-implanted boron concentration across the interfacial region represent by curve 4(a). Although the FWHM depth resolution is on the order of 10-15 nanometers, it is clear that no discontinuity exists at the interface of the two materials. Curve 4(b) shows the same region profiled again after annealing the sample for 30 minutes at 1000 °C (30). At the mean depth of the original implant, a residual peak remains.

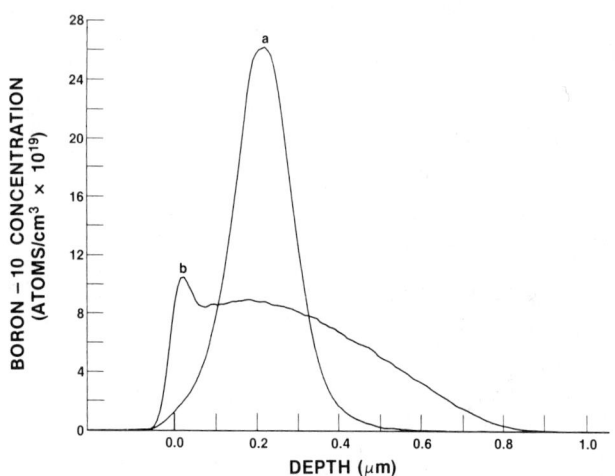

Figure 3. NDP depth profiles for a 70 keV ^{10}B implant in silicon at a dose of 4×10^{15} per cm^2. Depicted are (a) the as-implanted profile and (b) after a 30 minute anneal at 1000 °C.

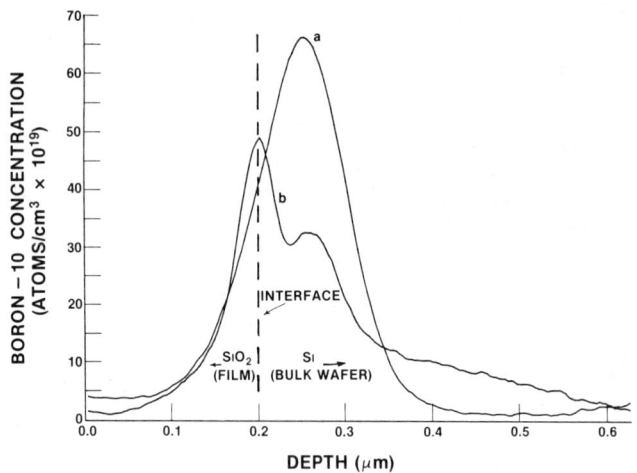

Figure 4. NDP depth profiles for a 70 keV ^{10}B implant to a dose of 1×10^{16} in an Si wafer covered with a 0.2 micrometer film of thermally grown SiO_2 (a) as deposited and (b) after a 30 minute anneal at 1000 °C.

The solid solubility of boron in silicon had been exceeded in the
original implant which is suspected (35) to give rise to Si-B compounds. Since the diffusivity of boron is much less in silicon
dioxide than in silicon, the boron on the silicon side of interface
migrates into the bulk silicon while the boron on the SiO_2 side of
the interface remains essentially immobile during the annealing. The
segregation coefficient of boron between Si and SiO_2 favors the SiO_2
which accounts for the increase in boron concentration at the
interface analogous to the effect seen in Figure 3(b).

A recent report by Matsumura, et al. (15,36) discusses the use
of the NDP method to investigate the diffusivity of boron in hydro-
genated amorphous silicon (a-Si:H), an important material in solar
cell production. Using a p-i-n, i.e., p-type/intrinsic/n-type,
layered amorphous silicon structure, the boron from the 60 nanometer
thick p-type layer was observed to diffuse into the underlying
undoped a-Si:H layer. From these measurements, they were able to
calculate the activation energy and diffusion coefficient for boron
in a-Si:H (the later being a dramatic twelve orders of magnitude
larger than for crystalline silicon) and estimate the deterioration
rate of boron-doped solar cells.

Channel Blocking. Matrix-charged particle detectors (26) are used
with the NDP technique to determine both the energy and lateral position of emitted particles in channel blocking experiments. The minor
damage incurred from thermal neutron induced reactions is negligible
when compared to RBS which bombards the sample with highly energetic
charged particles. It therefore seems appropriate that one of the
first applications of NDP was to establish the depth and lattice
position of dopants in single crystal materials (2,37).

Using NDP, Fink, et al. (38) have reported variations in the
lattice position of the dopant atoms with respect to the depth and
temperature treatment for boron implants in silicon. One example,
where a boron implant of 1 x 100 atoms per cm^2 was made at 120 keV
and annealed at 1000 °C for one hour, showed that two-thirds of the
boron atoms located near the average range of the implant remained
unordered. The remaining one-third atoms in that region were shown
to be interstitial. The further from the average range of the
implanted atoms, both above and below the plane, the more nearly substitutional the boron atoms were in the matrix. The largest component of the total boron implanted in these regions, however, remained
randomly located in the lattice.

In the past, researchers (39) have used etchable acetate foils
to map the channel blocking pattern, somewhat analogous to the
nuclear track technique (NTT) method of particle counting, however
quantitative analysis becomes tedious with this method and little
depth information is obtained. A review of channel blocking by NDP
for boron in silicon is presented by Fink, et. al. (38).

Thin Films and Leaching. Materials for optical waveguides and fiber
optics depend on uniform composition to prevent changes in the
refractive index of the material, which can reduce the intensity of
signal transmissions. Similar materials are used in thin, insulating
overcoats on electronic devices. The high solubility and mobility of

boron and lithium in these technologically important materials make them susceptible to leaching during wet processing, annealing at elevated temperatures, and during the cutting or polishing of surfaces. Riley, et al. (40) have studied some of the effects processing steps can have on boron in the near surface region of fiber-optic-grade glasses. Using NDP, SIMS, NTT, and prompt gamma activation analysis (PGAA) to quantify and map the boron distribution, they were able to show that a significant amount of leaching occurs within the first few micrometers of the samples.

Figure 5 shows boron depth profiles for a borosilicate glass which demonstrate the leaching effect. Curve 5(a) is from the surface of a freshly broken glass representing a homogeneous boron distribution. Curve 5(b) is the glass surface after a smooth saw cut was made using an aqueous coolant. The leaching of boron from the near surface is obvious and can be attributed to the action of the water during the cutting step. In their study, Riley, et al. demonstrated that leaching could be avoided by substituting a glycol based liquid for the water coolant during the cut.

For a sufficiently thin film, such as the BSG overcoat on a silicon water, a single NDP spectrum is capable of depicting the thickness, the distribution profile, and the total amount of boron present. After treatment to drive out trapped reaction products and voids in the glass, any boron or other mass losses can be quantified. Also, the effect of reflowing the glass film on the original boron profile can be identified (7).

Synergism

The major attributes of the NDP technique alone are illustrated through the preceding sections; however, the effectiveness of other surface analysis techniques can be enhanced through interaction with NDP.

SIMS. Secondary ion mass spectrometry has better sensitivity than NDP in most applications, however the SIMS concentration determination is on a relative scale and is not always linear. On the other hand, NDP is capable of independently measuring the absolute concentration for several of the key light elements, which in turn can be used to calibrate the SIMS signal. Presently, SIMS can achieve better spatial resolution (lateral and depth) as well, since it slowly creates a freshly formed surface to analyze. However, matrix effects can distort the depth and concentration scales. Such matrices require multiple corrections (41) and special experimental constraints to extract accurate information from the 'as-measured' spectrum. When analyzing insulating materials, especially for sodium, and where major chemical gradients occur, such as at interfacial boundaries, spurious effects are prominent for SIMS during the first nanometer or so of profiling. NDP is capable of identifying some of the problem areas and can offer quantitative solutions. Downing, et al. (42) coupled these techniques to correct the enhanced signal response observed by SIMS for both elements during an investigation of bismuth redistribution in tin.

An experiment by Ehrstein, et al. (30) compared the response of NDP, SIMS, and SRP for boron in silicon and across silicon/silicon oxide boundaries. Discrepancies were observed between the techniques

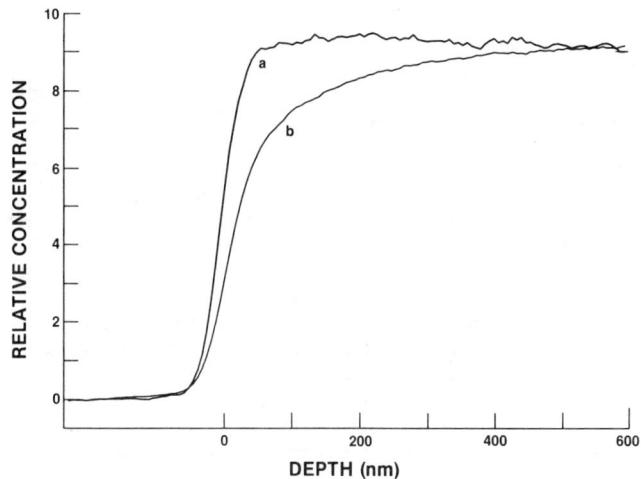

Figure 5. ^{10}B depth profiles by NDP depicting (a) the freshly broken borosilicate glass sample and (b) the near-surface of the glass after a smooth cut in an aqueous media.

in the depth scales, and the responses to the implantation features, which were only partially resolved and require further investigations to fully clarify.

SRP. Spreading resistance profiling measures the response of the dopant atoms that reside at electrically active sites in the lattice. Combining NDP and SRP allows one to distinguish dopants, such as boron, that are activated into electrically active sites from those located in nonactivated sites, such as in precipitates or interstitials. Therefore, the techniques can be used to select treatment methods that best activate the boron dopant and to provide information on the regions where non-electrically active dopant resides.

Another application of the combined techniques is in the study of overlapping dopants. When one element of a dopant pair is profilable by NDP, SRP would measure the combined response of both dopants, whereas NDP would map only the boron distribution. Subtracting the distribution, each dopant profile could be determined independent of the other.

One of the advantages of NDP over SRP characterization is the ability of NDP to profile dopants in infrared sensitive semiconductors such as CdHgTe (CMT) (31,32,39) and PbSnTe (PTT) (39). Intrinsic semiconductor lattices are not well suited for SRP measurements except under special conditions. Ehrstein, et al. (30) further discusses the attributes and the measurement discrepancies that exist between the techniques when profiling in silicon-based materials.

RBS. Rutherford backscattering and neutron depth profiling share several principals of measurement and approaches to spectral deconvolution. They both rely on the accuracy of the stopping power values for their depth scales and upon the energy resolution of charged particle energy analyzers for depth resolution. RBS is a more versatile technique since, in principle, all elements can be studied. It uses an external source of charged particles to probe the sample instead of depending on the limited number of neutron-induced charged particle reactions available to NDP. However, due to the physics of a particle scattering technique like RBS, light elements important to the semiconductor industry (i.e., B, Li, Be) are difficult to resolve, in the energy spectrum, from the signal produced by the heavier elements such as silicon. It follows that NDP can be used to complement RBS in analyses where both heavy and light dopants (or contaminants) are present. Studies can be envisioned to investigate the effect of distributions of heavy species on the diffusion, gettering or implantation of the lighter atoms, while using the same specimen repetitively.

RBS and NDP have used the Channel blocking technique to determine the presence of dislocated atoms in an ordered lattic, Biersack, et al. (2,37) describe the configuration of a NDP channel blocking system. In a comparison of channel blocking measurements between RBS and NDP, Fink, et al. (38) observed that radiation induced migration of boron can occur during the RBS measurement. The radiation damage is attributed to the recoil of the lattice atoms from the high energy protons used as the RBS probe. Using NDP, a more accurate understanding of the sample is obtained and measurement artifacts of the RBS technique can be investigated.

Summary

Neutron depth profiling has been applied in many areas of electronic materials research, as discussed here and in the references. The simplicity of the method and the interpretation of data are described. Major points to be made for NDP as an analytical technique include: i) it is nondestructive; ii) isotopic concentrations are determined quantitatively; iii) profiling measurements can be performed in essentially all solid materials, however depth resolution and depth of analysis are material dependent; iv) NDP is capable of profiling across interfacial boundaries; and v) there are few interferences.

Significant improvements in sensitivity will require more intense neutron sources. Systems using larger solid angles for particle collection will allow more efficient use of the present neutron fluences. Better algorithms for the deconvolution of system response from the energy spectrum will be necessary as well. Both NDP and RBS will benefit as the energy resolution of charged particle detectors improves with a corresponding gain in depth resolution. Fink, et al. (3) have designed a charged particle energy analyzer for an NDP system that should improve the energy resolution, while reducing the photon induced background levels.

Finally, the development of two- and three-dimensional neutron depth profiling should be possible through the use of position sensitive detectors and ion optics, providing an even more advanced tool for the further understanding of microelectronic materials.

Literature Cited

1. Ziegler, J. F.; Cole, G. W.; Baglin, J. E. E. Journal of Applied Physics, 1972, 43(9), 3809-3815.
2. Biersack, J. P.; Fink, D.; Lauch, J.; Henkelmann, R.; Muller, K. Nuclear Instruments and Methods, 1981, 188, 411-419.
3. Fink, D.; Biersack, J. P.; Liebl, H. In "Ion Implantation: Equipment and Techniques"; Ryssel, H.; Glawischnig, H., Eds.; Springer-Verlag: Berlin, 1983; pp. 318-326.
4. Myers, D. J. "Range Profiles of Helium in Copper After Thermal Anneals", University of Michigan-Ann Arbor: (Masters Thesis), 1979.
5. Myers, D. J.; Halsey, W. G.; King, J. S.; Vincent, D. H. Radiation Effects, 1980, 51, 251-252.
6. Halsey, W. G. "Concentration Dependent Thermal Release of Helium-3 Implanted in Molybdenum" University of Michigan-Ann Arbor: (Ph.D. Dissertation) 1980.
7. Downing, R. G.; Fleming, R. F.; Langland, J. K.; Vincent, D. H. Nuclear Instruments and Methods, 1983, 218, 47-51.
8. Deruytter, A. J.; Pelfer, P. Journal of Nuclear Energy, 1967, 21, 833-845.
9. Ziegler, J. F. "The Stopping and Ranges of Ions in Matter" Pergamon Press Inc.: New York, 1977.
10. Janni, J. F. Atomic Data and Nuclear Data Tables, 1982, 27, 147-529.
11. Thwaites, D. I. Radiation Research, 1983, 95, 495-518.

12. Bogancs, J.; Gyulai, J.; Hagy, A.; Nazarov, V. M.; Seres, Z.; Szabo, A. Joint Institute for Nuclear Research, 1979, 1, 59-64.
13. Bogancs, J.; Szabo, A.; Nagy, A. Z.; Csoke, A.; Pecznik, J.; Krakkai, I. Radiochemical and Radioanalytical Letters, 1979, 39(6), 393-403.
14. Crowder, B. L.; Ziegler, J. F.; Cole, G. W. In "Ion Implantation in Semiconductors and Other Materials"; Crowder, B. L., Ed.; Plenum Press: New York, 1973; pp. 257-266.
15. Matsumura, H.; Sakai, K.; Maeda, M.; Furukawa, S.; Horiuchi, K. Journal of Applied Physics, 1983, 54(6), 3106-3110.
16. Muller, K.; Henkelmann, R.; Bierseck, J. P.; Mertens, P. Journal of Radioanalytical Chemistry, 1977, 38, 9-17.
17. Nagy, A. Z.; Bogancs, J.; Gyulai, J.; Csoke, A.; Nazarov, V.; Seres, Z.; Szabo, A.; Yazvitsky, Y. Journal of Radioanalytical Chemistry, 1977, 38, 19-27.
18. Biersack, J. P.; Fink, D.; Henkelmann, R.; Muller, K. Nuclear Instruments and Methods, 1978, 149, 93-97.
19. Cervena, J.; Hnatowicz, V.; Hoffmann, J.; Kosina, Z.; Kvitek, J.; Onheiser, P. Nuclear Instruments and Methods, 1981, 188, 185-189.
20. Kvitek, J.; Hnatowicz, V.; Kotas, P. Radiochemical and Radioanalytical Letters, 1976, 24, 205-213.
21. Ryssel, H.; Haberger, K.; Hoffmann, K.; Prinke, G.; Dumcke, R.; Sachs, A. IEEE Transactions on Electron Devices, 1980, ED27(8), 1484-1492.
22. Nagy, A. Z.; Vasvari, B.; Duwez, P.; Bakos, L.; Seres, Z.; Bogancs, J.; Nazarov, V. M. In "KFKI Research Report 1979-91"; Ed.; Hungarian Academy of Sciences: Budapest, 1979; pp. 1-30.
23. Cervena, J.; Hnatowicz, V.; Hoffmann, J.; Kvitek, J.; Onheiser, P.; Rybka, V. Tesla Electronics, 1981, 14(1), 16-20.
24. Crowder, B. L.; Ziegler, J. F.; Morehead, F. F.; Cole, G. W. In "Ion Implantation in Semiconductors and Other Materials"; Crowder, B. L., Ed.; Plenum Press: New York, 1973; pp. 267-274.
25. Ziegler, J. F.; Crowder, B. L.; Cole, G. W.; Baglin, J. E. E.; Masters, B. J. Applied Physics Letters, 1972, 21, 16-17.
26. Muller, K.; Henkelmann, R.; Jahnel, F.; Ryssel, H.; Haberger, K.; Fink, D.; Biersack, J. Nuclear Instruments and Methods, 1980, 170, 151-1M5.
27. Ryssel, H.; Prinke, G.; Haberger, K.; Hoffmann, K.; Muller, K.; Henkelmann, R. Applied Physics, 1981, 24, 39-43.
28. Geissel, H.; Lennard, W. N.; Alexander, T. K.; Ball, G. C.; Forster, J. S.; Lone, M. A.; Milani, L.; Phillips, D. Nuclear Instruments and Methods, 1984, B2, 770-773.
29. Biersack, J. P.; Fink, D. In "Ion Implantation in Semiconductors"; Namba, S., Ed.; 1975b; pp. 211-218.
30. Ehrstein, J. R.; Downing, R. G.; Stallard, B. R.; Simons, D. S.; Fleming, R. F. In "Semiconductor Processing, ASTM STP 850"; Gupta, D., Ed.; American Society for Testing and Materials: Baltimore, 1984; pp. 409-425.
31. Ryssel, H.; Muller, K.; Biersack, J.P.; Kruger, W.; Lang, G.; Jahnel, F. Physica Status Solidi(a), 1980, 57, 619-624.
32. Vodopyanov, L. K.; Kozyrev, S. P. Physica Status Solidi(a), 1982, 72, K133-K136.

33. Jahnel, F.; Biersack, J.; Crowder, B. L.; d'Heurle, F. M.; Fink, D.; Isaac, R. D.; Lucchese, C. J.; Petrersson, C. S. Journal of Applied Physics, 1982, 53(11), 7372-7378.
34. Pelikan, L.; Rybka, V.; Krejci, P.; Hnatowicz, V.; Kvitek, J. Physica Status Solidi (a), 1982, 72, 369-373.
35. Ryssel, H.; Muller, K.; Haberger, K.; Henkelmann, R.; Jahnel, F. Applied Physics, 1980, 22, 35-38.
36. Matsumura, H.; Maeda, M.; Furukawa, S. Japanese Journal of Applied Physics, 1983, 22 No.5, 771-774.
37. Biersack, J. P.; Fink, D. In "Atomic Collisions in Solids"; Datz, S.; Appleton, B. R.; Moak, C. D., Eds.; Plenum Press: New York, 1975a; pp. 737-747.
38. Fink, D.; Biersack, J. P.; Carstanjen, H. D.; Jahnel, F.; Muller, K.; Ryssel, H.; Osei, A. Radiation Effects, 1983, 77, 11-33.
39. Biersack, J. P.; Fink, D. Nuclear Instruments and Methods, 1973, 108, 397-399.
40. Riley, J. E. Jr.; Mitchell, J. W.; Downing, R. G.; Fleming, R. F.; Lindstrom, R. M.; Vincent, D. M. Journal of Solid State Chemistry, 1984, to be published.
41. Deline, V. R. In "Thin Films and Interfaces II"; Baglin, J. E. E.; Campbell, D. R.; Chu, W. K., Eds.; North-Holland: New York, 19843; pp. 649-654.
42. Downing, R. G.; Fleming, R. F.; Simons, D. S.; Newbury, D. E. In "Microbeam Analysis"; Heinrich, K. F. J., Ed.; San Francisco Press: San Francisco, 1982; pp. 219-221.
43. Biersack, J. P. Radiation Effects, 1983, 78, 363.
44. Biersack, J. P.; Fink, D. Journal of Nuclear Materials, 1974, 53, 328-331.
45. Biersack, J. P.; Fink, D. In "Proceedings of the International Conference on Radiation Effects and Tritium Technology for Fusion Reactors"; Ed.; USERDA CONF-750989: Gatlinburg, 1975c; pp. II362-II371.
46. Biersack, J. P.; Fink, D.; Henkelmann, R. A.; Muller, K. Journal of Nuclear Materials, 1979, 85-86, 1165-1171. 47. Biersack, J. P.; Fink, D.; Mertens, P.; Henkelmann, R. A.; Muller, K. In "Plasma Wall Interactions"; Ed.; Pergamon Press: Oxford, 1976; pp. 421- 430.
48. Downing, R. G.; Fleming, R. F.; Maki, J. T.; Simons, D. S.; Stallard, B. R. In "Thin Films and Interfaces II"; Baglin, J. E. E.; Campbell, D. R.; Chu, W. K., Eds.; North-Holland: New York, 1984; pp. 655-656.
49. Fink, D. Nuclear Instruments and Methods, 1983, 218, 456-462.
50. Fink, D.; Biersack, J. P.; Grawe, H.; Riederer, J.; Muller, K.; Henkelmann, R. Nuclear Instruments and Methods, 1980, 168, 453-457.
51. Fink, D.; Biersack, J. P.; Jahnel, F.; Henkelmann, R. In "Analysis of Nonmetals in Metals"; Ed.; Walter de Gruyter & Co.: Berlin, 1981; pp. 163-171.
52. Fink, D.; Biersack, J. P.; Stadele, M.; Tjan, K.; Cheng, V. Nuclear Instruments and Methods, 1983, 218, 171-175.
53. Fink, D.; Biersack, J. P.; Stadele, M.; Tjan, K.; Haring, R. A.; De Vries, R. A. Nuclear Instruments and Methods, 1984, B1, 275-281.

54. Fink, D.; Biersack, J. P.; Tjan, K.; Cheng, V. K. Nuclear Instruments and Methods, 1982, 194, 105-111.
55. Fink, D.; Riederer, J. Nuclear Instruments and Methods, 1981, 191, 408-413.
56. Jahnel, F.; Ryssel, H.; Prinke, G.; Hoffmann, K.; Muller, K.; Biersack, J.; Henkelmann, R. Nuclear Instruments and Methods, 1981, 182/183, 223-229.
57. Kotas, P.; Obrusnik, J.; Kvitck, J.; Hnatowicz, V. Journal of Radioanalytical Chemistry, 1976, 30, 475-488.
58. Kristiakova, K.; Kristiak, J.; Kvitek, J.; Cervena, J. Nuclear Instruments and Methods, 1982, 199, 371.
59. Mezey, G.; Szokefalvi-Nagy, Z.; Badinka, C. S. Thin Solid Films, 1973, 19, 173-175.
60. Muller, K.; Henkelmann, R.; Boroffka, H. Nuclear, Instruments and Methods, 1975, 129, 557-559.
61. Nagy, A. Z.; Vasvari, B.; Duwez, P.; Bakos, L.; Boganos, J.; Nazarov, V. M. Physica Status Solidi (a), 1980, 61, 689-692.
62. Ryssel, H.; Kranz, H.; Muller, K.; Henkelmann, R. A.; Biersack, J. Applied Physics Letters, 1977, 30(8), 399-401.

RECEIVED August 8, 1985

10

Thermal-Wave Measurement of Thin-Film Thickness

Allan Rosencwaig

Therma-Wave, Inc., Fremont, CA 94539

> We have developed a method for measuring the
> thickness of semiconductor thin films that is
> nondestructive, noncontact and that can make
> measurements with 2-µm spatial resolution on
> both optically opaque and optically
> transparent films. This method is based on
> the use of high-frequency thermal waves.

Thin-film thickness measurements are of considerable importance
in semiconductor device processing. Direct measurements of the
geometric thickness of a thin film require either microscopic
examination of a cross-sectioned sample, a destructive procedure,
or a profilometer (stylus) measurement of a step height, a
contact procedure. Several noncontact, nondestructive methods
are also available, although these are all indirect measurements
of the geometric film thickness and therefore require either
knowledge of several material parameters in conjunction with a
comprehensive model, or more commonly, careful calibration with
known standards. The most common of these noncontact indirect
methods include optical interferometry and optical ellipsometry
for optically transparent films, eddy current and resistivity
probes for metallic films, and beta backscatter and x-ray
fluorescence for certain metallic and other optically opaque
films. None of the above noncontact methods is widely
applicable, and in particular none is suitable for both the
optically opaque and optically transparent films used in
semiconductor processing. We have developed a new method,
utilizing high-frequency thermal waves, that is noncontact,
nondestructive and that can be used for most, if not all,
semiconductor thin films, including optically opaque and
optically transparent films. In addition, the thin-film
thickness measurements can be performed with high spatial
resolution, that is, down to a 2-µm spot size.

It is well known from photoacoustic theory (1) that one can,
with thermal waves, obtain information about the thermal

0097-6156/86/0295-0181$06.00/0
© 1986 American Chemical Society

characteristics of a sample as a function of depth beneath its surface. Although there has been some experimentation in thermal wave depth profiling, (2-3) this capability has not been extensively exploited, primarily because of the lack of adequate theoretical models. The recent model of Opsal and Rosencwaig (4) (OR model) shows how depth profiling and multilayer thickness analysis can be performed from thermal-wave measurements using either surface temperature or thermoacoustic probes, and allows for a fuller exploitation of this depth-profiling capability. There have also been experimental impediments to thermal-wave profiling. For example, in many cases one would like to operate outside of a photoacoustic cell, to employ a completely contactless method for thermal-wave generation and detection, and to couple thickness measurements with high spatial resolution, this last requirement necessitating the use of high frequency (>100 kHz) thermal waves.

To date, only the thermoacoustic probe, which detects the thermoacoustic signals arising from the periodic stress-strain variations set up within the volume of the sample where the thermal waves are propagating, has been used routinely for detecting high-frequency (i.e., megahertz regime) thermal waves. (5) The thermoacoustic detection methodology has found important applications in thermal-wave imaging at high spatial resolution, where micron-sized thermal waves are needed, as in the study of semiconductor materials and devices. (5-7) The use of a thermoacoustic probe to detect the reflection and scattering of the thermal waves from the thermal features suffers, however, from the major drawback of requiring acoustic coupling between the sample and an ultrasonic transducer. In the analysis of semiconductor materials and devices, one would like to operate in an open environment, employ completely contactless methods for thermal-wave generation and detection, and be able to make measurements or obtain images at high spatial resolution.

To satisfy all the above requirements one needs to utilize lasers for both the generation and the detection of the thermal waves. The generation is, of course, straightforward. The detection is more involved, performed either by interferometric detection of the thermoelastic displacements of the sample surface or by laser detection of the local thermoelastic deformations of the surface. All the other methods for thermal-wave detection suffer from either being limited to low modulation frequencies or from needing contact to the sample.

There have been some initial studies of thermal-wave detection using the techniques described above. Ash and his colleagues have performed some imaging experiments with the laser interferometric technique, (8-11) while Amer and his colleagues have used both the laser interferometric and a laser deflection (surface deformation) technique for spectroscopic studies on amorphous silicon. (12-13) These various investigations were all performed at low to moderate modulation frequencies (<100 kHz) only.

We have developed a laser beam deflection technique which, although similar to the method employed by Amer's group, differs in several important respects. (14-15) In particular, our method employs highly focused heating and probe laser beams, both incident normal to the sample surface, and the experiments are performed at high modulation frequencies of up to 10 MHz.

Figure 1 depicts the experimental arrangement that we use. The 488-nm beam of a 100-mW Ar+ ion laser is intensity-modulated with an acousto-optic modulator, directed through a beam expander, and then focused to a 2-4μm diam spot on the sample. This is the heating beam, and it has a sample incident power of ~30 mW. The 633-nm beam of a 5-mW He-Ne laser, the probe beam, is directed through a beam expander, a polarizing beam splitter and quarterwave plate, reflected off a dichroic mirror, and then focused onto a 2-4μm diam spot on the sample with an incident power of ~2mW. The two laser spots are displaced ~2μm from each other at the sample surface. The 488-nm heating beam reflects back on itself, while the 633-nm probe beam undergoes a periodic deflection arising from the periodic change in the local slope of the sample surface. The reflected probe beam passes through the quarterwave plate again, and since it is 90° out of phase from the beam leaving the He-Ne laser, it is directed by the polarizing beam splitter to the bicell photodetector, which measures the periodic deflections of the probe beam. Unlike the laser interferometric technique, which measures local surface displacements in the vertical direction, the laser probe method measures changes in the local slope of the surface as depicted in Figure 2.

With the apparatus depicted in Figure 1, we are able to detect, at a 1-MHz modulation frequency, changes in the local surface slope in Al that result from local surface displacements of $\sim 10^{-4}$ Å/$(H_z)^{\frac{1}{2}}$. We have presented the basic theory underlying the thermoelastic deflection probe technique elsewhere. (15) Here we shall discuss some of the results that we have obtained with this technique for the measurement of thin film thickness.

First let us consider the signals obtained on bulk Al and Si with a view to understanding the roles of the thermal expansion coefficient and the thermal lens effect. Since Al and Si have comparable thermal diffusivities ($k/\rho C$), the surface temperatures in these two materials will be comparable under the same heating conditions. Still the beam deflection from an Al surface is approximately ten times greater than from a Si surface because of the difference in thermal expansion coefficient. Using a more complete model that accounts for optical reflectivities, finite absorption depths, and finite probe beam diameters, we have calculated the relative laser beam deflections from Al and Si as a function of modulation frequency. The results shown by the dashed curves in Figure 3 are found to be in excellent agreement with the experimental results plotted as open circles on the same figure.

Thermal lens effects occur in the air above the sample surface and within any layer of the sample that is not optically

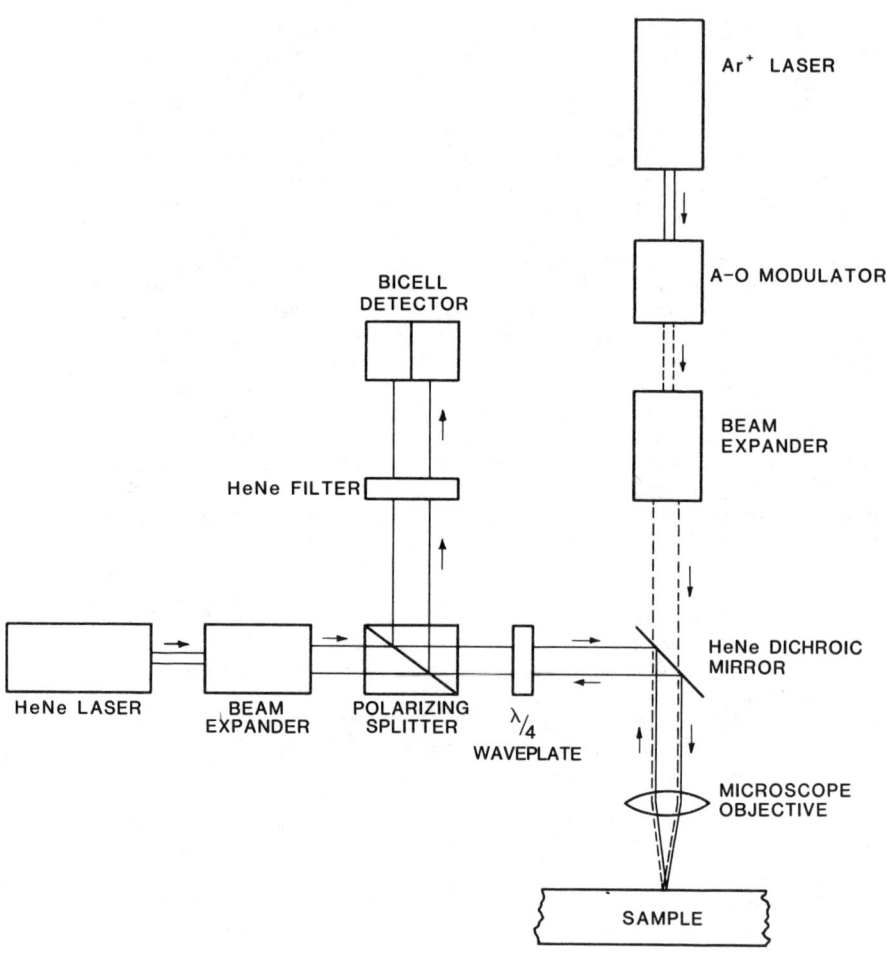

Figure 1. Schematic depiction of laser beam deflection technique used for the thin-film thickness measurement experiments.

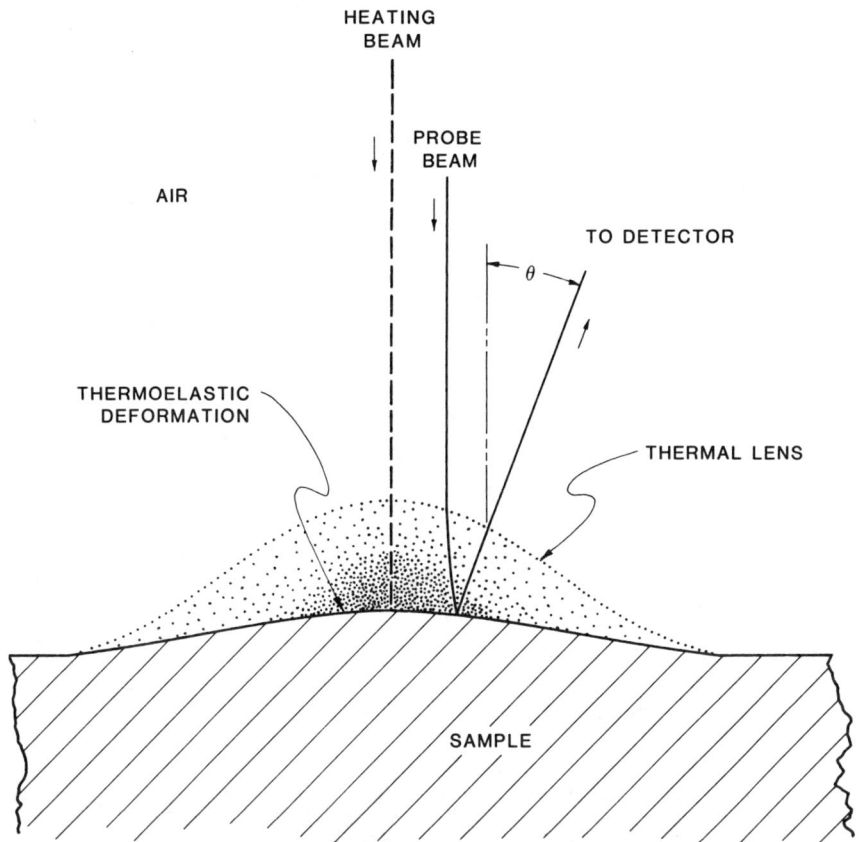

Figure 2. Schematic depiction of physical processes affecting the laser probe beam for an opaque homogeneous sample including thermoelastic deformation of the air-sample interface and thermal lens effects in the air above the sample.

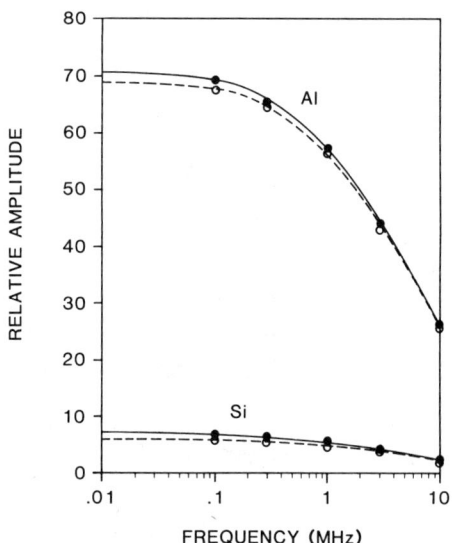

Figure 3. Relative amplitude of laser beam deflection signal as a function of thermal-wave (modulation) frequency for Al and Si under air (with thermal lens) and vacuum (no thermal lens) conditions. Experimental data are plotted as open (vacuum) and closed (air) circles and theoretical results as dashed (vacuum) and solid (air) curves.

opaque. Even though these thermal lenses have only micron-sized dimensions at the high modulation frequencies employed, their refractive power is still considerable since the normalized refractive-index gradient $n^{-1}(dn/dx) = -\epsilon(dT/dx)$ across the lens is now quite high, ϵ being of the same order as the thermal expansion coefficient of a solid. Also, even though the probe laser beam is incident normal to the sample surface, it strikes the thermal lens off-axis and thus undergoes refraction in both incident and reflected directions. Consequently, the theory predicts, and we find experimentally, that the thermal lens effect can be appreciable for some materials such as Si.

For air at 1 atm and 0°C, $\epsilon = 1.1 \times 10^{-6}/$ C, and thus the ratio of thermal lens effect to surface deflection effect for Si is ~0.6 and for Al is ~0.05. However, nonlinear effects due to higher temperatures (to be discussed later) will tend to reduce these ratios. Figure 3 presents comparisons with experiments for a complete calculation (that includes nonlinearities) under vacuum, where there is no thermal lens effect (dashed curves), and in air, where there is a thermal lens effect (solid curves). The agreement between theory and experiment is excellent. Note that the somewhat stronger dependence on frequency predicted for the thermal lens effect is observed experimentally in that its contribution to the total measurement decreases with increasing frequency.

In these thermal-wave experiments dc and ac temperature excursions can range from 30°C to several hundred degrees, depending on the sample's thermal characteristics. With such temperature of the various excursions, the dependence on temperature of the various thermal, optical, and elastic parameters has to be considered as well. In general, the most critical parameters appear to be the refractive index and the thermal conductivity. The index of refraction of air is given by, (16)

$$n = 1 + \frac{n_0-1}{1+\alpha_v T} P \qquad (1)$$

where $n_0 = 1.003$ is the index of refraction of air at 0°C, T is the temperature above 0°C, $\alpha_v = 3.66 \times 10^{-3}/$°C is the volume thermal expansion coefficient, and P is the normalized pressure of the air. Thus for temperature excursions in air of 10-100°C, ϵ will decrease by 30-50%. In addition, for most solids the temperature dependence of the thermal conductivity at our operating temperatures is given by, (17)

$$k = \frac{k_0}{1+\beta_k T} \qquad (2)$$

where k_0 is the thermal conductivity at T = 0°C, T is the temperature above 0°C and, β_k is a temperature coefficient. Temperature excursions of 50-100°C in Si where $\beta_k = 7.1 \times 10^{-3}/$°C, decrease k by 30-60%. These temperature effects on ε and k introduce appreciable nonlinearities in the model that cannot be neglected.

Optical effects will, of course, play an important role in these experiments as well. For example, in Si we have to take into account the optical absorption length (~1μm) for the 488-nm Ar+ ion laser light. Optical reflectivities must also be included. In addition, when dealing with optically transparent films such as SiO_2, optical interference effects within the film have to be considered as well. Figure 4 schematically depicts the situation encountered for an SiO_2 film on Si. Here we see the thermoelastic deformations of both the Si-SiO_2 and the SiO_2-air surfaces, the thermal lenses in both the SiO_2 and the air, and the optical interference effects on the probe beam in the SiO_2 film. Note that the thermal lenses have opposite signs in air and SiO_2 because of the opposite signs of their respective ε's.

When all the thermal lens, optical, and nonlinear effects are properly included in a 3-D O-R model, (15) we have a quantitative tool for measuring the thickness of thin films. This is illustrated in Figure 5, where we show theoretical curves and data obtained for single films of Al on Si and for double films of Al and SiO_2 on Si. We have used the magnitude of the thermal-wave signal rather than the phase in these measurements, since the magnitude has a greater dynamic range and can be measured more precisely. The data in Figure 5 are in excellent agreement with the theory both for the single and double films. The precision of the readings obtained with a 1-sec averaging time and 1-MHz modulation frequency translates to a thickness sensitivity of $\pm 2\%$ over the thickness range of 500-25,000 Å for these films.

In Figure 6 we show the theoretical curves and the data for a series of transparent SiO_2 on Si. Although Si-SiO_2 on Si is only a single film problem, the theory in this case must include thermoelastic deformations at both the Si-SiO_2 and the SiO_2-air interfaces, thermal lens effects in both the SiO_2 and the air, and optical interference effects in the SiO_2 (see Figure 4). The fit between theory and experiment is, with all this complexity, quite good, indicating that transparent as well as opaque films can be measured with this thermal-wave technique. The thickness sensitivity for SiO_2 films on Si appears to be $\pm 2\%$ over the 500-15,000 Å range.

This sensitivity is, of course, the precision of the measurement based on signal:noise considerations, and it does not reflect the absolute accuracy of the measurement. As with other noncontact, nondestructive methods, the thermal-wave technique provides an indirect measure of the geometric film thickness, and absolute accuracy must rely on either an accurate knowledge of the relevant physical parameters, or, as is common with the other methods, the use of calibration standards. In analyzing the data presented here we have used a rather complete (and complex) theoretical model to explain our experimental data, and thereby

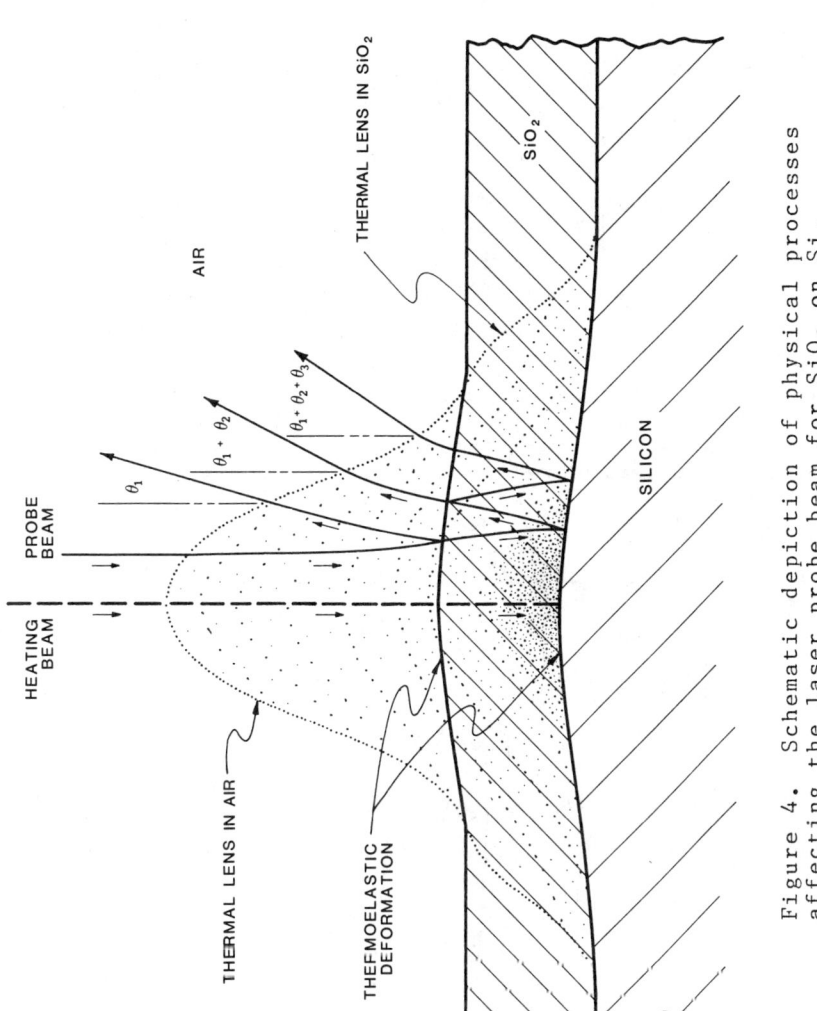

Figure 4. Schematic depiction of physical processes affecting the laser probe beam for SiO_2 on Si, including thermoelastic deformations of $Si-SiO_2$ and SiO_2-air interfaces, thermal lenses of opposite sign in air and SiO_2, and optical interference effects in the SiO_2 film.

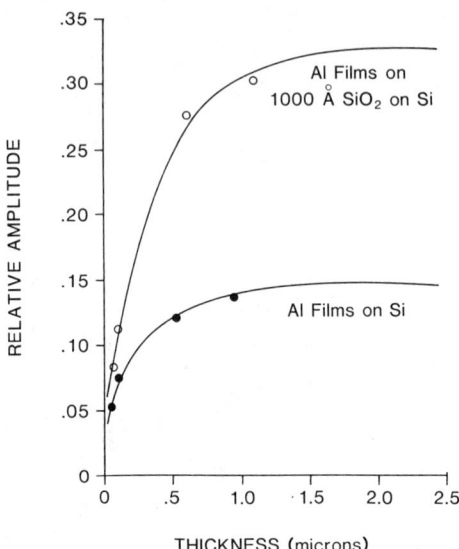

Figure 5. Relative amplitude at 1 MHz of laser beam deflection signal as a function of Al film thickness for a series of Al-on-Si and Al-on-SiO$_2$-on-Si films. Circles are experimental data, and curves are from the extended Opsal-Rosencwaig model.

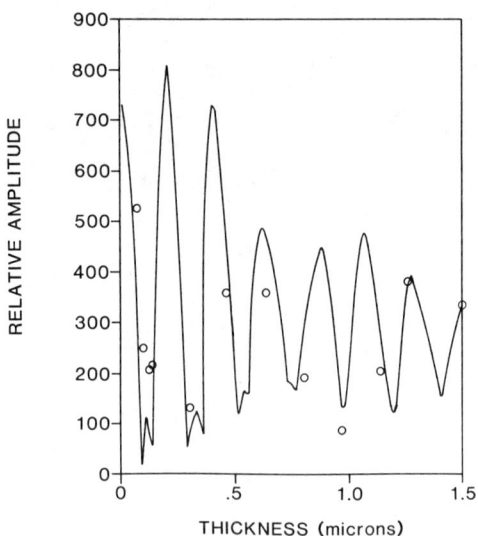

Figure 6. Relative amplitude at 1 MHz of laser beam deflection signal as a function of SiO$_2$ film thickness for a series of SiO$_2$-on-Si films. Circles are experimental data, and curves are from the extended Opsal-Rosencwaig model.

provide a firm theoretical basis for our method. Similarly, complex theoretical models are needed to account for all of the effects that play a role in any of the other indirect measurement techniques. In practice, however, the use of suitable calibration standards greatly simplifies the modeling requirements for all of the methods including the thermal wave technique. Finally, it should be noted that when the thickness of a thin film is known, then the thermal-wave signal can be used to characterize the composition or uniformity of the thin film material.

Literature Cited

1. Rosencwaig, A.; Gersho, A. J. Appl. Phys. 1976, 47, 64.
2. Adams, M.J.; Kirkbright, G.F. Analyst 1977, 102, 678.
3. Rosencwaig, A. J. Appl. Phys. 1978, 49, 2905.
4. Rosencwaig, A. Science 1982, 218, 223.
5. Opsal, J.; Rosencwaig, A. J. Appl. Phys. 1982, 4240, 53.
6. Cargill, G.S. Physics Today, 1981, 34, 27.
7. Rosencwaig, A. In "Thermal-Wave Measurement of Thin Film Thickness"; ACS SYMPOSIUM SERIES, this volume.
8. Ameri, S.; Ash, E.A.; Neuman, V.; Petts, C.R. Electron Lett. 1981, 17, 337.
9. Martin, Y.; Ash, E.A. Electron Lett. 1982, 18, 763.
10. Wickramasinghe, H.K.; Ameri, S.; See, C.W. Electron Lett. 1982, 18, 973.
11. Wickramasinghe, H.K.; Martin, Y.; Spear, D.A.H.; Ash, E.A. J de Physique, 1983, Coll. C6, Suppl. FASC. 10, C6-191.
12. Olmstead, M.A.; Amer, N.M.; J. Vac. Sci. Technol. 1983, 751, B1.
13. Amer, N.M.; J de Physique, 1983, Coll. C6, Suppl. FASC. 10, C6-185.
14. Rosencwaig, A.; Opsal, J.; Willenborg, D.L. Appl. Phys. Lett. 1983, 43, 166.
15. Opsal, J.; Rosencwaig, A.; Willenborg, D.L. Appl. Optics, 1983, 22, 3169.
16. Gray, D.E., Ed. "American Institute of Physics Handbook" McGraw-Hill: New York; 1972; Table 6e-5.
17. Kwong, D.L.; Kim, D.M. J. Appl. Phys. 1983, 54, 366.

RECEIVED October 31, 1985

11

Characterization of Materials, Thin Films, and Interfaces by Optical Reflectance and Ellipsometric Techniques

D. E. Aspnes

Bell Communications Research, Inc., Murray Hill, NJ 07974

> Reflectance-based optical characterization techniques offer the advantages of high energy resolution and sensitivity to both macrostructural and microstructural effects while nondestructively providing real-time information with the sample in any transparent ambient. Experimental and analytical methods are discussed, and examples are given to illustrate representative applications to problems of current interest in semiconductor technology.

Reflectance-based optical characterization techniques such as polarimetry, ellipsometry, and specular, modulated, and scattered-light reflectometry derive information about a sample from the component of incident flux returned by the optical impedance mismatch between sample and ambient. Despite an historical head start, fundamental limitations on resolving power and instrument-imposed limitations on wavelength tuning ranges cause reflectance-based optical techniques to be intrinsically less powerful than the more sophisticated electron- and ion-scattering spectroscopies that have been developed over the last few years. However, optical capabilities have also advanced. Significant progress has been made in a number of areas including automation, spectral scanning capabilities, and data analysis. The properties of rather complicated laminar and microscopically inhomogeneous samples can now be determined routinely from their measured optical behavior. At the same time, optical techniques retain their standard advantages of the generality of being able to obtain data in any transparent ambient, the convenience of being able to obtain these data in air or air-pressure environments, the energy resolution to determine alloy compositions to within fractions of a percent, the submonolayer surface sensitivity necessary to optimize cleaning procedure and to monitor film growth even in the initial stages of film formation, and finally the capability of being able to obtain these data nondestructively. The last advantage is particularly important, because the sample or material is not affected by the measurement and can be used for further purposes.

This paper is intended to give a brief overview of reflectance-based optical characterization techniques and their applications to determining sample properties. The next section deals with general principles, and includes comments about instrumentation and analytic methods. The rest of the paper consists of representative examples. Other applications can be found in several recent reviews and symposium proceedings (1-5). Length limitations preclude extensive discussions; references should be consulted for further details.

General Principles

The kind of information provided by reflectance-based optical data depends on whether the measurement is specular or nonspecular and on the spectral range involved. Nonspecular (scattered light) data carry information about macroscopic, spatially resolvable extrinsic features such as pits, scratches, or particulate contamination. The characteristic dimensions of these features must be of the order of or larger than the Rayleigh length $l = 0.61\lambda$, where λ is the wavelength of light.

Specular data carry information about intrinsic properties such as the atomic and electronic structure of materials as well as extrinsic information about microstructure and the thicknesses of films and interfaces. Throughout this paper, structure, microstructure, and macrostructure will refer to local atomic configurations, grains whose sizes are small compared to the wavelength of light yet sufficiently large to retain their own dielectric identity, and features larger than the Rayleigh length, respectively. Structure determines the vibrational lines in the infrared (ir), while microstructure and electronic energy levels determine the dielectric response in the visible-near ultraviolet (v-uv). Film thicknesses affect phase shifts and can cause interference oscillations in all regions of the optical spectrum. This information is encoded most clearly in the wavelength dependence of the reflected signal.

Instrumentation limitations create a natural division between the ir and the v-uv. Ir photodetectors have low sensitivities and high intrinsic noise levels. Fourier transform infrared (FTIR) spectrometers overcome these limitations by allowing all wavelengths to impinge on the detector simultaneously and separating them in the data reduction process. V-uv photodetectors have high sensitivities and very low intrinsic noise levels and typically operate in the shot-noise limit where the noise is determined by the light flux itself. Here, wavelength multiplexing provides no advantages and can saturate detectors. Therefore, v-uv systems are typically quasimonochromatic with the wavelength tunable over a finite range. The speed advantage of multiple wavelength detection can be realized in the v-uv by coupling the dispersing element (prism or grating) to a linear detector array.

Numerous methods have been used to measure and analyze specularly reflected light. A brief review of general concepts will place these in better perspective. Information in a monochromatic beam is contained in the complex amplitudes E_x and E_y of the projections of the electric field E parallel and perpendicular, respectively, to the plane of incidence. The incident components E_p^i and E_s^i are transformed by reflection to the reflected components $E_p^o = r_p E_p^i$ and $E_s^o = r_s E_s^i$, where r_p and r_s are the complex reflectance coefficients. At any given wavelength, these two coefficients completely describe the reflectance properties of a specular sample that is laterally isotropic no matter how complicated the structure normal to the surface may be. Polarimetry is the optical technique that deals with the amplitudes and the relative phase of r_p and r_s. Ellipsometry deals only with the relative amplitude and relative phase, usually represented as the complex reflectance ratio $\rho = r_p/r_s = (\tan\psi)e^{i\Delta}$. Reflectometry deals only with the amplitudes $R_p = |r_p|^2$, $R_s = |r_s|^2$, while modulated reflectometry deals with changes induced in these amplitudes by externally applied perturbations such as electric fields.

Some of the many different instrument configurations that have been proposed to obtain these data have been discussed by Hauge (6). The technical difficulties associated with accurately measuring amplitudes and relative phase simultaneously are formidable, so polarimetry is not presently a viable technique. Spectroscopic ellipsometry (SE) is rapidly becoming the dominant specular technique in the v-uv because two pieces of information are obtained in each measurement, and accuracy is not difficult to achieve. Ellipsometric data are relatively unaffected by source intensity fluctuations or by artifacts caused by light scattering from the surfaces of macroscopically rough samples. Accurate data are more difficult to obtain in normal-or near-normal-incidence reflectometry than in ellipsometry, but the instrumentation is much simpler, the computational burden is much less, and the results are relatively unaffected by focusing the probe beam to small spot sizes. Consequently, reflectometry and its modulated and scattered-light variations are ideally suited for obtaining spatially resolved information over large sample areas in a short time. Reflectometry and modulated reflectometry are also dominant in ir spectroscopy, where the presence or absence of a particular vibrational line is more important than its detailed shape.

<u>Analysis of Laminar and Microstructurally Inhomogeneous Samples</u>. Much recent attention has been directed to the problem of determining the characteristics of laminar samples and microscopically inhomogeneous materials from their optical properties. The science (or art) of optical data analysis is that of constructing models that accurately reproduce the measured functions of r_p and r_s for given samples for as many different conditions as possible. Independent variables may include wavelength, angle of incidence, index of refraction of the ambient medium, or combinations thereof. Models are idealized representations of actual sample configurations. Their properties are mathematically evaluated using the Fresnel reflectance expressions (7), which

describe wave propagation in laminar configurations, effective medium theories (8), which describe the dielectric responses of microscopically inhomogeneous materials, least-squares regression analysis (9), which allows the best-fit values, correlations, and confidence limits of the free parameters of the model to be determined systematically, and accurate reference data for the dielectric functions of the model constituents, which provide the foundation from which the model properties are evaluated. Typical free parameters include film and interface thicknesses, and compositions, densities, and degrees of crystallinity of microscopically inhomogeneous materials. The dielectric functions of these materials also depend on microstructural details such as grain shapes that cannot be measured directly, but the uncertainties associated with these ambiguities can be estimated with recently derived limit theorems (10).

Data analysis is usually performed in two distinct ways. In the first, the measured function of r_p and r_s serves as the template to which the model is forced to conform. The form that the model takes and the values of its free parameters are adjusted until the best fit is obtained. Success (credibility) is measured by the size of the confidence limits and by the accuracy to which the model actually represents the data. In the second, the equations representing the model are solved using the optical data for one (reflectometry) or two (ellipsometry) of the free parameters for each combination of independent variables, and the form of the model and the remaining free parameters are adjusted until artifacts such as wavelength-dependent film thicknesses or interference-related structure in refractive indices vanish. Here, success is measured by the extent to which such artifacts can be eliminated.

Both approaches require a model, that is, assumptions to be made about the nature of the sample. While independent knowledge or the data themselves can often provide the information needed to establish the gross features of a model, a general weakness of optical characterization methods is that considerable physical insight may be required to properly construct a model that accurately represents the finer details of a sample and its spectrum. It should be obvious that the more data, the better; only a very limited amount of information can be obtained from a single combination of independent variables, and except possibly for extremely simple or otherwise well characterized samples, such information should be considered suspect because no cross-checks are provided to establish its reliability. For these reasons, a spectroscopic capability is virtually essential for the analysis of complicated samples. However, single-wavelength measurements may be adequate for measurements on growing films where film thickness is the independent parameter, although the recent development of rapid-scan instrumentation may soon allow spectroscopic measurements to be performed under dynamic conditions (11).

Examples

Surface and Sample Quality. A focused-spot reflectometer capable of determining macroscopic imperfections and particulate contamination on a 4 in. semiconductor wafer in 10 s has recently been described (12). The sample is rotated and translated to scan a focused laser beam over its surface. The light scattered by macroscopic particles and defects is collected by a large-aperture lens. The detector threshold may be set to discriminate among different scattering strengths. A typical series of scans of the same wafer at different threshold settings is shown in Figure 1. Particle sizes down to fractions of a micron can be detected with this technique.

By using a near-uv laser, the same instrument can also determine microstructural properties such as the degree of crystallinity or the concentration of twin boundaries in amorphous (a-) Si (13) and silicon-on-sapphire (14) films, respectively. These parameters correlate with the uv reflectance because the dielectric properties of crystalline (c-), polycrystalline (p-), and a-Si are different.

Alloy Compositions and Carrier Concentrations. Modulated reflectometry can be used to accurately determine the uniformity of carrier concentrations and alloy compositions over a sample surface. The principle of the method can be understood by considering the electroreflectance (ER) spectrum of the E_1 transition of $Hg_{1-x}Cd_xTe$, $x = 0.2$, shown in Figure 2 (15). This spectrum was obtained by applying a small audio-frequency voltage between an ohmic contact on the back of the sample and the KCl-water electrolyte in which the sample was immersed. The modulating electric field is developed in the space charge region adjacent to the semiconductor-electrolyte interface. The threshold energy of the transition is given by the zero crossing of the structure with respect to the baseline, here near 5200Å. Since the threshold energy changes by about 1 eV across the alloy series, the sensitivity of the crossing wavelength, λ_o, to x at x = 0.2 is $d\lambda_o/dx = 2300$Å. If the wavelength were set at the zero crossing at a reference point on the surface, then a positive or negative signal would be present if the value of x at a second point were greater or less than that of the reference, respectively. A signal amplitude equal to 1% of the peak value near 5300Å would correspond to a shift by x by 0.0007. A contour map of the sample of Figure 2 with compositional variations in increments of 0.2% is shown in Figure 3 (15).

The maximum value itself is proportional to the local carrier concentration if the modulating field does not exceed the low-field limit. Therefore, the spatial variation of the carrier concentration can also be mapped. The percentage deviation from a reference value obtained in this way for a GaAs wafer is shown in Figure 4 (16). The penetration depth of light in this case was 150Å, so that the spatial variation of the near-surface carrier concentration was measured. Information about depth uniformity can be obtained by choosing wavelengths where the penetration depths are different. This method is applicable for carrier concentrations in the range of $10^{16}-10^{18}$ cm^{-3}.

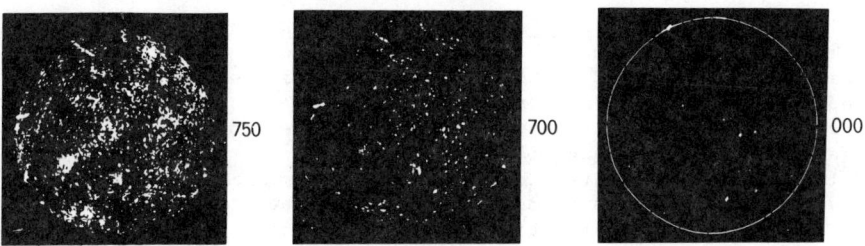

Figure 1. Particulate-contamination scans of the same wafer with different threshold sensitivities. (Reproduced with permission from Ref. 12. Copyright 1983 RCA Review.)

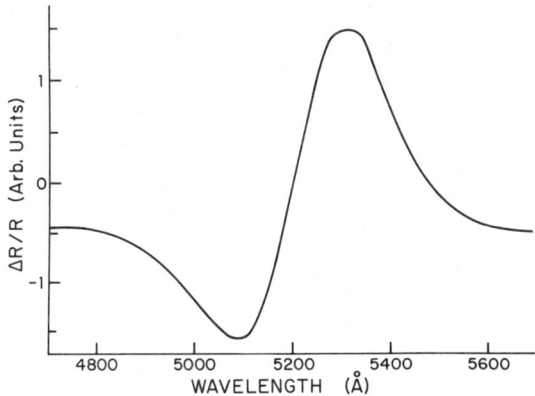

Figure 2. Electrolyte ER spectrum of $Hg_{0.8}Cd_{0.2}Te$ near the E_1 transition. (Reproduced with permission from Ref. 15. Copyright 1978 American Institute of Physics.)

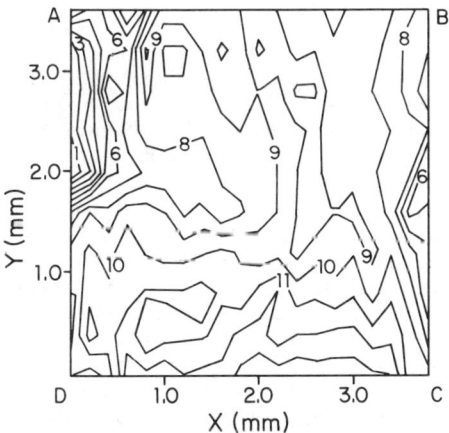

Figure 3. Variation in composition of a $Hg_xCd_{1-x}Te$ sample. Contours differ by 0.2%. (Reproduced with permission from Ref. 15. Copyright 1978 American Institute of Physics.)

Thickness Determinations in Laminar Systems. From now on, all samples will be assumed to be laterally uniform. The simplest characterization task is to determine the thickness of films where the dielectric properties of the films and substrates are already known. Data from single-wavelength ellipsometers can be used to solve model equations for thicknesses of transparent films, modulo a repeat thickness $\lambda/2n$, with submonolayer sensitivities (7). If the films are more than about 100Å thick, then both thickness and refractive index can be obtained. However, too many unknowns are involved if the films are optically absorbing, and for very thin films the results are extremely sensitive to the exact values of the dielectric properties of the substrate and interface. Also, no cross-check is obtained on the reliability of the solution.

Comparative instruments using white light sources offer a simple way to circumvent some of these difficulties if modest ($\sim$10Å) resolution is adequate (17-19). Comparative instruments rely on two reflections, one from a reference and one from the sample. In a recently described comparison ellipsometer (17), the sample and reference are oriented orthogonally so that the p-polarized component of the first becomes the s-polarized component of the second, and vice versa. Reflectance effects are exactly cancelled at all wavelengths if the film thicknesses of reference and sample are identical. By positioning cross polarizers at the source and detector, regions on the sample where this condition is met appear dark. If the reference film thickness is tapered, then a dark band appears where the film thicknesses are equal. This instrument provides direct readouts of thicknesses with no moving parts. An imaging capability limited only by depth of field is also possible for uniformity measurements over a wafer surface (18).

In one form of comparison reflectometer, a similar double-reflectance geometry is used but the reference is translated while the intensity is monitored (19). When the film thicknesses are equal, two of the four optical paths that result when all possible combinations of front- and back-surface reflections are taken into account are identical and therefore add in phase. This gives rise to an intensity maximum. Accuracies are comparable to those attainable with comparison ellipsometers.

Other alternatives involve the analysis of interference oscillations in data taken as a continuous function of wavelength, as shown for reflectance in Figure 5 (20). One spectrum is calculated for a 5000Å p-Si film deposited on a 1000Å SiO_2 layer thermally grown on a c-Si substrate, and the other for a 3000Å oxide overlayer on the p-Si film. A significant change in wavelength dependence is produced by the extra layer, indicating that multilayer as well as single layer analysis is possible with this approach. A scanning reflectometer using optical fibers for flux transmission and determining film thicknesses by the template approach is described in ref. (21). A microminiaturized version of this system, with all electronic and optical components — including an optical prism — mounted on a single card for insertion into the back plane of a minicomputer, has recently been described (22).

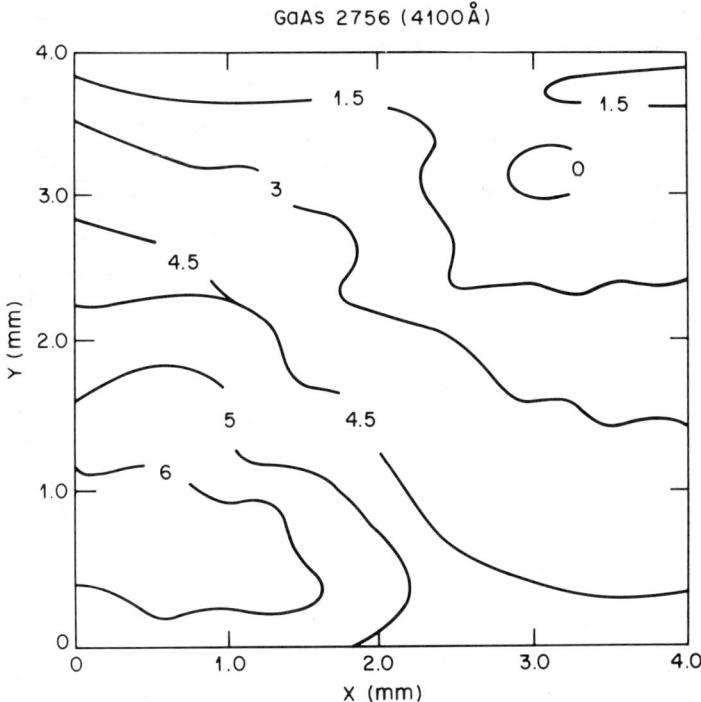

Figure 4. Variation in impurity concentration for a GaAs sample as determined by ER. Contours represent 1% increments. (Reproduced with permission from Ref. 16. Copyright 1979 American Institute of Physics.)

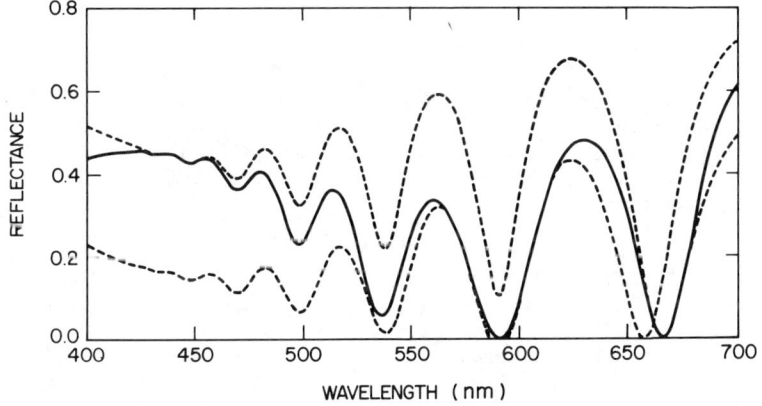

Figure 5. Calculated reflectance of a polysilicon multilayer stack with and without a surface oxide. (Reproduced with permission from Ref. 20. Copyright 1979 Journal of the Optical Society of America.)

Materials Analysis. The next level of complexity involves the measurement of dielectric properties for the determination of composition and microstructure as well as thicknesses. Thin films are typically microscopically inhomogeneous with substantial fractions of grain boundaries and voids, so their dielectric properties are rarely equal to those of the corresponding materials in bulk form. As an example, the pseudodielectric function $<\epsilon> = <\epsilon_1> + i<\epsilon_2>$ for a p-Si film deposited by low pressure chemical vapor deposition (LPCVD) is shown in Figure 6 (23). A pseudodielectric function is an apparent dielectric function calculated in the two-phase model, that is, without taking into account the possible existence of overlayers. For comparison, the imaginary part, ϵ_2, of the complex dielectric function $\epsilon = \epsilon_1 + i\epsilon_2$ for c-Si is shown in Figure 7 (24). It is clear that a significant difference exists, and that a model must be developed to describe the apparent dielectric response of the p-Si film in terms of its constituents and microstructure.

This model can be constructed with the help of Figure 7. The structures at 3.4 and 4.2 eV clearly show that the p-Si film contains c-Si. The broad background of the $<\epsilon_2>$ peak centered near 3.5 eV shows that the p-Si film also contains a-Si. The small amplitude of the $<\epsilon>$ spectrum shows that the density of the p-Si film must be less than that of c- and a-Si, i.e., the film contains an appreciable fraction of voids. Finally, the low value of the 4.2 eV peak in $<\epsilon_2>$ relative to that at 3.4 eV shows that the surface is microscopically rough. With the microstructure qualitatively established, the model can be cast in mathematical form using effective-medium theory and the equations describing propagation in laminar media. Then the free parameters can be evaluated using the template approach and the reference data of Figure 7. The best-fit spectra obtained are shown as the dashed curves. The agreement between calculation and experiment over the entire range for both real and imaginary parts of $<\epsilon>$ demonstrates that the model indeed accurately represents the sample. The film is found to consist of approximately 14 ± 2 vol. % c-Si and 25 ± 6 vol. % voids (leaving 61 ± 8 vol. % a-Si), and to be covered with a microscopically rough overlayer represented by 12 ± 3Å of SiO_2. Analysis of data for similar samples heat treated at 950C for 1/2 h in a PBr_3 atmosphere showed that the crystalline fraction increased to 57 ± 4 vol. % and the void fraction decreased to 8 ± 1 vol. %.

Figure 8 shows the results of an analysis of SE data for a p-Si layer deposited on a 1000Å thermally grown oxide on a c-Si substrate and subsequently annealed at 1000C for 1/2 h (25). Here, the model equations were solved for the dielectric function of the p-Si layer using the SE data, reference data for the thermal oxide and the c-Si substrate, and estimated values of the free parameters (film thicknesses). The top curve shows that the free parameters can be chosen so that all interference-related artifacts can be completely eliminated to the accuracy of the measurement. This demonstrates that the model accurately represents the sample. The lower curves show the effects of small changes in the values of the free parameters, allowing the

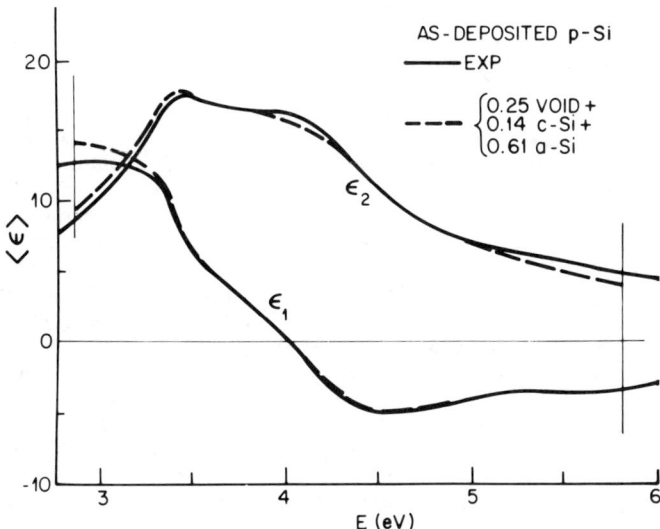

Figure 6. Ellipsometric <ε> spectrum for a sample, compared to model calculations. (Reproduced with permission from Ref. 23. Copyright 1981 American Institute of Physics.)

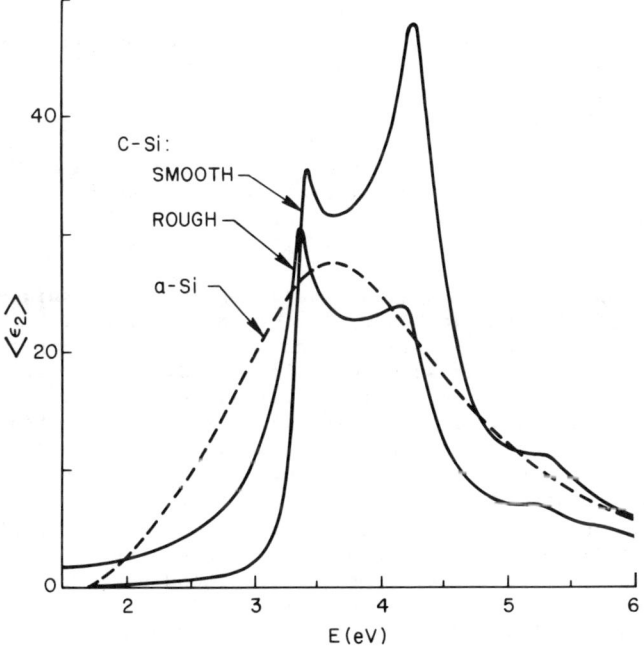

Figure 7. Ellipsometric <ε> spectra for Si samples. (Reproduced with permission from Ref. 24. Copyright 1982 Thin Solid Films.)

sensitivity to be estimated directly. The bottom curve shows that a fifth phase, representing microscopic roughness, must be included. The thickness of this microscopically rough overlayer is 6.5 ± 1Å.

An example illustrating the nondestructive analysis of a locked-in interface is shown in Figure 9 (26). Here, a similar computation is performed to yield the dielectric function of an electrochemically grown anodic oxide on a CdTe substrate where one of the free parameters is the thickness of an assumed interfacial layer of a-Te. Interference-related artifacts are minimized for 1.5 ± 2Å of a-Te between substrate and oxide. Similar calculations for HgTe, $Hg_{0.8}Cd_{0.2}Te$ and $Hg_{0.29}Cd_{0.71}Te$ showed 0 ± 2Å a-Te at these interfaces. The results demonstrated that the relatively large (~30Å) amounts of a-Te typically found at the interfaces between these materials and their electrochemically grown anodic oxides by destructive analyses are artifacts of ion milling.

The final example shows that under certain conditions optical characterization techniques can rival cross-sectional transmission electron microscopy (XTEM) as a means of determining sample properties. Figure 10 shows an XTEM cross section of the surface region of a c-Si wafer that had been implanted with 200 keV Si ions to a fluence of 1×10^{16} cm^{-2} (27). Implantation yielded a multilayer structure consisting of a surface oxide, a mixed c- and a-Si subsurface region, a relatively unperturbed c-Si channeling region, and a substantially amorphized region where the penetrating ions came to rest. The sample properties deduced from an analysis of 2.0-4.0 eV SE data and independently from XTEM micrographs compare as follows (XTEM values given first): oxide thickness: 25Å, 24 ± 3Å; subsurface region: thickness 120 ± 20Å, thickness and composition 119 ± 19Å, 18 ± 3% a-Si; channeling region: thickness 550 ± 50Å, 511 ± 21Å; termination region: thickness 250 ± 50Å, thickness and composition 270 ± 30Å, 79 ± 3% a-Si. The thickness results are in excellent agreement. However, the optical analysis goes further, providing additional information about the relative fractions of crystalline and amorphous material in the different layers.

Real Time Applications. Optical characterization techniques can provide unique information in real-time applications involving deposition or etching. Line-of-sight access to the sample by both incident and reflected beams is required. The use of reflectance techniques to monitor material removal, e.g., in plasma reactors, is well known (28); the examples selected here will illustrate less well explored possibilities in deposition.

Figure 11 shows a $\Delta-\psi$ trajectory measured at 5471Å for an a-Si film deposited by CVD on a Si_3N_4 substrate (29). After initial heating to 580C, deposition begins as indicated. As the thickness increases, optical contact to the substrate is lost and the trajectory approaches the point corresponding to a-Si. The dashed curve shows the trajectory expected if the film were being deposited uniformly. While the initial and final points agree, a large discrepancy occurs for intermediate coverages and particularly during the initial stage of deposition. This discrepancy

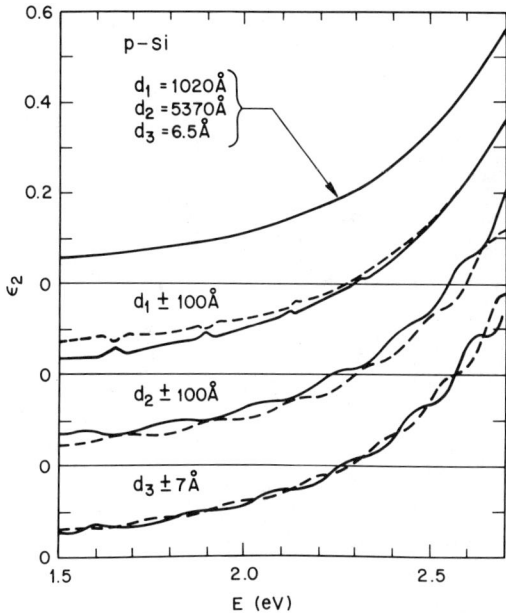

Figure 8. Calculated ε_2 values for the p-Si layer in a multilayer stack, calculated as described in text. (Reproduced with permission from Ref. 25. Copyright 1984 American Physical Society.)

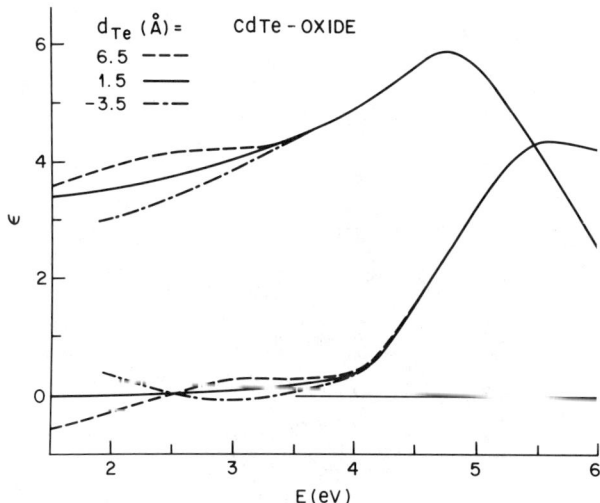

Figure 9. Calculated dielectric properties of oxide on CdTe for different interface thicknesses of a-Te. (Reproduced with permission from Ref. 26. Copyright 1984 American Institute of Physics.)

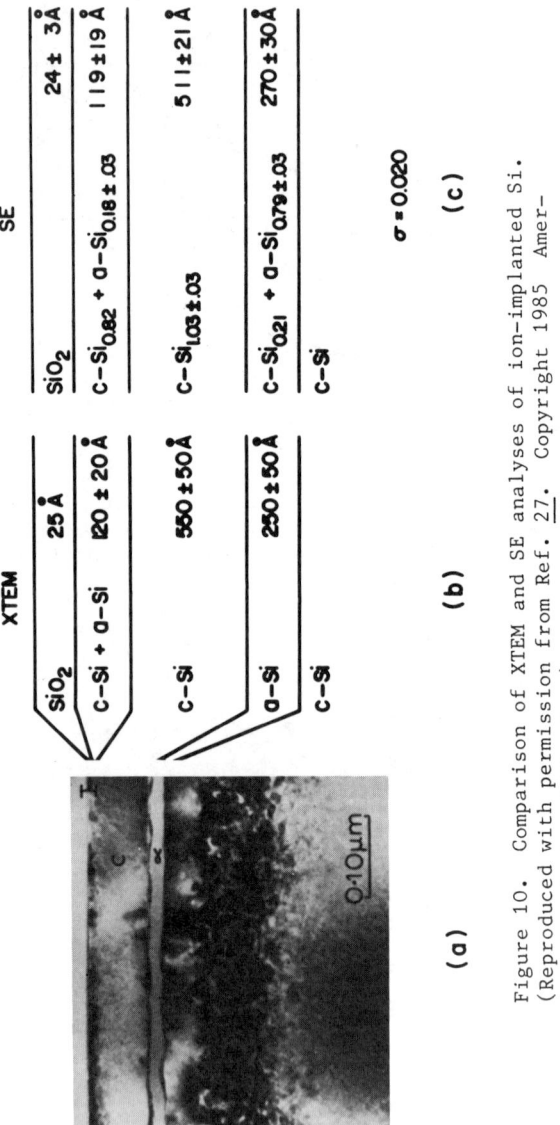

Figure 10. Comparison of XTEM and SE analyses of ion-implanted Si. (Reproduced with permission from Ref. 27. Copyright 1985 American Institute of Physics.)

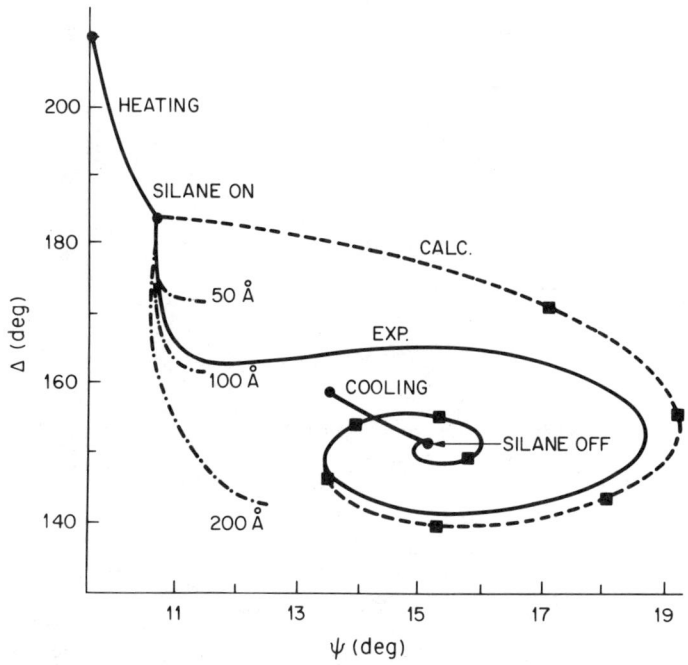

Figure 11. Evolution of Δ-ψ trajectory for deposited Si film. (Reproduced with permission from Ref. 29. Copyright 1980 Journal of Crystal Growth.)

indicates that the film does not grow uniformly. If one now assumes that growth proceeds by accretion of material as hemispherical bumps centered at nucleation sites scattered with some mean separation over the surface, the dielectric response of the film can be calculated by separations are shown as the dot-dashed curves. The results accurately represent the data until the hemispheres coalesce and the film begins to acquire the characteristics of a uniform layer. Thus not only is the growth mechanism identified, but the mean distance between nucleation centers is established to be of the order of 90Å. Note that the growth mechanism can already be identified after only two monolayers of material have been deposited.

A variation of the above procedure has been used to obtain information about the alloy compositions of growing films that are laterally microscopically homogeneous but inhomogeneous in the direction of film growth. The growing film is modeled as a multilayer stack with the properties of a given layer being determined from the ellipsometric data and the previously determined solutions for the underlying layers. The results of an analysis for an epitaxial $Al_xGa_{1-x}As$ film prepared by metalorganic chemical vapor deposition (MOCVD) on a GaAs substrate are shown in Figure 12 (30). The nominal composition of x = 0.25 was not reached until nearly 700Å of material had been deposited.

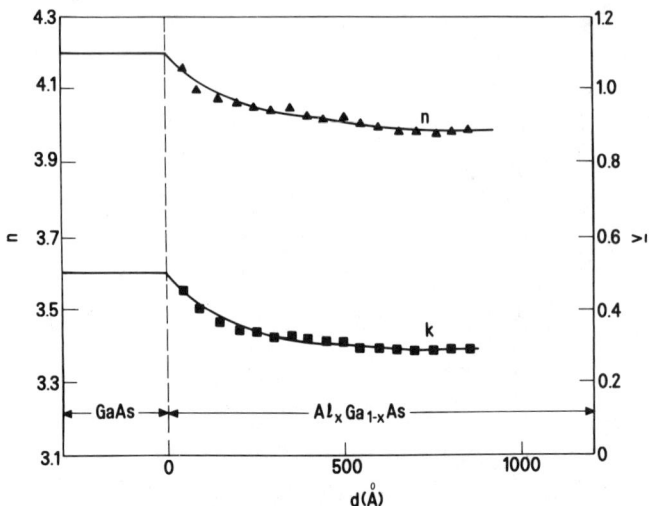

Figure 12. Evolution of complex refractive index for $Al_xGa_{1-x}As$ film grown by MOCVD. Reproduced with permission from Ref. 30. Copyright 1981 Revue de Physique Appliquee.)

When GaAs was deposited on $Al_{0.25}Ga_{0.75}As$, the proper composition was found to be reached in 100Å. The results were interpreted as evidence of loss of the reactants due to gettering on the chamber walls. The gas did not reach its nominal composition at the growing surface until the walls of the chamber had been saturated. The real-time control possibilities evident from this example are obvious.

Conclusion

The above examples are representative of the present capabilities of reflectance-based optical characterization techniques. Other applications can be found in the general references given in the introductory paragraphs. Ir reflectance has not been discussed, not from lack of examples (31, 32), but because the major fraction of reflectance characterization has been done in the v-uv. Additional progress and new applications can be expressed in all areas.

Literature Cited

1. "Spectroscopic Characterization Techniques for Semiconductor Technology"; Pollak, F. H.; Bauer, R. S. Eds.; SPIE, 1983, 452.
2. "Optical Characterization Techniques for Semiconductor Technology"; Aspnes, D. E.; So, S.; Potter, R. F. Eds.; SPIE, 1981, 276.
3. "Ellipsometry and Other Optical Methods for Surface and Thin Film Analysis"; Abeles, F. Ed.; J. de Physique-Colloque, 1983, C10.

4. "Optical Methods for the Characterization of Thin Films"; Acta Electronica 1981/1982, 24, 118-286.
5. Surface Science; Gijzeman, O. L. J.; van Silfhout, A. Eds.; 1983, 135.
6. Hauge, P. S. Surface Science 1979, 96, 803.
7. Azzam, R. M. A.; Bashara, N. M. "Ellipsometry and Polarized Light"; North-Holland: Amsterdam, 1977.
8. Aspnes, D. E.; Theeten, J. B.; Hottier, F. Phys. Rev. 1979, B20, 3292.
9. Keeping, E. S. "Introduction to Statistical Inference"; Van Nostrand: Princeton, 1962, Chapter 12.
10. Bergman, D. J. Ann. Phys. 1982, 138, 78.
11. Muller, R. H.; Farmer, J. C. J. de Physique-Colloque 1983, C10, 57.
12. Steigmeier, E. F.; Auderset, H. RCA Review 1983, 44, 5.
13. Harbeke, G.; Meier, E.; Sandercock, J. R.; Tgetgel, M.; Duffy, T. T.; Soltis, R. A. RCA Review 1983, 44, 19.
14. Duffy, M. T.; Corboy, J. F.; Cullen, G. W.; Smith, R. T.; Soltis, R. A.; Harbeke, G.; Sandercock, J. R.; Blumfeld, M. J. Cryst. Growth 1982, 58, 10.
15. Pollak, F. H.; Okeke, C. E.; Vanier, P. E.; Raccah, P. M. J. Appl. Phys. 1978, 49, 4216.
16. Pollak, F. H.; Okeke, C. E.; Vanier, P. E.; Raccah, P. M. J. Appl. Phys. 1979, 50, 5375.
17. Stenberg, M.; Sandstrom, T.; Stiblert, L. Mat. Sci. Eng. 1980, 42, 65.
18. Stiblert, L.; Sandstrom, T. J. de Physique-Colloque 1983, C10, 790.
19. Sandercock, J. R. J. Phys. E 1983, 16, 866.
20. Hauge, P. S. J. Opt. Soc. Am. 1979, 69, 1143.
21. Kennerth, K. L.; Dill, F. H. Solid State Electron. 1972, 15, 371.
22. Korth, H.-E. J. de Physique-Colloque 1983, C10, 101.
23. Bagley, B. G.; Aspnes, D. E.; Adams, A. C.; Mogab, C. J. Appl. Phys. Lett. 1981, 38, 56.
24. Aspnes, D. E. Thin Solid Films 1982, 89, 249.
25. Aspnes, D. E.; Studna, A. A.; Kinsbron, E. Phys. Rev. 1984, B29, 768.
26. Arwin, H.; Aspnes, D. E. J. Vac. Sci. Technol. 1984 (in press).
27. McMarr, P. J.; Vedam, K.; Narayan, J. in "Ion Implantation and Ion Beam Processing of Semiconductors"; Hubler, G. K., Clayton, C. R., Holland, O. W., and White, C. W. Eds.; Mater. Res. Society 1984 (in press).
28. Marcoux, P. J.; Foo, P. D. SPIE 1981, 276, 170.
29. Hottier, F.; Theeton, J. B. J. Cryst. Growth 1980, 48, 644.
30. Laurence, G.; Hottier, F.; Hallais, J. Revue Phys. Appl. 1981, 16, 579.
31. Durschlag, M. S.; DeTemple, T. A. Sol. State Commun. 1981, 40, 307.
32. Schaefer, R. R. J. de Physique-Colloque 1983, C10, 87.

RECEIVED December 1, 1985

12

Measurement of the Oxygen and Carbon Content of Silicon Wafers by Fourier Transform IR Spectrophotometry

Aslan Baghdadi

Semiconductor Materials and Processes Division, National Bureau of Standards, Gaithersburg, MD 20899

> Fourier transform infrared (FT-IR) spectrophotometry is a rapid, nondestructive characterization technique which is being increasingly applied on a large scale to the routine measurement of the oxygen and carbon content of silicon wafers used for the fabrication of microelectronic devices. Control of the oxygen content is needed to achieve acceptable yields in modern device processing, particularly for those processes which utilize oxide precipitates to protect active regions of devices from contamination by metallic impurities during high-temperature processing. The interlaboratory reproducibility of the measurement is not adequate considering the degree of control of the oxygen that is required. This review focuses primarily on the measurement of oxygen and carbon in silicon and on methods for improving quantitative FT-IR absorption measurements on semiconductor wafers.

The evolution of silicon processing technologies for producing very large-scale, high-density integrated circuits has resulted in more demanding starting material specifications, and has increased the need for nondestructive and inexpensive characterization techniques (1) which can be applied routinely to measure the properties of silicon wafers in production quantities. Infrared absorption spectrophotometry, using either dispersive or Fourier transform (FT-IR) instruments, is used for the characterization of both neutral (oxygen and carbon) and electrically active impurities (e.g., group III or group V elements, as well as some oxygen complexes) in silicon. Oxygen and carbon content measurements, which can be made at room temperature without damaging the wafers, are being widely used to control material properties prior to high-temperature processing.

This paper describes the configuration of oxygen and carbon impurities in silicon and techniques which can be used to improve the quality and quantitative accuracy of IR absorption spectra obtained on FT-IR instruments. It also discusses instrument-to-instrument

This chapter not subject to U.S. copyright.
Published 1986, American Chemical Society

variations, as well as differences in the silicon wafers themselves, which may affect the interlaboratory reproducibility of oxygen content measurements on as-received silicon wafers.

Oxygen and Carbon in Silicon

The most common use of infrared absorption for the characterization of silicon wafers for microelectronics processing is for the determination of the interstitial oxygen and, to a much lesser extent, substitutional carbon content. Close to the melting point of silicon, it is possible to dissolve as much as 30 parts per million atomic (ppma) oxygen in silicon (2). Well below the melting point, however, the equilibrium solid solubility is considerably lower so that, during high-temperature processing, the supersaturated oxygen can condense into a variety of complexes and precipitates, as shown in Table I. Since most semiconductor silicon crystals are pulled from crucibles lined with fused silica using the Czochralski crystal growth method, the silicon melt contains significant amounts of dissolved oxygen. Moreover, the distribution of oxygen in the melt is not uniform, since it depends upon a number of crystal growth parameters including the rotation rates of the crucible and the crystal, the melt aspect ratio (i.e., the ratio of the melt height to the crucible diameter), the pull rate of the crystal, temperature fluctuations in the melt, the crucible dissolution rate and the evaporation of silicon monoxide from the exposed melt surface (3). Fortunately, the oxygen incorporation has been empirically found to be fairly reproducible for a given apparatus and set of growth conditions, so that silicon suppliers have been able to produce crystals with specified oxygen contents (over a substantial portion of the ingot) to meet specific process requirements (2,4). Typical Czochralski crystals contain 10 to 20 ppma of oxygen, with both radial and axial macroscopic variations. There are also variations on a submillimeter scale caused by changes in the instantaneous crystal growth rate (5-7). These microscopic variations in the oxygen concentration can introduce systematic errors in the measurement of the oxygen distribution across a silicon wafer (8,9). Czochralski crystals grown using a double-crucible technique (3), magnetic fields (10), or nitride-lined crucibles (11) have lower and more uniform oxygen concentrations, although these techniques have not yet been adopted on a commercial scale.

Oxygen occupies interstitial sites in the silicon lattice. This was demonstrated by showing that the lattice parameter of silicon containing high concentrations of oxygen was greater than that of low-oxygen silicon (12). Since isolated oxygen in silicon is electrically neutral, it was proposed that the oxygen atom forms bonds with two nearest-neighbor silicon atoms which give up their original bonds with each other (13-15). At room temperature two absorption bands are observed, a strong, broad band at 1107 cm^{-1} and a much weaker band at 515 cm^{-1}. The 1107 cm^{-1} band has been attributed to an anti-symmetric stretching mode of the triatomic Si-O-Si defect "molecule" (15), and the 515 cm^{-1} mode has been recently proposed as a symmetric stretching motion, similar to vibrations occurring in disiloxane (16). Near liquid nitrogen temperatures, another band appears at 1205 cm^{-1} (17), as well as some far-infrared bands. At liquid helium temperatures, the broad band at 1107 cm^{-1} splits into a

Table I. Examples of the Effects of Annealing on the Configuration of Oxygen in Silicon

Anneal Time & Temperature	Initial Oxygen Configuration	Resulting Oxygen Configuration & its Electrical Activity	Reference
450°C, 1-10 hrs	Interstitial oxygen	SiO_x complexes, double donors known as thermal donors (TD's)	27
650°C, 30 mins	SiO_x complexes formed at 450°C	Electrically inactive oxygen-silicon complexes or very small microprecipitates	27
700-800°C up to 100 hrs	Nucleation centers, such as SiO_x complexes or carbon-oxygen complexes	SiO_2 microprecipitates, 100 Å - 1000 Å, donors (ND's)	73
1000-1150°C, 2-10 hrs	Microprecipitates or other nucleation centers	SiO_2 precipitates, ~1 μm, no electrical activity	74
1350°C, hrs	Any of the above defects	Interstitial oxygen no electrical activity	20

number of sharp lines at 1128 to 1137 cm^{-1}. These lines could be subdivided into sets of lines whose intensities were in the ratio of the relative abundance of ^{28}Si-O-^{28}Si, ^{28}Si-O-^{29}Si, ^{28}Si-O-^{30}Si, ^{29}Si-O-^{29}Si, ^{29}Si-O-^{30}Si, and ^{30}Si-O-^{30}Si. This reduces the number of observed lines for a single silicon isotope to three or possibly four (18). The remaining lines have been explained as being due to additional levels in the far-infrared, due to oscillations of the oxygen atom as it tunnels between its two neigboring silicon atoms to equivalent positions around the <111> axis (17). Figure 1 is a schematic energy-level diagram, showing the allowed infrared optical transitions of a model of the ^{28}Si-O-^{28}Si complex, which results in the multiple-line spectrum (17).

Electrically active defects can also exhibit absorption in the infrared due to changes in their electronic states. For example, acceptor or donor impurities, at cryogenic temperatures, can have absorption lines due to transitions from the ground state of the neutral impurity to shallow hydrogen-like or helium-like levels lying just below their respective band edges (19). Thus IR absorption can be used to detect the presence of electrically active silicon-oxygen complexes which may form during high-temperature processing (20).

Silicon-oxygen complexes can be formed by annealing the wafers at temperatures such as 450°C (20), as was shown in Table I. The configuration of these complexes has not been clearly determined, but from the kinetics of their formation they involve a relatively small number of oxygen atoms, possibly three to five atoms, forming the silicon-oxygen complex. The 450°C complex is electrically active, and it has been shown to form donor (21) levels just below the conduction band edge. At cryogenic temperatures, the carriers "freeze out" onto the oxide complex, so that an IR spectrum of the specimen exhibits a series of sharp intense bands (22). The absorption process is shown schematically in Figure 2. The details of the actual formation and configuration of these defects are still the subject of active debate (23-26), but their IR spectra have been identified as being due to neutral hydrogen-like or singly-ionized helium-like donors (22). The spectra have not all been assigned to a single oxygen-silicon complex, but rather to a series of complexes which come into existence one after the other, and which exhibit different annealing behaviors (22). These defect complexes which are known for historical reasons as thermal donors (TD) can be destroyed by a relatively short (~30 minutes) heat treatment at about 650°C (27).

Further long-term heat treatments, at temperatures ranging from 550 to 950°C, create a new class of donors, which have been termed new donors (ND) (28). These donors are also not fully understood, but they are presumably a type of oxide microprecipitate. Oxide microprecipitates ranging in size from 100 Å to 1000 Å have been detected in silicon by transmission electron microscopy (29,30) following heat treatments in that temperature range. Long-term annealing at temperatures above 1000°C promotes the formation of larger (>1 μm) oxide precipitates. The growth of the oxide precipitates results in a reduction in the intensity of the interstitial oxygen band at 1107 cm^{-1}, and in the growth of broad new bands, the most prominent of which is centered at about 1225-1230 cm^{-1} (31-33). The size and shape of these precipitates, as well as their composition and crystallography, can have important effects on their IR spectra (31). Formation of the precipitates generally also results in the creation

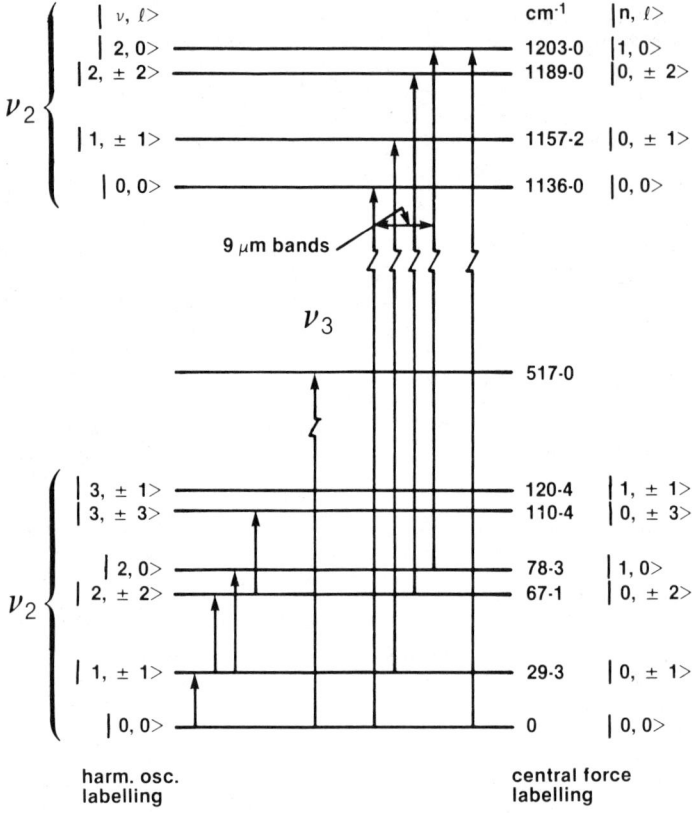

Figure 1. Energy-level diagram for interstitial oxygen in silicon, showing observed vibrational transitions in the near and far infrared for a $^{28}Si-O-^{28}Si$ complex. (Reproduced with permission from Ref. 17. Copyright 1970 The Royal Society.)

of networks of dislocations around the precipitates. These dislocations can serve to trap fast-diffusing metallic impurities. This is the basic mechanism which is used in the intrinsic gettering (IG) of undesirable impurities to improve the yield of semiconductor fabrication processes (21).

Carbon in silicon is a much simpler and better understood defect. Since, at room temperature, it occupies substitutional sites and since it is isoelectronic with the silicon atoms in the host lattice, isolated carbon atoms have only subtle or indirect effects on the properties of the silicon. Its infrared spectrum is straightforward, consisting of a single local vibrational mode for each isotope. The ^{12}C absorbance band is completely dominated by a two-phonon silicon lattice band at 610 cm^{-1} (see Infrared Absorption Measurements). The intensity of the two-phonon band is very strongly dependent on temperature, as shown in Figure 3. The open triangles represent the intensity of the silicon lattice band at 610 cm^{-1}, while the closed triangle represents the absorption coefficient due to substitutional carbon at 300 K for a nominal carbon concentration of 1 ppma. Thus the temperature of the reference and specimen wafers must be kept within 1°C of each other in order to maintain a good match between the specimen and reference spectra. This degree of temperature control is not necessary for the measurement of the interstitial oxygen peak height, since the absorption due to the silicon lattice band at 1120 cm^{-1} (shown as the open circles in Figure 3) is much smaller than the absorption due to interstitial oxygen for a wafer containing 20 ppma oxygen (shown as the closed circle in Figure 3). However, when the ratio of the absorptivity of the oxygen band to the absorptivity of the 610 cm^{-1} silicon band (34) is used for the computation of the oxygen content, the temperature dependence of the silicon lattice band must be taken into account.

The great reduction in the intensity of the interfering 610 cm^{-1} silicon lattice band at low temperatures might be expected to lead to increased sensitivity and accuracy in carbon determinations based on cryogenic IR measurements. However, the reflection losses at each of the cryostat windows almost outweigh this advantage (35). The remaining advantage is usually too slight to justify the additional time for cooling the sample and for preparing special small specimens necessary for most low-temperature measurements.

The distribution of carbon in single-crystal silicon is not uniform (36). Since the carbon atom is smaller than the silicon atom it replaces, high local concentrations of carbon can produce a localized strain in the lattice, which can serve as nucleation centers for the formation of structural defects such as "swirl" patterns (36). For this reason, float-zoned silicon intended for the fabrication of semiconductor power devices is often specified to contain low carbon concentration, e.g., less than 1 ppma. Carbon has been reported to play a role in the heterogeneous nucleation of oxide precipitates (37-40). In principle, carbon nucleation of oxide microprecipitates could be used to advantage in designing IC processes. In practice, however, the nonuniformity of the carbon distribution renders this an impractical approach. Thus in order to avoid uncontrolled variations in the oxide precipitation, low-carbon silicon is often specified for processes which use intrinsic gettering.

Low-carbon silicon must also be used for the fabrication of extrinsic infrared detectors to prevent the formation of carbon-

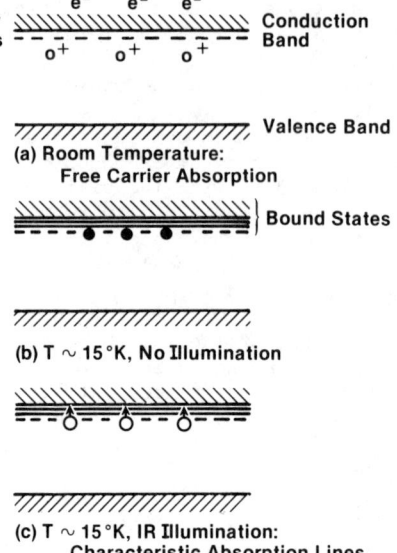

Figure 2. Energy-level diagram showing the absorption of radiation in n-type silicon under various conditions: (a) room temperature, (b) same specimen at cryogenic temperatures, without any illumination, and (c) same specimen at cryogenic temperatures, with sub-bandgap illumination, showing characteristic absorption of the IR photons from the ground state of the donor to bound states just below the conduction band.

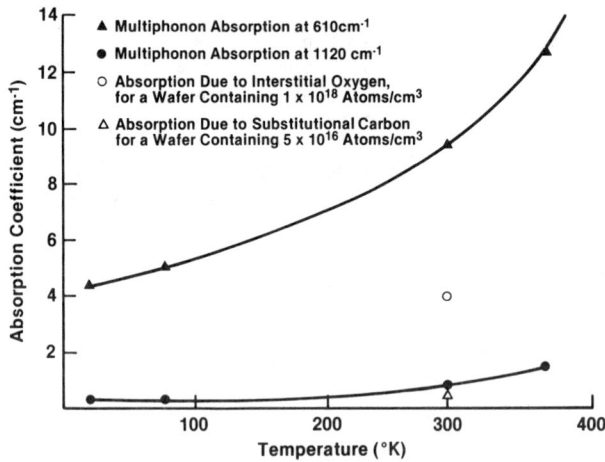

Figure 3. Temperature dependence of the silicon lattice absorption at 610 cm^{-1} (▲) and at 1120 cm^{-1} (●), compared to the room-temperature absorption due to substitutional carbon at 605 cm^{-1} (△) and interstitial oxygen at 1107 cm^{-1} (○). (Adapted from Ref. 75.)

acceptor complexes (41). These electrically active defect complexes consist of a carbon atom paired with a group IIIA acceptor atom (B, Al, Ga, In).

Infrared Absorption Measurements

Infrared absorption resulting from localized modes of vibration of impurities in the semiconductor lattice can be used to determine the concentrations of certain impurities in the semiconductor, as shown in Table II. Localized modes of vibration generally occur when the impurity atom is lighter than the host atom. However, heavier atoms can sometimes exhibit characteristic absorption because of the removal of the translational symmetry of the lattice due to the presence of the impurity. This occurs when there are no silicon lattice bands with any significant amplitude at that characteristic frequency (42). In these cases, in-band resonance modes characteristic of the impurity can then be observed (43,44) Note that nominally neutral impurities, such as substitutional carbon or interstitial oxygen in silicon, can give rise to infrared absorption due to localized modes since charge redistribution occurs (44), creating a dipole moment at the defect.

Figures 4a and 4b show the infrared transmission spectra of a high-purity float-zoned silicon wafer and of a Czochralski wafer, respectively (13). The spectral features due to the presence of oxygen and carbon in the Czochralski wafer are clearly shown in Figure 4c, which is the difference spectrum of the Czochralski wafer relative to that of the float-zoned wafer. This difference spectrum was obtained using a double beam dispersive spectrometer, with the float-zoned wafer in the reference beam and the Czochralski wafer in the sample beam. The broad band at 1107 cm^{-1} and the smaller band at 515 cm^{-1} are due to interstitial oxygen, and the band at 605 cm^{-1} is due to substitutional carbon.

The raw data in a Fourier transform spectrophotometer are obtained in the form of an interferogram, i.e., a plot of the detector response versus the optical path difference between the fixed and moving mirrors in the interferometer. In either dispersive or FT-IR instruments, multiple reflections within the semiconductor specimen can result in Fabry-Perot fringes in the infrared spectrum obtained for that wafer. In FT-IR instruments, multiple reflections between the specimen surfaces and the interferometer can also result in interference fringes (45). These fringes can obscure weak features in the spectra, as well as reduce the accuracy of quantitative measurements. In Fourier transform spectrometry, the effects of these multiple reflections are evident, upon examination of the interferogram prior to transformation, as a series of secondary and tertiary (45) interferograms, the secondary interferograms being due to multiple passes within the specimen, and the tertiary interferograms being due to multiple passes between the specimen surfaces and the interferometer mirrors. These are superimposed on the main interferogram, as shown in Figure 5 (46). Several methods have been developed for eliminating these fringes in FT-IR instruments:

1. If the data are obtained at a sufficiently low resolution, for example at 6 cm^{-1} (47), the interference fringes are not observed for silicon wafers thicker than 0.3 mm because the fringes are

Table II. Localized Vibrational Modes of Isolated Impurities in Semiconductors

Host Semiconductor	Impurity Species	Mode Frequency (temperature, K) cm^{-1}	Reference
Silicon	^{10}B (s)	644(300), 646(80)	44
	^{11}B (s)	620(300), 622(80)	44
	^{12}C (s)	605(300), 608(80)	44
	^{13}C (s)	586(300), 589(80)	44
	^{14}C (s)	570(300), 573(80)	44
	O (i)	515(300), 1107(300) 517(80), 1136(80), 1205(80)	44 44
	As (s) resonance mode	366(80)	43
	P (s) resonance mode	491(80)	43
Germanium	Si (s)	389(300)	43
	^{10}B (s)	571(80)	43
	^{11}B (s)	547(80)	43
	^{16}O (i)	855(300), 862(4)	43
Gallium Arsenide	Si (s) gallium site	384(80)	43
	Si (s) arsenic site	399(80)	43

(s) - substitutional
(i) - interstitial

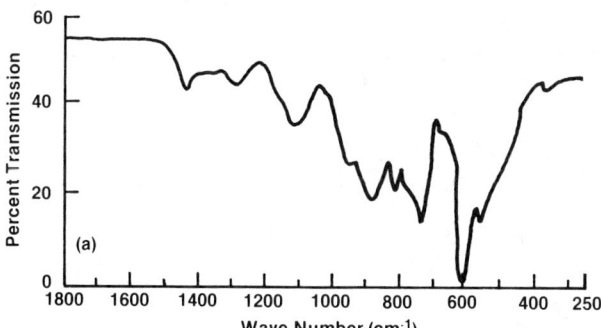

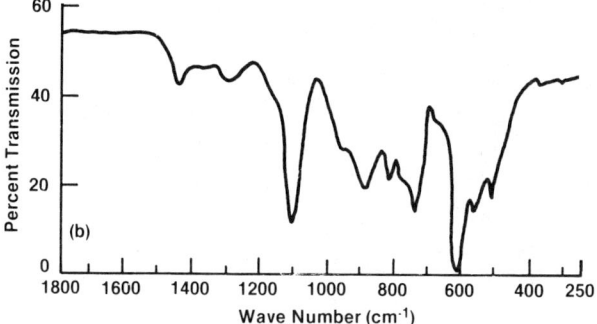

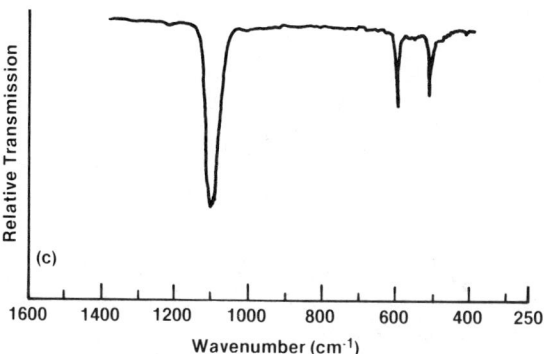

Figure 4. (a) Room-temperature, air-reference, transmission spectrum of a high purity 4.5 mm thick float-zoned silicon specimen.
(b) Room-temperature, air-reference, transmission spectrum of a 150 Ω·cm 4.5 mm thick Czochralski silicon specimen.
(c) Room-temperature difference spectrum of the Czochralski specimen used for (b) obtained in a double-beam dispersive spectrometer with the float-zoned specimen used for (a) in the reference beam. All these spectra were obtained using a resolution of 1 cm^{-1} above 600 cm^{-1}, and 2 cm^{-1} below 600 cm^{-1}. (Reproduced with permission from Ref. 13. Copyright 1977 Masson S. A., Paris.)

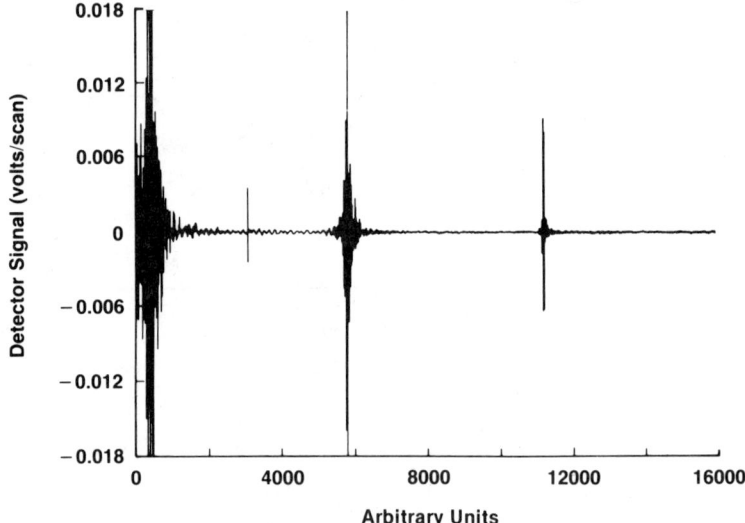

Figure 5. Interferogram of a 0.5-mm thick silicon wafer. The units on the x-axis are directly proportional to the mirror position, with 10,000 units ~0.633 mm. The signal at about 3000 units is a tertiary interferogram, and the signals at about 5,700 and 11,200 units are secondary interferograms (44).

closer together than the instrumental resolution. This is an attractive approach when measuring a broad absorption band, such as the interstitial oxygen in silicon peak at 1107 cm^{-1}, which has a full-width-at-half-maximum (FWHM) of 32 cm^{-1} (48). On the other hand, if the absorption band of interest is not much broader than the instrumental resolution, this may not be the best approach. This is especially true when working with a region of a spectrum containing adjacent or overlapping lines such as the substitutional carbon line in silicon at 605 cm^{-1}, which is dominated by the two-phonon lattice absorption band at 610 cm^{-1} (13).

2. Another method, which is best suited to FT-IR measurements, is to replace the regions of the interferogram which contain the extraneous secondary and tertiary interferograms with zero amplitude (49). This method is equivalent to multiplying the interferogram with a function which is zero in certain regions and unity elsewhere. In order to minimize the resulting distortions to the transformed spectrum, the background interferogram should also be multiplied by the same function prior to transformation (46).

3. A third method, closely related to the previous one, is to replace the regions of the interferogram containing the secondary and tertiary interferograms with a section from another interferogram, taken with a thicker sample, which does not contain the offending interferograms in the same regions (49). The spectrum obtained by this method may not be fully satisfactory because, even if the two samples are well matched in their other properties, the difference in thickness itself makes it impossible for the new sections of the interferogram to be fair representations of the ones they are replacing.

4. The amplitude of the Fabry-Perot fringes can be reduced by mounting the semiconductor specimen at Brewster's angle (50). In the case of silicon, Brewster's angle is 73° 41'. At this angle, the reflectance for incident light polarized parallel to the plane of incidence is zero, while the reflectance for light polarized perpendicular to the plane of incidence, is 0.71. Thus the parallel component does not contribute to the fringes. On the other hand, the perpendicular component, which does contain interference fringes, is greatly reduced in intensity since its reflectance is so large that only 17% of the incident perpendicular component is transmitted. In practice this method, for the case of circularly polarized light, has reduced the amplitude of the fringes from 0.28, for a 0.4 mm thick silicon wafer mounted normal to the IR beam, to 0.05 when the wafer is mounted at Brewster's angle (50). However, mounting the specimen at such a large angle may cause further problems: the light transmitted through the wafer is shifted in position significantly compared to the background light. The change in the distribution of radiation at the detector caused by this shift could introduce an unnecessary element of uncertainty in the measurement.

Multiple reflections within the specimen can also reduce the accuracy of quantitative measurements (51,52). The intensity of light transmitted through a plane parallel plate, after the interference fringes have been eliminated, is given by (53):

$$I(\bar{\nu}) = \frac{I_o(\bar{\nu}) T_f T_b e^{-\alpha(\bar{\nu})x}}{1 - R_f R_b e^{-2\alpha(\bar{\nu})x}} \qquad (1)$$

where $I(\bar{\nu})$ is the intensity of the transmitted light at wavenumber $\bar{\nu}$ reaching the detector, x is the thickness of the parallel plate, T and R are the nonscattered transmittance and the nonscattered (specular) reflectance, respectively, with the subscripts f and b specifying the front and back surfaces of the specimen, $I_o(\bar{\nu})$ is the intensity of the incident radiation and $\alpha(\bar{\nu})$ is the absorption coefficient of the specimen at wavenumber $\bar{\nu}$. Equation 1 has been rewritten from its usual form to account for the effects of unpolished front or back surfaces which may scatter the radiation in addition to transmitting or reflecting it. For the case of a silicon wafer with both front and back surfaces polished, T = 0.7 and R = 0.3, so that almost 10% of the transmitted radiation in transparent regions of the spectrum passes through the specimen more than once.

Although specimens prepared especially for quantitative IR measurements would normally have both surfaces polished, typical production silicon wafers commonly have only a single polished surface. The opposite surface is generally lapped and etched, either with an acid or alkaline etch (54), or sometimes deliberately damaged, either mechanically or with a high-power laser. The deliberately damaged back surfaces are used to create dislocation networks in the silicon wafer, away from the active regions of the devices, which can trap or "getter" heavy-metal impurities, and thereby improve the overall yield of the fabrication process (55). Figure 6 is the absorption spectrum of a silicon wafer with a rough back surface. The strongly sloping baseline is due to the wavenumber dependence of the scattering at the back surface. In these cases, the absorption coefficient due to an impurity, in a region without silicon lattice bands, is given by (53):

$$\alpha_{imp} = \frac{2.303}{x} \left[\Delta A + \log(1-S) - \log\left(1 - S e^{-2\alpha_{imp} x}\right) \right] \qquad (2)$$

where

$$S = R_f R_b e^{(-2\alpha_{free} x)}$$

and where ΔA is the net height of the absorbance band, S is the multiple-reflection correction factor, and α_{free} is the absorption due to free carriers. When silicon lattice bands overlap the impurity band, their contribution must be subtracted from the observed absorption in order to obtain α_{imp}. Note that Equation 2 contains the factor α_{imp} on both sides of the equation, so that it must be solved iteratively. However, it converges very rapidly, and generally needs less than four iterations.

The correction factor S is close to zero for a wafer with a very rough back surface, such as the one whose spectrum is shown in Figure 6, and is 0.09 for a lightly-doped (e.g., resistivity 5 Ω·cm or greater) wafer with both surfaces polished. The back surfaces of most wafers, however, fall between these two extremes, so that S must be determined in order to measure accurately the concentration of an impurity. Several methods have been proposed to obtain the multiple-reflection correction factor S (53,56-58). If the wafer has a uniform thickness, the ratio of the amplitudes of the secondary to primary interferograms can be used to obtain S (53). Alternatively, the value of the baseline in a transparent region of the spectrum can also be used to determine the value of S (56). This requires that the detector system linearity be well-established, the reflections between the specimen surfaces and the spectrometer components be understood quantitatively, and there be no other problems, such as a shift of the position of the radiation at the detector when the background interferogram is being obtained. A third method uses the magnitude of the apparent absorbance obtained for a multiphonon silicon lattice absorbance band of known value as a calibration to calculate the true value of the absorbance due to the impurity (57). This method, however, can be quantitative only when the scattering is not dependent upon the wavelength of the radiation. This is not generally the case, as is illustrated in Figure 6. Once the value of S has been determined by one of these methods, Equation 2 can be used to obtain the absorption coefficient of the specimen.

Measurements of the Oxygen Content of Silicon Wafers

The short-term reproducibility of infrared absorption spectroscopy in the determination of the oxygen content of a typical production silicon wafer can be good, as shown in Figure 7. These data represent twenty-four consecutive measurements of a single spot on a silicon wafer with an etched back surface. The data were obtained at 4 cm^{-1} resolution on an FT-IR spectrophotometer equipped with a mercury cadmium telluride detector, a KBr/Ge beamsplitter and a Globar (59) IR source. A 1 mm aperture was placed in front of the silicon wafer, and 200 scans were taken for each data point. Unfortunately, this degree of reproducibility only demonstrates the potential precision of FT-IR measurements. In fact, variations of up to 20% have been reported (60-62). These variations can occur in the measurement of the same specimen by different laboratories or in the analysis of a given specimen before and after it has undergone different back-surface treatments. Experiences of this kind have caused a lack of confidence in the precision of the IR method for the determination of the oxygen content of production silicon wafers. This problem has been ascribed to differences among the wide variety of instrument models and computer software packages used in the measurements (61). Variations in the specimen condition, i.e., dopant level, back-surface treatment or oxygen distribution have also been a source of poor reproducibility (8,9,52,53,63).

Shive and Schulte (61) have described a reference materials approach to improve the precison of the IR analysis method. The reference silicon wafers were given one of the standard back surface treatments: acid etched (AE), back-surface damaged (BSD), or enhanced gettered (EG). The EG wafers were acid-etched wafers with a polycrystalline silicon layer deposited on their rough side. Each wafer was quartered and then one of the quarter sections was polished on the back surface. An oxygen content was assigned to each wafer by analyzing this double-side polished (DSP) section. The oxygen contents were determined automatically from the transformed spectra using computer programs written by the instrument's manufacturer.

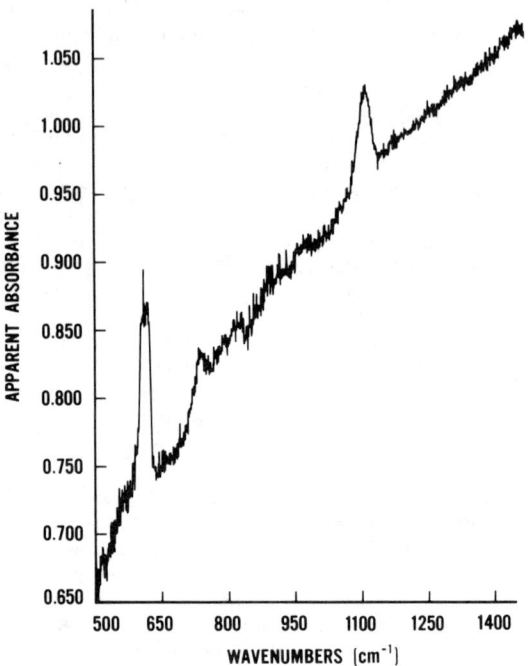

Figure 6. Spectrum of a silicon wafer with a back surface which was mechanically damaged to provide gettering sites.

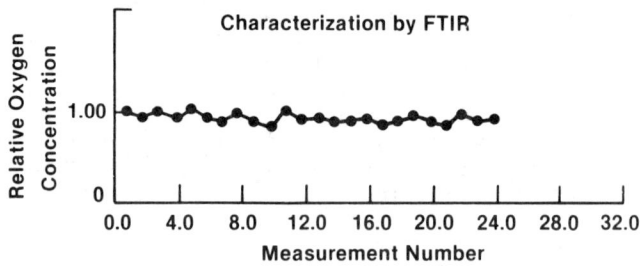

Figure 7. Reproducibility of consecutive measurements of the same spot on a Czochralski silicon wafer containing approximately 7.3×10^{17} interstitial oxygen atoms/cm^3.

The bias of the data, which was defined as the difference between the oxygen content determined for the DSP section and the oxygen content determined for the single-side polished (SSP) sections, was determined for each type of back surface condition. It should be noted that the oxygen concentrations quoted in Shive and Shulte's paper were based upon the "old" ASTM calibration coefficient (64), and have been converted here to values using the most current ASTM coefficient (48). This method turned out to be successful with the BSD wafers, but not with the AE or EG wafers (61). The reason for the failure with the AE and EG back surface conditions was attributed to the dramatic variability of the scattering of these wafers. In these cases, a method for obtaining the multiple reflection correction directly from the specimen interferogram or spectrum, such as one of the methods described in the section on Infrared Absorption Measurements, should be used. The calibration method also worked well with DSP wafers. Thus the calibration method was successful either when the full multiple reflection correction could be applied, i.e., in DSP wafers, or when the back surface was so rough that the magnitude of the multiple-reflection correction was very small, for example, in the case for BSD wafers. Huber and Stallhofer reached substantially similar conclusions in a recent article (65).

High concentrations of free carriers have important effects on the precision of the oxygen content determination in silicon wafers by IR spectrophotometry (63). Absorption of the IR beam by the free carriers (see Figure 8) can reduce the signal-to-noise ratio of the measurement until, for low resistivity wafers, meaningful measurements can no longer be made. The free carriers also affect the reflectance of the silicon, thus changing the magnitude of the multiple-reflection coefficient. However, since wafers with resistivities much below 0.1 $\Omega \cdot$cm cannot be measured, the error due to this effect is less than 1%. Figure 9 shows the contribution of the free carriers to the baseline for the case of n-type silicon, as calculated by Weeks (63). These calculated baselines proved to be an excellent simulation of the baselines of transmission spectra obtained for a series of n-type silicon samples with free carrier concentrations ranging from 5×10^{13} donors/cm^3 to 5×10^{18} donors/cm^3. The effect of this curving baseline must be taken into account if the quantitative analysis of the spectrum relies on the ratio of the oxygen peak at 1107 cm^{-1} to one of the strong silicon lattice absorbance bands at lower wavenumbers.

Some of the lack of interlaboratory reproducibility may be caused by the use of different instruments. Variations in the oxygen content determination can be due to the use of different apodization functions in different FT-IR instruments, to different designs of their optical benches or simply to differences in the performance of their components. Apodization functions are used to reduce the amplitude of artifacts in the transformed spectra, caused by the truncation of the interferograms by the instrument. Four of the most commonly used apodization functions are boxcar, cosine, Happ-Genzel, and triangular (66). The effect of these different apodization functions on the magnitude of the oxygen absorbance is shown in Figure 10 (66). Note that these data are not intended to be used to compare the peak height at one resolution to the peak height at another resolution. They should only be used at a given resolution, to compare the effects of different apodization functions at that resolution.

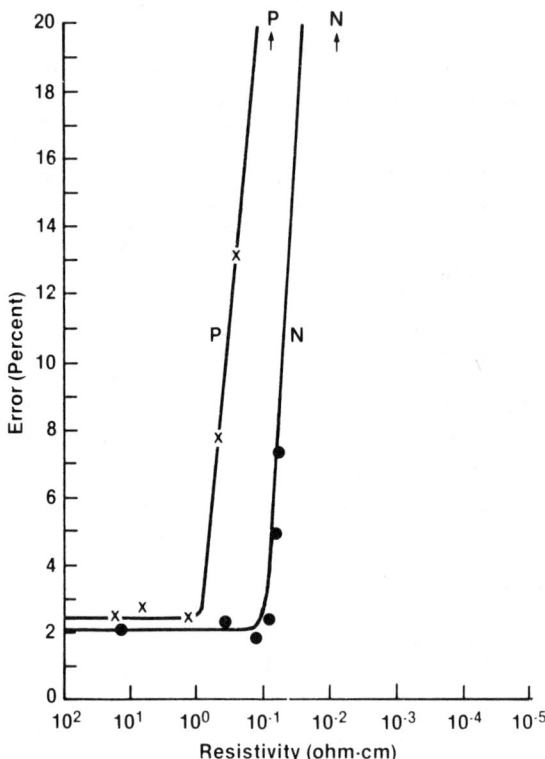

Figure 8. Error, based upon two standard deviations of a set of 10 measurements, in the determination of the interstitial oxygen content of silicon as a function of resistivity for both n- and p-type silicon, expressed as a percentage of the oxygen concentration. (Reproduced with permission from Ref. 62. Copyright 1983 ASTM.)

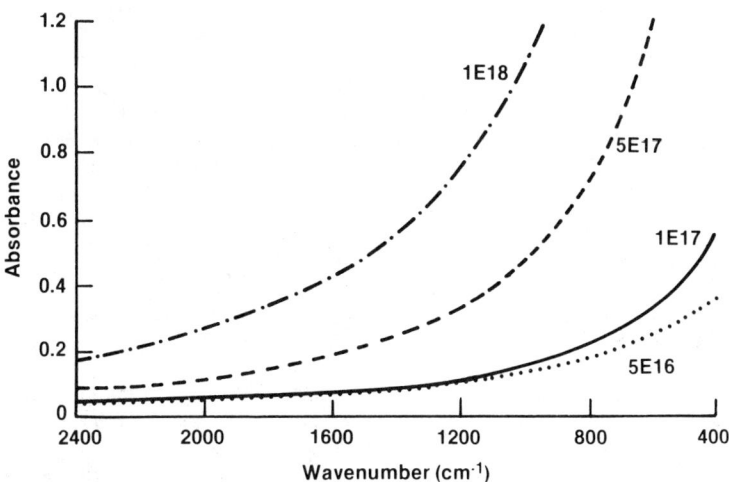

Figure 9. Calculated absorption due to free-carriers in 500-μm thick n-type silicon containing 5×10^{16} to 1×10^{18} donors/cm^3 (Reproduced with permission from Ref. 62. Copyright 1983 ASTM.)

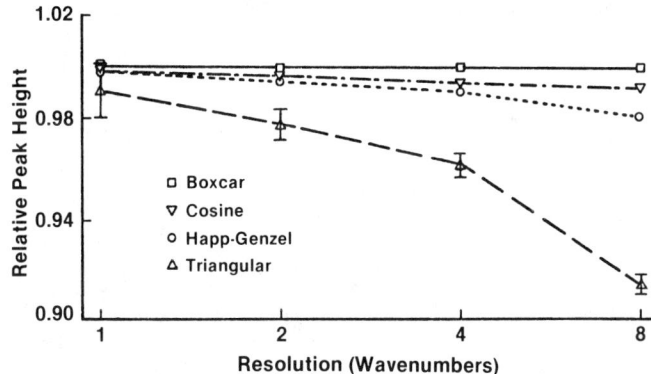

Figure 10. The effect of the apodization function on the relative magnitude of the absorbance band at 1107 cm^{-1} due to interstitial oxygen in silicon. These data were obtained on a 0.87-mm thick silicon wafer polished on both sides. The error bars plotted on the triangular apodization curve can also be applied to the other curves. They represent the greatest standard deviation for any of the data sets (cosine, Happ-Genzel, or triangular) obtained at that resolution (65).

The oxygen determination is most often carried out at a resolution of 4 cm^{-1}; at this resolution, the use of triangular apodization results in a reduction of the peak height by about 4%, compared to the peak height obtained by the use of boxcar apodization. The effects of beam geometry and polarization have also been calculated (66), and found not to be significant, except in the most extreme cases. For example, the change in the reflectance of a silicon wafer for a vertical, columnar IR beam with a maximum angle of incidence at the sample of 10 deg, polarized horizontally, is only 0.6%, compared to the case of an unpolarized beam. On the other hand, detector system nonlinearity in one instrument was found to be in part responsible for the lack of precision in the oxygen content determination (66). Therefore, the lack of interlaboratory reproducibility is apparently not due to different optical bench designs in the different instruments, but rather is attributed to differences in the software or to the performance of the detector systems.

The American Society for Testing and Materials (ASTM) recommends obtaining the interstitial oxygen concentration in atoms/cm^3 by multiplying the peak absorptivity at 1107 cm^{-1} by 2.45×10^{17} (48); the same conversion coefficient is also used by the German standards organization, Deutsches Institut für Normung (DIN) (48,67). This value was obtained by calibrating the infrared absorptivity using vacuum fusion analysis. However, the Japan Electronics Development Association (JEIDA), using charged-particle activation analysis, has obtained a significantly different value, 3.0×10^{17} (68). The ASTM and DIN conversion coefficient for obtaining the substitutional carbon content from the peak absorptivity at 605 cm^{-1} is 1.0×10^{17} (69). This coefficient was determined using deuteron activation analysis as the absolute analytical method (70).

Conclusions

Fourier transform infrared spectrophotometry is used widely in the semiconductor industry for the routine determination of the interstitial oxygen content of production silicon wafers. However, the lack of interlaboratory reproducibility in this method has forced the use of ad-hoc calibration methods. The sources of this lack of reproducibility are just beginning to be understood. As investigation of this problem continues and wider acceptance is gained for improved experimental and analytical techniques, a greater degree of reproducibility should be achieved. Furthermore, new standard test methods and standard reference materials being developed by the ASTM (71), DIN (72), JEIDA (73), and the Bureau Centrale de Materiaux References (73) should also significantly improve the precision of the impurity content determination.

Literature Cited

1. Miller, R. L.; Robinson, A. H.; Ferris, S. D. In "Semiconductor Characterization" edited by Barnes, P. A., and Rozgonyi, G. A., The Electrochemical Society, Pennington, N.J., 1978, pp. 1-17.
2. Craven, R. A.; Korb, H. W. Solid State Technology 1981, 24, 55-61.

3. Benson, K. E.; Lin, W.; Martin, E. P. In "Semiconductor Silicon 1981", Huff, H. R.; Kriegler, R. J.; and Takeishi, Y., Eds.; The Electrochemical Society: Pennington, N.J., 1981; pp. 33-48.
4. Murgai, A.; Patrick, W. J.; Combronde, J. C.; Felix, J. C. IBM J. Res. Develop. 1982, 26, 546-552.
5. Abe, T.; Kikuchi, K.; Shirai, S.; Muraoka, S. In "Semiconductor Silicon 1981", Huff, H. R.; Kriegler, R. J.; and Takeishi, Y., Eds.; The Electrochemical Society: Pennington, N.J., 1981; pp. 54-71.
6. Rava, P.; Gatos, H. C.; Lagowski, J. In "Semiconductor Silicon 1981", Huff, H. R.; Kriegler, R. J.; and Takeishi, Y., Eds.; The Electrochemical Society: Pennington, N.J., 1981; pp. 232-243.
7. Ohsawa, A.; Honda, K.; Yoshikawa, M. Fujitsu Scientific and Technical J. 1980, pp. 123-134.
8. Forman, R. A.; Bell, M. I.; Mayo, S.; Kahn, A. H. J. Appl. Phys. 1984, 55, 547-554.
9. Forman, R. A.; Bell, M. I.; Baghdadi, A.; Mayo, S. Proc. Symp. on Defects in Silicon, Bullis, W. M. and Kimerling, L. C., Eds.; The Electrochemical Society: Pennington, N.J., 1983; pp. 303-312.
10. Suzuki, T.; Isawa, N.; Okuba, Y.; Hoshi, K., ibid, pp. 90-100.
11. Watanabe, M.; Usami, T.; Muraoka, H.; Matsuo, S.; Imanishi, Y.; Nagashima, H., ibid, pp. 126-137.
12. Bond, W. L.; Kaiser, W. J. Phys. Chem. Solids 1960, 16, 44-45.
13. Pajot, B. Analusis 1977, 5, 293-303.
14. Hrostowski, H. J.; Adler, B. J. J. Chem. Phys. 1960, 33, 980-990.
15. Corbett, J. W.; McDonald, R. S.; Watkins, G. D. J. Phys. Chem. Solids 1964, 25, 873-879.
16. Stavola, M. Appl. Phys. Lett. 1984, 44, 514-516.
17. Bosomworth, D. R.; Hayes, W.; Spray, A. R. L.; Watkins, G. D. Proc. Royal Society of London 1970, A317, pp. 133-152.
18. Pajot, B.; Deltour, J. P. Infrared Phys. 1967, 7, 195-200.
19. Baber, S. C. Thin Solid Films 1980, 72, 201-210.
20. Patel, J. R. In "Semiconductor Silicon 1977", Huff, H. R. and Sirtl, E., Eds.; The Electrochemical Society: Pennington, N.J., 1977; pp. 521-545.
21. Jastrzebski, L. IEEE Trans. Electron Devices 1982, ED-29, 475-487.
22. Oeder, R.; Wagner, P. In "Defects in Semiconductors II", Mahajan, S. and Corbett, J. W., Eds.; North-Holland, N.Y., 1983; pp. 171-175.
23. Kaiser, W.; Frisch, H. L.; Reiss, H. Phys. Rev. 1958, 112, 1546-1554.
24. Helmreich, D.; Sirtl, E. In "Semiconductor Silicon 1977", Huff, H. R. and Sirtl, E., Eds.; The Electrochemical Society: Pennington, N.J., 1977; pp. 626-636.
25. Gosele, U.; Tan, T. Y. Appl. Phys. 1982, 28, 79-92.
26. Oehrlein, G. S.; Corbett, J. W. In "Defects in Semiconductors II", Mahajan, S. and Corbett, J. W., Eds.; North-Holland, N.Y., 1983; p. 107-123.
27. Capper, P.; Jones, A. W.; Wallhouse, E. J.; Wilkes, J. G. J. Appl. Phys. 1977, 48, 1646-1655.
28. Craven, R. A. Technical Digest, International Electron Devices Meeting, Washington, DC, 1981, IEEE, N.Y., pp. 228-231.

29. Shimura, F.; Tsuya, H. J. Electrochem. Soc. 1982, 129, 2089-2095.
30. Patel, J. R. In "Semiconductor Silicon 1981", Huff, H. R.; Kriegler, R. J.; and Takeishi, Y., Eds.; The Electrochemical Society: Pennington, N.J., 1981; pp. 189-207.
31. Hu, S. M. J. Appl. Phys. 1980, 51, 5945-5948.
32. Patrick, W. J., NBS Symposium on Silicon Processing, National Bureau of Standards Special Publication Number 337, pp. 442-449, 1970.
33. Tempelhoff, K.; Hahn, B.; Gleichman, R. In "Semiconductor Silicon 1981", Huff, H. R., Kriegler, R. J., and Takeishi, Y., Eds., The Electrochemical Society: Pennington, N.J., 1981; pp. 244-253.
34. Vidrine, D. W. Anal. Chem. 1980, 52, 92-96.
35. Kolbesen, B. O.; Kladenovic (sic), T. Kristall und Technik 1980, 15, K1-K3.
36. Kolbesen, B. O.; Muhlbauer, A. Solid State Electronics 1982, 25, 759-775.
37. Kishino, S.; Matsushita, Y.; Kanamori, M. Appl. Phys. Lett. 1979, 35, 213-215.
38. Ogino, M. Appl. Phys. Lett. 1982, 41, 847-849.
39. Oehrlein, G. S.; Lindstrom, J. L.; Corbett, J. W. Appl. Phys. Lett. 1982, 40, 241-243.
40. Ohsawa, A.; Takizawa, R.; Honda, K.; Shibatomi, A.; Ohkawa, S. J. Appl. Phys. 1982, 53, 5733-5737.
41. Jones, C. E.; Schafer, D.; Scott, W.; Mazer, R. J. J. Appl. Phys. 1981, 52, no. 8, pp. 5148-5158.
42. Donovan, B.; Angress, J. F. "Lattice Vibrations"; Chapman and Hall: London, 1971; p. 167.
43. Barker, Jr., A. S.; Sievers, A. J. "Reviews of Modern Physics"; 47, Suppl. No. 2, 1975; S39-S40.
44. Newman, R. C. "Infrared Studies of Defects in Crystals"; Taylor and Francis: London, 1973; ch. 6.
45. Baghdadi, A. Appl. Spectros. 1981, 35, 473-475.
46. Baghdadi, A. Appl. Spectros. 1983, 37, 520-523.
47. Mead, D. G.; Lowry, S. R. Appl. Spectros. 1980, 34, 167-172.
48. Standard Test Method F121-80, Annual Book of ASTM Standards, Part 43, 1982, pp. 538-540.
49. Hirschfeld, T.; Mantz, A. W. Appl. Spectros. 1976, 30, 552-553.
50. Leroueille, J. Appl. Spectros. 1982, 36, 153-155.
51. Thurber, W. R., NBS Technical Note 529, National Bureau of Standards, Washington, DC, 1970.
52. Graupner, R. K., in Silicon Processing, ASTM STP 804, Gupta, D. C., Ed.; American Society for Testing and Materials, 1983; pp. 459-468.
53. Baghdadi, A. Proc. Symp. Defects in Silicon, Bullis, W. M. and Kimerling, L. C., Eds.; The Electrochemical Society: Pennington, N.J., 1983; pp. 293-302.
54. Pearce, C. W. "VLSI Technology"; Sze, S. M., Ed.; McGraw-Hill: N.Y., 1983; p. 39.
55. Monkowski, J. R. Solid State Technology 1981, 24, 44-51.
56. Kulkarni, M. V., IBM Corporation, private communication.

57. Krishnan, K. Proc. Symp. on Defects in Silicon, Bullis, W. M. and Kimerling, L. C., Eds.; The Electrochemical Society: Pennington, N.J., 1983; pp. 285-292.
58. Graff, K. J. Electrochem. Soc. 1983, 130, 1378-1381.
59. Certain commercial equipment, instruments, or materials are identified in this paper in order to adequately specify the experimental procedure. Such identification does not imply recommendation or endorsement by the National Bureau of Standards, nor does it imply that the materials or equipment identified are necessarily the best available for the purposes.
60. Matlock, J. H., SEH America, Inc., private communication.
61. Shive, L. W.; Shulte, B. K. Third ASTM Symposium on Semiconductor Processing, Gupta, D. C., Ed.; 1984 (to be published).
62. Bullis, W. M.; O'Mara, W. C. Proc. Symp. on Defects in Silicon, Bullis, W. M. and Kimerling, L. C., Eds.; The Electrochemical Society: Pennington, N.J., 1983; pp. 275-284.
63. Weeks, S. P. Third ASTM Symp. on Semiconductor Processing, Gupta, D. C., Ed.; 1984 (to be published).
64. Standard Test Method F121-79, Annual Book of ASTM Standards, Part 43, 1979, pp. 519-521.
65. Stallhofer, P.; Huber, D. Solid State Technology 1983, 26, 233-237.
66. Baghdadi, A. Third ASTM Symp. on Semiconductor Processing, Gupta, D. C., Ed.; 1984 (to be published).
67. Graff, K.; Grallath, E.; Ades, S.; Goldbach, G.; Tolg, G. Solid State Electronics 1973, 16, 887-893.
68. Iizuka, T.; Takasu, Shin.; Tajima, M.; Arai, T.; Nozaki, M.; Inoue, N; Watanabe, M. Proc. Symp. on Defects in Silicon, Bullis, W. M. and Kimerling, L. C., Eds.; The Electrochemical Society: Pennington, N.J., 1983; pp. 265-274.
69. Standard Test Method F123-81, Annual Book of ASTM Standards, Part 43, pp. 543-548.
70. Haas, E.; Kolbesen, B. O., Contract No. NT 74 Dk 546.28.05:621.382.2/.3 Technological Program of the Federal Department of Research and Technology of the FRG, 1973; p. 42.
71. Baghdadi, A., verbal report at the meeting of the ASTM Subcommittee on Semiconductor Physical Properties, F1.04, San Jose, California, 1984.
72. Vieweg-Gutberlet, F., verbal report at the meeting of the ASTM Subcommittee on Semiconductor Physical Properties, F1.04, San Jose, California, 1984.
73. Scace, R. I., National Bureau of Standards, private communication.
74. Yasutake, K.; Umeno, M.; Kawabe, H.; Nakayama, N.; Nishino, T.; Hamakawa, Y. Jpn. J. Appl. Phys. 1982, 21, 28-32.
75. Patrick, W.; Hearn, E.; Westdorp, W.; Dohg, A. J. Appl. Phys. 1979, 50, 7156-7164.
76. Johnson, F. A. Proc. Phys. Soc., 1959, 73, 265-272.

RECEIVED June 18, 1985

13

Application of the Raman Microprobe to Analytical Problems of Microelectronics

Fran Adar

Instruments SA, Inc., Metuchen, NJ 08840

> The Raman microprobe MOLE consisting of a conventional research microscope that is optically and mechanically coupled to a Raman spectrometer, is described. Microanalysis for quality control and materials development is performed non-destructively and requires minimum sample modification. Spatial resolution, determined by optical wavelengths and available microscope optics is approximately 1μm laterally and 10μm axially in transparent materials; in opaque materials the axial resolution, determined by the optical skin depth, will be ca. 300-3000Å. The minimum Raman detectable volume is dependent on the intrinsic Raman intensity. Examples of successful identification of organic contaminants on Si wafers include silicone oil, PET, EDTA, PTFE and cellulose. Elemental silicon has been identified in the vicinity of an Al wire bonded to Au. Stress in laser-annealed laterally seeded Si on silicon oxide has been measured quantitatively and stress-relief at the surface has been detected. The crystalline phase of a ZrO_2 particle precipitating a break in an optical fiber has been identified; since two possible sources of ZrO_2 contaminants exist in different crystalline polymorphs, only phase identification enabled elimination of this problem. Other potential uses of the Raman microprobe in a microelectronics analytical laboratory will be discussed.

Raman scattering is a vibrational spectroscopic technique that can fingerprint both organic and inorganic species and identify polymorphs of crystalline materials. By coupling an optical microscope to a conventional spectrometer, the technique becomes a microprobe with spatial resolution of 1μm, determined by the wavelength of the radiation (ca. .5μm) and the numerical aperature of the microscope objective (n.a ≤ .9 on dry objectives).([1])

Since Raman microprobe (RMP) spectra are identical to spectra acquired in a classical (macro) geometry, literature spectra and spectral files can be used for interpretation. Applications in the microelectronics industry include fingerprint identification of microscopic contaminants and inclusions, and characterization of technologically important new materials.

General Instrument Design

The coupling between micro-sampling optics and a Raman spectrometer was accomplished simultaneously in France and in the United States. Delhaye and co-workers coupled a commerical research grade microscope to a Raman monochromator (**2,3**) while Rosasco coupled an elliptical reflector (**4**) to a Raman monochromator.(**5,6**) The work described here was accomplished on a second-generation instrument of the French design.

The instrument is shown in Figure 1. The coupling of laser to microscope to sample to monochromator is achieved by the beam splitter in the nose piece of a classical metallographic microscope. The beam splitter enables the separation of illuminating and scattered radiation with high numerical aperture at the sample. Standard microscope objectives serve to focus the laser irradiation and to collect the scattered light. Sample preparation is identical to that required for standard optical microscopy. Sample alignment is achieved by positioning the region of interest under the focussed laser beam. Subsequently the focal plane is adjusted to maximize the signal when samples are large and high spatial resolution is not essential. When maximum spatial resolution is required, especially in the axial direction, an iris diaphragm behind the microscope (not shown in figure) can be employed as a spatial filter. The iris is mounted at an image plane conjugated to the sample, and has the effect of eliminating radiation originating outside of the laser focal volume.(**4,7,8**)

Microanalysis - Organic Contamination

Over the past years we have been asked to identify organic contaminants that appear on silicon wafers during processing operations. As the scale of integration of the circuits increases, and the size of the smallest features decreases, the size of contaminants that can effect a device's operation becomes more critical and the ability to identify foreign materials becomes increasingly important. The amount of material present is often too small for analysis by infrared absorption or X-ray diffraction. The Auger and electron microprobes are incapable of yielding chemical (i.e., molecular) identification. The Raman microprobe is unique in its ability to identify organic contaminants that appear as particles as small as $1\mu m$, or as films as thin as $1\mu m$.

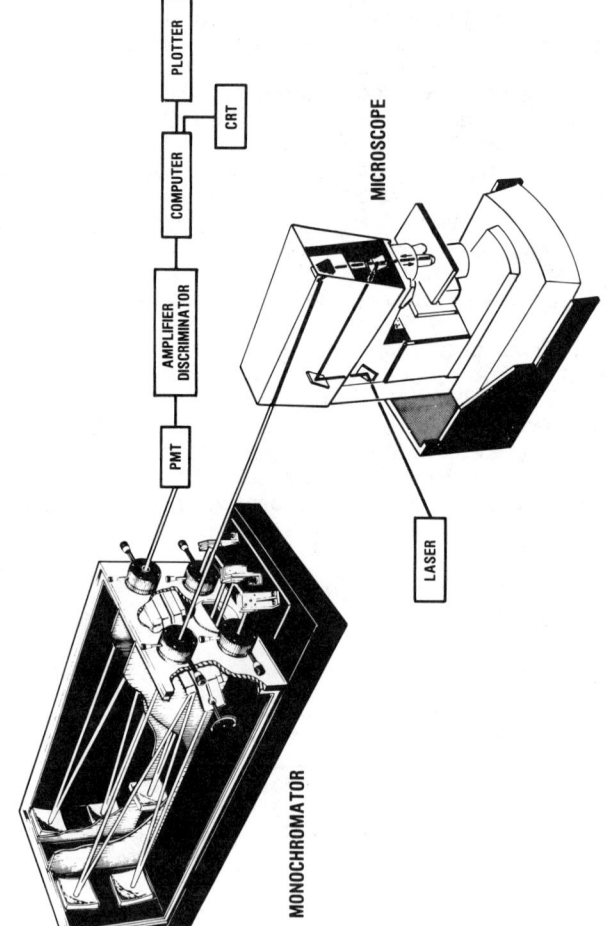

Figure 1 - Instrument schematic showing the light path through microscope and monochromator. Digital data collection was achieved by computer-controlled scanning of the monochromator usually with 1 wavenumber increments and then counting individual photon pulses from an amplifier discriminator. Spectra are manipulated by the computer and plotted on a digital plotter.

EDTA. A very large particle (>100μm) on a silicon wafer was identified as EDTA (ethylene diamine tetraacetic acid). Because of the size of the particle, it was apparent that the EDTA had not precipitated on this wafer after inadequate rinsing (precipitated crystals would exhibit morphology of growth on a substrate). It was inferred that this particle had inadvertently settled on the wafer when clean room gloves had not been appropriately changed.

SILICONE. A silicone film about 5μm thick was identified on a silicon substrate. It would have been difficult by X-ray microanalysis to detect the additional silicon due to the film above the silicon X-ray signal from the substrate. Moreover, the X-ray spectrum would not have revealed the presence of surface silicon as organically-bound silicon, i.e., as a form of poly(dimethyl siloxane). However the Raman spectrum, by fingerprinting the polymer, provides clues to the source of the contaminant. In this case, a degraded silicone-rubber gasket was suspected of introducing material that had settled on this sample.

PET CHIP. A microscopic particle on a silicon wafer was fingerprinted as a piece of polyethylene terephthalate. The source of this contaminant was identified as a wafer handling basket. This result was especially surprising because these baskets had been "guaranteed" not to chip and produce debris.

Fluorinated Hydrocarbon. The bottom of Figure 2 shows a spectrum of material that had settled on another wafer. The features at 521 and 950cm^{-1} are one- and two phonon bands of the silicon substrate. The other features come from the contaminant. The top of the figure shows a spectrum of teflon (polytetrafluoroethylene). While the two spectra in the figure are not a perfect match, it is significant to note that there are no spectral features between 1400 and 1500cm^{-1}, which is where most hydrocarbons exhibit a strong band due to the CH_2 deformation. The lack of this feature implies the absence of any alkane regions longer than two carbon atoms.

Interpretation of this spectrum requires information on the history of the wafer. This specimen had been polished with a slurry containing organic solvents and then etched. The contaminant could be attributed to two sources. Low molecular weight polymer could have been carried from a teflon container. Or polymer could have been deposited on the wafer during etching in a CF_4/H_2 plasma if the plasma contained excess H_2.

CELLULOSE. The upper spectrum of Figure 3 was recorded from a hazy film (<5μm thick) on a Si wafer. In addition to the silicon phonon at 521cm^{-1}, other bands appeared which could be matched to literature spectra of native cellulose (polymorphy I), whose spectrum is shown in the lower portion of the figure.

It was speculated that the source of this contaminant could be cellulose filter material used in one of the processing operations. However, there is a problem in reconciling the assignment to cellulose I and the film morphology. It has usually

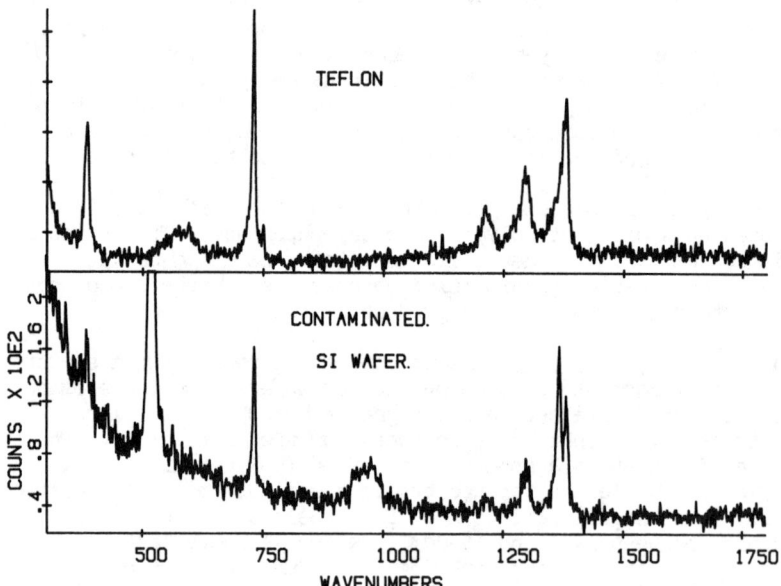

Figure 2 - Raman microprobe spectrum of a contaminated silicon wafer (lower trace) compared with a spectrum of teflon (upper trace). Acquisition conditions are displayed at the bottom of the figure. Note that laser power at the sample is 10% of the value displayed.

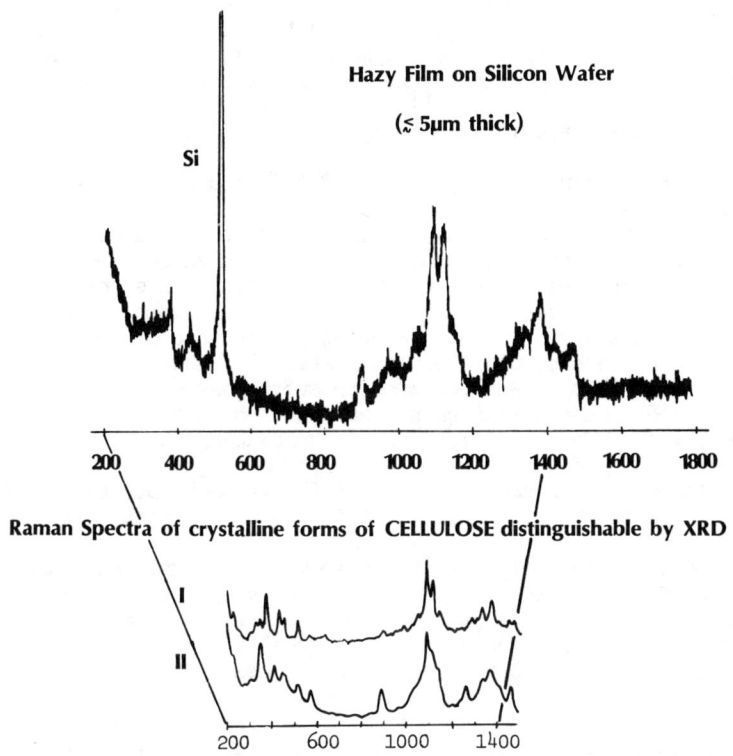

Figure 3 – Raman microprobe spectrum of a hazy film contaminating a silicon wafer (upper trace). Reference spectra of Celluloses I and II shown below. (Reproduced with permission from Ref. 9b. Copyright 1976 John Wiley.) Conditions used to acquire the microprobe spectrum: Laser – 514.5nm, 8.5mw at sample; Slits, 300 m, Scan rate 100cm^{-1}/min; Coaddition of 3 scans; Objective 80x.

been observed that cellulose I cannot be recovered from cellulose solutions. Chemically processed cellulose shows X-ray diffraction and Raman spectra of cellulose II. Thus one would not expect to observe cellulose I from a film of re-precipitated cellulose. However, Attala has been able to show that cellulose I can be recovered from solutions of cellulose by precipitation at elevated tempertures (>150°C).(9) Thus, this Raman spectrum enables one not only to identify the chemical composition of the contaminant, but also allows determination of the processing step during which the contaminant precipitated. It is useful to note here that identification of crystalline polymorphs of many other polymers provides information on their thermal and/or stress history.

Microanalysis - Break in Wire Connect

An integrated circuit package that had failed during use was submitted for analysis. The ceramic package had been opened and it was found that some of the aluminum wire connects had broken not far from the gold bonding pads. It was requested that material deposited on the gold around the broken aluminum wire be identified - the material was suspected to be organic.

Numerous scans taken of the deposited residue did not indicate the presence of an organic material. However, a band at approximately 519cm^{-1} was observed on every pass - and it was not observed when the instrument was focussed on the gold pad or aluminum wire. We were forced to the conclusion that the "contaminating" material was elemental, semi-crystalline silicon. (The silicon phonon frequency down shifts when the crystallite size is reduced(10) or when the lattice is under tensile stress (11)).

From infrared analysis, it was known that significant amounts of silica were present. It was subsequently argued that silica gel had contaminated the package and acted as a carrier for corroding chemicals. The Raman microprobe did not find evidence for silica in significant amounts. We argued that the failure could have been caused by the aluminum wires themselves. These often have a small percentage of micro-crystalline silicon added as a hardening agent. If the silicon is inadequately dispersed there might be local heating during use which would promote electron-migration of the silicon and could end ultimately in a break in the wire.

In order to reconcile the Raman microprobe results with those of infrared, it is important to recognize the limitations of the techniques. The Raman-active phonon mode in silicon is infrared inactive. However, all silicon is coated to some extent with a thin film of a thermal oxide to which infrared absorption is very sensitive. While the oxide is also active in the Raman effect, its signal is much weaker than that of the silicon itself.

Stress in Laser Annealed Silicon on Silicon Oxide

In collaboration with Hewlett Packard Laboratories lateral-epitaxially regrown films of laser-annealed silicon on silicon oxide islands were examined.(10) This study is motivated by the attempt to produce high quality single crystal silicon films isolated by oxide from the silicon crystal wafer. Such materials could then be used to manufacture devices with increased component density per package. It is known, however, that mismatch at the silicon/oxide interface will usually result in stress in the annealed material that causes dislocations and grain boundaries which will effect the silicon's conductivity. The RMP study ([10]) discribed here was an effort to combine the ability of the Raman signal to monitor stress with the high spatial resolution of the technique.([11]) The measured stress could then be correlated with the geometry of the sample.

The samples were prepared by depositing .55µm of polysilicon over a (001) silicon wafer patterned with 480µm square islands of SiO_2 that were 1.4µm thick. The edges of the island were oriented along (110) directions. The polysilicon was encapsulated with 60Å of silicon nitride. The samples were held at 500°C during recrystallization which was accomplished by scanning an 11W, 80µm diameter Gaussian argon ion laser beam across the sample at a rate of 25cm/sec, raster stepping 10 m between scans.

The Raman active phonon of high quality single crystal silicon occurs at 520.7cm^{-1}. Under tensile stress the band shifts to lower energies; under compressive stress the band shifts to higher energies.([12]) Because of the values of the thermal expansion coefficients of silicon and its oxide, the silicon film re-crystallized over the oxide will experience tensile stress. In order to maximize the accuracy with which the stress could be measured, the Raman spectra were recorded digitally with 0.1cm^{-1} between data points. (Instrumental repeatability was also 0.1cm^{-1}).

Figure 4 shows a plot of the Raman phonon frequency as a function of distance from the edge of the oxide island. The figure also shows the dependence of the stress (as extrapolated from the Raman frequency) on the position of the probe.

Two sets of data are illustrated in the figure; Raman spectra were excited at wavelengths 514.5 and 457.9nm. Because the optical penetration depth of silicon is different at these two wavelengths, the information generated reflects different thicknesses of samples. Both sets of data indicate an increase in stress as the distance from the oxide edge increases. A striking difference is the indication of a drop in stress at approximately 25µm from the edge that is detected in the 457.9nm-excited data. This is to be attributed to the shallower penetration depth of this radiation; the conclusion to be drawn is that after the sample breaks into polycrystalline grains approximately 11µm from the edge (as measured by Nomarski contrast micrographs), there is relief of stress at the surface of the recrystallized material.

Other Areas of Applicability of the RMP

Optical Fibers. Several manufacturers of optical fibers have suggested that a Raman microprobe could be a useful tool in characterizing fibers. The literature shows that it is possible to monitor the concentration of additives to silica down to the 1 mole percent level.([13-17]) Polished sections of preforms or drawn fibers have been monitored in the microprobe in this laboratory.

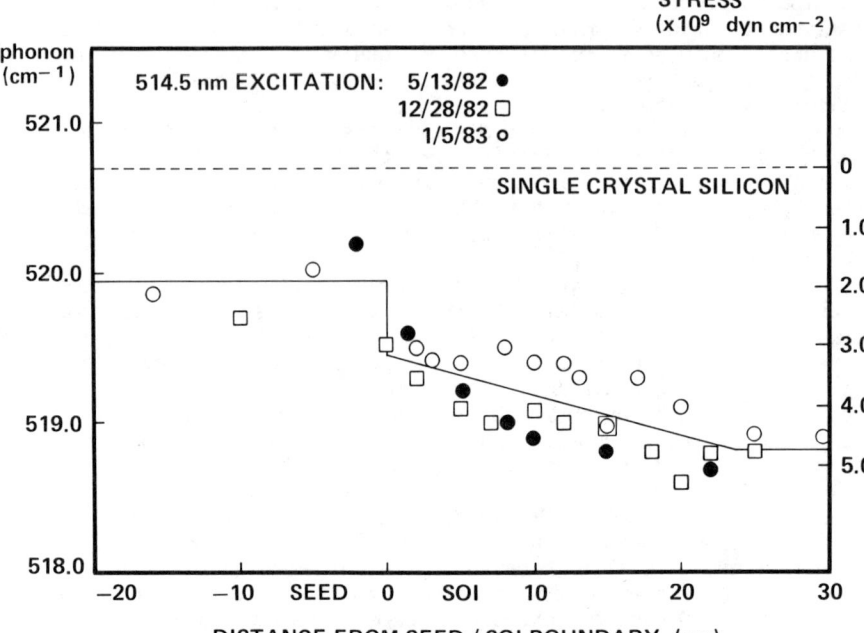

Figure 4 - Raman phonon frequency of laser-recrystallized lateral epitaxially annealed silion on insulating oxide (SOI) and the derived stress, as a function of distance from the seed/SOI interface: Spectra were acquired with laser wavelengths 514.5 and 457.9nm.

In collaboration with L. Soto at Bell Laboratories, we have recently succeeded in identifying the cause of a break in a communications fiber.([18]) By X-ray fluorescence zirconium was known to be present in the 2μm particle precipitating the break - the Raman spectrum showed clear evidence for the monoclinic polymorph. Because the furnace used to pull the fibers contained monclinic and tetragonal zirconias in different locations, it was only the polymorph identification by the Raman microprobe which provided the information necessary to refine the manufacturing process.

III-V Semiconductors. The use of Raman spectroscopy to characterize crystalline films of compound semiconductors has been reviewed.([19]) Effects that can be monitored are orientation, carrier concentration, charge carriers' scattering times, mixed-crystal composition ([19]) and Group V deposits in the native oxides.([20])

Literature Cited
1. Adar, F. In "Microbeam Analysis - 1981", Geiss, R.H., Ed.; San Francisco Press 1981; pp. 67-72.
2. Delhaye, M. and Dhamelincourt, P. J. Raman Spectrosc. 1975, 3, 33.

3. Dhamlincourt, P. In "Microbeam Analysis - 1982:; Heinrich, K.F.J., Ed.; San Francisco Press, 1982; pp. 261-269.
4. Hirschfeld, T. *J. Opt. Soc. Am.* 1973, 63, 476-7.
5. Rosasco, G.J., Etz, E.S., Cassatt, W.A. *Appl. Spectrosc.* 1975, 29, 396.
6. Rosasco, G.J. In "Advances in Infrared and Raman Spectroscopy Volume 7"; Clark, R.J.H. and Hester, R.J., Eds.; Heyden: London 1980; Chap. 4.
7. Dhamelincourt, P. PhD. Thesis, L'Universite des Sciences et Techniques de Lille, Lille 1978.
8. Adar, F. and Clarke, D.R. In "Microbeam Analysis - 1982"; Heinrich, K.F.J., Ed.; San Francisco Press 1982; pp. 307-310.
9a. Atalla, R.H., Dimick, B.E., Nagel, S.C. In "Cellulose Chemistry and Technology"; Arthur, J.C. Jr., Ed.; ACS Symposium Series No. 48, American Chemical Society.
9b. Atalla, R.H., *Appl. Polym. Symp.*, 1976, No. 28, 659-669
10. Iqbal, Z. and Veprek, S., J. Phys. C. Solid State Phys. 14, 0000(1981)
11. Zorabedian, P., Adar, F. *Appl. Phys. Lett.* 1983, 43, 177-179.
12. Anastassakis, E, Pinczuk, A, Burstein, E., Pollack, F.H., Cardona, M. *Solid State Commun.* 1970, 8, 133.
13. Walfrafen, G.E. and Stone, J., *Appl. Spectroscopy* 1975, 29, 337-344
14. Lan, G.-L., Banerjee, R.K., and Mitra, S.S., *J. Raman Spectrosc.* 1981, 11, 416-423
15. Sproson, W.A., Lyons, K.B, Fleming, J.W. *J. Non-Crystal. Solids* 1981, 45, 69-81
16. Shibata, U., Horiguchi, M., Edahiro, T. *J. Non-Crystal. Solids* 1981, 45, 115-126.
17. Noguchi, K., Murakami, Y., Uesugi, N. and Ishihara, K., *Appl. Phys. Lett.* 1984, 44, 491-493
18. Soto, L, and Adar, F. In "Microbeam analysis 1984" Romig, A.D. Jr. and Goldstein, J.I., Eds.; San Francisco Press, 1984, pp. 121-124.
19. Abstreiter, G, Bauser, E. Fischer, A., Ploog, K. *Appl. Phys.* (Springer Verlag) 1978, 16, 345-352.
20. Schwartz, G.P., Gualtieri, G.J., Griffiths, J.E., Thurmond, C.D., Schwartz, B. *J. Electrochem. Soc.* 1980, 127, 2488.

RECEIVED June 28, 1985

14

Characterization of Gallium Arsenide by Magneto-optical Photoluminescent Spectroscopy

D. C. Reynolds

Air Force Wright Aeronautical Laboratories, AADR, Wright-Patterson Air Force Base, OH 45433

> Identification of residual and doped donors have been identified in expitaxial GaAs using the photoluminescence technique in the presence of applied magnetic fields. Transitions occur between excited initial and final states of the neutral-donor-bound-exciton complexes. The magnetic field compresses the wave function which sharpens the optical transitions. The magnetic field also separates the different donors when viewed from the neutral-donor-bound-exciton transitions. These two effects make possible the identification of donors when the donor concentration is in the mid 10^{15}cm^{-3} range.

Reflection, emission and absorption in solids has long been studied. Intense photoluminescence is observed in many semiconductors at low temperatures. When spectrally analyzed, this photoluminescence provides an extensive source of experimental data which contributes to the ultimate identification of the electronic states of impurities and defects in these semiconductors. Many sharp lines appear in such spectra, particularly from bound excitons, which provide a "finger print" of the impurities and defects which are present in the semiconductor lattice.

The exciton is the probe in this case, becoming bound to various impurities, defects, and complexes and the subsequent decay from the bound state yields information concerning the center to which it was bound. The effective mass like donors in many III-V binary and in several of the III-V ternary systems are shallow. The chemical shifts and central cell corrections are small, therefore the energy separation of donors resulting from different impurities or host defects is small. This requires low residual concentrations to prevent concentration broadening and merging of the impurity levels with the conduction band. Controlled doping experiments with known donors must also be in the low concentration range to permit the identification of specific donors. The binding energies of acceptors are in general larger than donors. This makes the experimental characterization of acceptors easier than that for donors.

This chapter not subject to U.S. copyright.
Published 1986, American Chemical Society

The behavior of these sharp line transitions in perturbing magnetic and strain fields make it possible to differentiate between simple substitutional donors and acceptors and complexes composed of combinations of impurities and or defects.

Exciton Spectra

Intrinsic Excitons. The intrinsic fundamental gap exciton in semiconductors is a hydrogenically bound hole-electron pair, the hole being derived from the top valence band and the electron from the bottom conduction band. It is a normal mode of the crystal created by an optical excitation wave, and its wave functions are analogous to those of the Block wave states of free electrons and holes. When most semiconductors are optically excited at low temperatures it is the intrinsic excitons that are excited. The energy of the ground and excited states of the exciton lie below the band gap energy of the semiconductor. Hence, the exciton structure must first be determined in order to determine the band gap energy. The exciton binding energy can be determined from spectral analysis of its hydrogenic ground and excited state transitions. Precise bandgap energies can be determined by adding the exciton binding energy to the experimentally measured photon energy of the ground state transition.

Extrinsic Excitons. Bound exciton complexes or impurity exciton complexes are extrinsic properties of materials. These complexes are observed as sharp line optical transitions in both photoluminescence and absorption. The bound complex is formed by binding a free exciton to a chemical impurity atom (ion), complex, or a host lattice defect. The binding energy of the exciton to the impurity or defect is generally weak compared to the free exciton binding energy. The resulting complex is molecular-like (analogous to the hydrogen molecule or molecule-ion) and bound excitons have many spectral properties which are analogous to those of simple diatomic molecules. The center to which the free excitons are bound can be either neutral donor and acceptor centers or ionized donor and acceptor centers. The emission or absorption energies of these bound exciton transitions are always below those of the corresponding free excitons, due to the molecular binding energy.

The sharp spectral lines of bound exciton complexes can be very intense (large oscillator strength). The line intensities will, in general, depend on the concentrations of impurities and/or defects present in the sample.

The theory of "impurity" or defect absorption intensities in semiconductors has been studied by Rashba (1). By use of the Fredholm method, he finds that if the absorption transition occurs at k=0 and if the discrete level associated with the impurity approaches the conduction band, the intensity of the absorption line increases. The explanation offered for this intensity behavior is that the optical excitation is not localized in the impurity but encompasses a number of neighboring lattice points of the host crystal. Hence, in the absorption process, light is absorbed by the entire region of the crystal consisting of the impurity and its surroundings.

In an attack on the particular problem of excitons which are weakly bound to localized "impurities", Rashba and Gurgenishvili ($\underline{2}$) derived the following relation between the oscillator strength of the bound exciton F_d and the oscillator strength of the intrinsic excitons f_{ex}, using the effective-mass approximation

$$F_d = (E_o/|E|)^{3/2} f_{ex} \qquad (1)$$

where $E_o = (2h^2/m)(\pi/\Omega_o)^{2/3}$, E is the binding energy of the exciton to the impurity, m is the effective mass of the intrinsic exciton, and Ω_o is the volume of the unit cell.

It has been shown in some materials that F_d exceeds f_{ex} by more than four orders of magnitude. An inspection of Equation 1 reveals that, as the intrinsic exciton becomes more tightly bound to the associated center, the oscillator strength, and hence the intensity of the exciton complex line, should decrease as $(1/E)^{3/2}$.

In magnetic fields, bound excitons have unique Zeeman spectral characteristics, from which it is possible to identify the types of centers to which the free excitons are bound. Bound exciton spectroscopy is a very powerful analytical tool for the study and identification of impurities and defects in semiconductor materials.

Magneto-Optical Spectroscopy Techniques

Magneto-optical spectroscopy techniques have been applied to the characterization of GaAs which is a potentially important material for many technological applications. High quality material is required for many of these applications. To improve the quality requires a knowledge of the residual impurities in undoped material. The acceptors, having relatively large binding energies as compared to the donors (~30 vs. 5.7 meV), can be identified quite easily. The shallow hydrogenic donors, on the other hand, have small binding energies and also have small central-cell corrections. This makes the resolution of different donors resulting from different chemical impurities difficult to achieve. The early experiments from which different chemical donors were identified employed high-resolution Fourier-transform infrared magnetospectroscopy (FTIR) which used the modulated photoconductivity detection technique to monitor the $1S-2P_{-1}$ transition in a fixed magnetic field.

FTIR Studies. These studies were motivated by both fundamental and technological interests. Fundamentally there was a desire to determine how effectively these impurity states could be treated by effective mass theory. Also the effect of local potentials due to the different core configurations of different chemical impurities was of particular interest. It is these chemical shifts that require corrections to be made to the effective mass theory. These shifts also make possible the identification of different chemical species which is of great technological interest. Considering III-V semiconductors, donors are introduced by substituting group IV atoms on the group III site or substituting group VI atoms on the group V site. In a similar manner acceptors are introduced by substituting group II atoms on the group III site or by substituting group IV atoms on the group V site. The slight energy shift (central cell correction) resulting from introducing different chemical donors or acceptors makes possible the identification of the substituted atom.

The FTIR experiment can be either a transmission experiment or a photoconductive experiment. Photoconductive detection is much more sensitive than the transmission experiment while still retaining the high resolution feature of the technique. It was shown by Stillman (3) that photoconductive detection allows one to observe transitions between bound impurity states. The excited carrier can absorb a phonon and make a transition to a conduction band state.

Effective Mass Model. The effective mass description of an impurity (donor) state requires that the electron orbit extend over many lattice distances. This permits the use of the macroscopic dielectric constant to describe the electronic motion. With these conditions satisfied, the impurity energy state can be approximated by the hydrogenic form with the electron mass replaced by an effective mass m^*

$$E_m = -R^*/m^2 \tag{2}$$

where R^* is an effective Rydoerg which is related to the hydrogen Rydberg by the following expression

$$R^* = \tfrac{1}{2} m^* e^4/\varepsilon_o^2 \hbar^2 \tag{3}$$

It is clear that the effective mass model must be modified in order to account for different chemical donors. The FTIR experiment monitors the $n = 1$ to $n = 2$ transition. For shallow donors in many of the III-V compounds the 2s and 2p states are degenerate for all practical purposes. The resolution is improved by applying a magnetic field. The magnetic field separates the orbital angular momentum states. The magnetic field also compresses the wave function which sharpens the transitions, it also separates transitions due to different chemical impurities (4).

High Resolution Photoluminescence Studies. More recently, shallow residual donors have been identified in high-purity VPE GaAs using high-resolution photoluminescence spectroscopy (5, 6). The optical transitions that were used to identify the residual donors result from the collapse of neutral-donor-bound excitons. The decay of an exciton bound to a donor (acceptor) may leave the donor (acceptor) in an excited state. This was first pointed out by Thomas and Hopfield (7). They observed transitions in CdS that were characterized by large magnetic field splittings and negative diamagnetic shifts which they tentatively identified with transitions of this type. Excited-state transitions of this type were later identified in GaP, (8), CdSe, (9, 10) CdS, (11) ZnO, (12) and ZnSe (13). Residual donors have subsequently been resolved in GaAs by Almassy et al. (5) from optical transitions resulting from the collapse of an exciton bound to an excited donor state, leaving an excited terminal state. The terminal state of this transition, from which chemical identifications are made, is the $n = 2$ state. When the terminal state is a 2s state, then the central-cell correction to this state is assumed to be 1/8 of its value for the 1s state. In transitions of this type it is possible to observe 7/8 of the total central-cell correction. When the terminal state is a 2p state the full central-cell correction is observed.

Emission lines have been observed on the high-energy side of the neutral-donor – bound-exciton transition (D^0,X) in many materials, CdTe, (14) GaAs, (15, 16) CdS, (17) and ZnSe (13). These transitions were interpreted as excited states of the D^0,X but with very little detail as to the nature of the excited states. Guillaume and Lavallard (18) proposed a rigid-rotator model to explain these excited states in CdTe. In this model the hole is excited to rotate around the fixed donor analogous to rotation of diatomic molecules. This model had difficulty in predicting the observed energies for the excited-state transitions. A non-rigid-rotator model was subsequently proposed by Ruhle and Klingenstein, (19) which was successful in predicting the excited-state energies in InP and GaAs. It was the collapse of excitons bound to the n = 2 rotational state of the donors from which the residual donors were resolved in Reference 5. It has been observed that in many crystals the D^0,X state as well as the first non-rigid-rotational state are broadened while other bound-exciton states are not. Herzberg and Spinks (20) have observed broadening of rotational lines from diatomic molecules and have also observed that this broadening decreased with increasing rotational quantum number. It has been proposed (21) that an analogous broadening mechanism is responsible for the broadening of the D^0,X transition as well as the transition from the first non-rigid-rotational state. It was observed that when the ground state was broadened the excited- (n = 2) state transitions were also broadened. The energy separation between the D,X state and the first rotational state in GaAs in 50μV. If one attempts to resolve residual donors from the excited-state (n = 2) transitions of these two states, the combined line broadening and small energy separation render it impossible in most cases. Almassy et al. (5) circumvented this problem by observing the n = 2 states resulting from the collapse of the exciton bound to the second non-rigid-rotational state of the donor, which is not broadened. This has the advantage that the experiment is done in zero magnetic field and therefore the donor energies are directly measured. The scheme has the disadvantage that the intensity of the n = 2 state associated with the second non-rigid-rotational state is considerably less than the intensity of the n = 2 state associated with the first non-rigid-rotational state.

It would be desirable to identify the residual donors in GaAs from the transition involving an exciton bound to the first non-rigid-rotational state. The terminal state consists of the excited state (n = 2) of the electron on the donor. The observation of different residual donor species from this transition is made possible by performing the experiment in a magnetic field. The magnetic field produces two effects: (a) It separates out states with different orbital angular momentum and (b) it compresses the wave function which sharpens the lines and separates the donors. In the final state the transition can terminate in either the 2s or 2p state. From parity arguments it can be shown that the initial state of the D^0,X transition has odd parity. The 2s final state in this transition will have even parity whereas the 2p final state will have odd parity. The preferential transition, therefore, from the D^0,X initial state will be to the 2s final state. By similar

argument it can be shown that the initial state of the exciton bound
to the first rotational state of the neutral donor will have even
parity. The preferential transition for this state then will be to
the 2p final state having odd parity. The intensities of these
transitions were observed by Dean et al. (22) for the case of ZnTe.
They showed that when the exciton collapsed from the first rotational
state the terminal state intensity ratio 2p/2s = 20. They further
showed that when the exciton collapsed from the D^o,X state the
terminal state intensity 2p/2s = 1/5. It can therefore be concluded
that in the case of GaAs the 2p final states are predominately
associated with excitons bound to the first non-rigid-rotational
state, and the 2s final states are predominately associated with the
D^o,X state.

Transition from Neutral Donor States. When an exciton decays from
the D^o,X state in a magnetic field the excited 2s and 2p terminal
states are separated. The 2p angular momentum states are also
separated. The initial state of the complex consists of two paired
electrons and one unpaired hole as shown in Figure (1). The
unpaired hole will split in a magnetic field. The final state of
the complex consists of one unpaired electron either in the ground
state or in an excited state. The transitions of interest in this
paper are those associated with the terminal electron in the n = 2
state. The inset in Figure (1) is a densitometer trace of the D^o,X
state and its associated non-rigid-rotational states. The initial
state of the transition can originate from D^o,X or any of its rotational states. The transition of interest is the one whose
initial state is the first non-rigid-rotational state. The specific
transition results from the collapse of an exciton bound to the
first non-rigid-rotational state and terminating in the 2p state of
the electron on the neutral donor. The initial state of the complex
splits into a quartet and the final 2p state splits into three
widely separated states $2p_{+1}$, $2p_0$, and $2p_{-1}$. In this transition a
negative diamagnetic shift occurs so that the lowest-energy 2p
state results in the highest-energy optical transition. This is the
$2p_{-1}$ state which has the highest intensity of the p-state transitions. The magnetic field splitting of this state at 40 kG is shown
in Figure (2). The inset of Figure (2) shows the zero-field trace
in the n = 2 spectral region of the neutral-donor - bound exciton.
Here it is seen from the collapse of the exciton bound to the
second, third, and fourth non-rigid-rotational states that the
crystal contains residual Si and S donors. The splitting of the
$2p_{-1}$ state shows components of these two donors as marked in Figure
(2). The highest-intensity transitions are the spin-conserving
transitions. The lines are sharp and well resolved. The donors
could not be resolved from this transition in zero magnetic field
due to broadened lines and the near proximity of the principal-
donor - bound exciton. Much can be gained by using a magnetic
field with the photoluminescence identification of shallow donors.

It was shown by Fetterman et al. (4) that the separation between different chemical donors increased monotonically with magnetic field strength. In their case they were analyzing the donors
by the FTIR technique. Since the exciton is very loosely bound to
the donor in the photoluminescence scheme, a similar response might

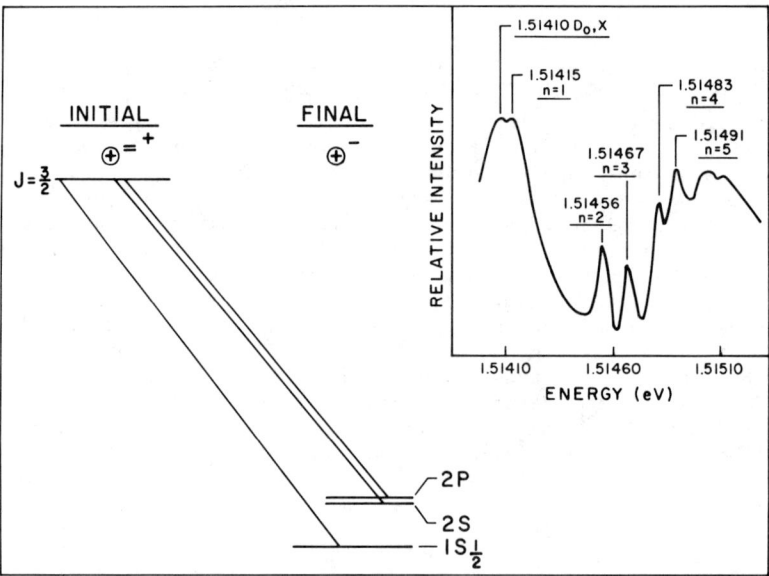

Figure 1. Schematic representation of radiative recombination of an exciton bound to a neutral donor where the final state is the donor in the ground or in the excited configuration. The inset shows the initial state of the neutral-donor-bound exciton in the ground and several excited rotational states. (Reproduced with permission from Ref. 24. Copyright 1983 American Physical Society.)

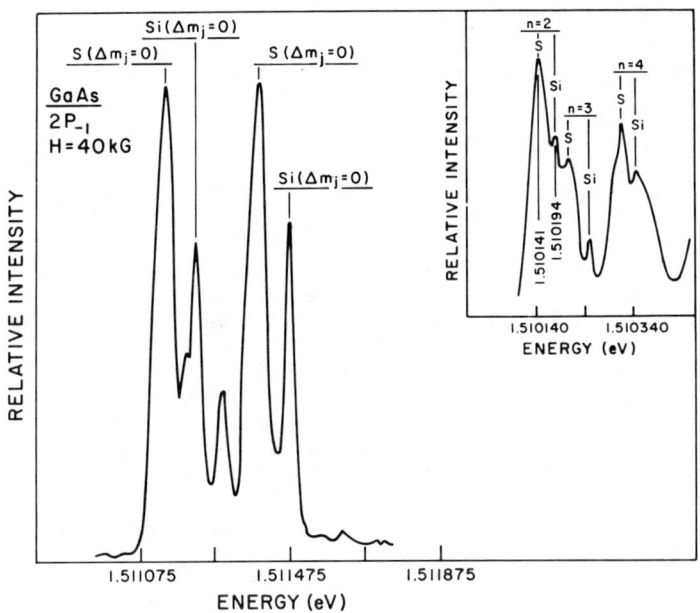

Figure 2. Magnetic field components of the $2p_{-1}$ state at 40 kG resulting from the collapse of an exciton bound to the first-rotational state of the neutral donor for both sulfur and silicon. The inset shows the same two donors in zero magnetic field resulting from the exciton bound to the second, third, and fourth rotational states of these neutral donors. (Reproduced with permission from Ref. 24. Copyright 1983 American Physical Society.)

be expected. The perturbation theory of Fetterman et al. resulted
in the following expression for the magnetic field separation of
donors 1 and 2:

$$\Delta_{1,2}(B) = K_{1,2}|\chi_{1s}(0)|^2 \tag{4}$$

where $K_{1,2}$ is an adjustable parameter independent of magnetic field.
$\chi_{1s}(0)$ is the value of the effective-mass envelope function for an
electron in the donor ground state at the origin in the presence of
a magnetic field. This function is magnetic field dependent. χ_{1s}
can be calculated from effective-mass theory. The solid line in
Figure (3) is a theoretical plot of $|\chi_{1s}(0)|^2$ as a function of
magnetic field taken from Cabib et al.(23). In the same figure is
plotted $\Delta_{Si,S}$ (silicon, sulfur donor separation) by adjusting the
right-hand scale to place the point at 40 kG close to the theoretical curve. With the use of unit K the remaining experimental points
fall as shown in Figure (3). The fit is reasonably good and shows
that the perturbation theory is also applicable when an exciton is
loosely bound to the donor. It is evident that at higher magnetic
fields increased donor separation will occur.

A geometric construction was used to analyze the magnetic
field splitting of the $2p_{-1}$ states of the Si and S donors whose
densitometer traces are shown in Figure (2). Since the excitons
are very loosely bound to the donors it is reasonable to assume
that the electron g value in these bound states will be essentially
the same as the free-electron g value. Using an electron g value
of $g_e = -0.55$, the nomogram at 40 kG shown in Figure (4) was produced. From the nomogram a hole κ value of 0.65 is measured.

In Figure 5 the deviation from the center of mass of the $M_{J=0}$
transition for the silicon donor as a function of magnetic field
is plotted. An effective g value of 1.1 is obtained from this
splitting. This is close to the sum of the magnitudes of the electron and hole g values which is expected for this splitting. An
identical splitting is obtained for the sulfur donor.

Identification of Donors in Doped GaAs

Magneto-optical photoluminescent spectroscopy was used to identify
the donors in two samples of Si doped GaAs. The Si concentration
in sample A as determined by C-V profiling was $N_D = 5X10^{15} cm^{-3}$.
The total donor concentration in sample B was $N_D = 3X10^{15} cm^{-3}$ with a
carrier mobility = 45,000 cm^2/V-sec at 77 K determined from Hall
effect measurements.

The donors were identified from the collapse of excitons bound
to neutral donors. The optical transitions involve the collapse
of exciton complexes whose initial state consists of an exciton
bound to the first nonrigid rotational state of the neutral donor
and whose terminal state consists of the $2p_{-1}$ state of the neutral
donor. The most intense transitions in this process are the spin
conserving transitions. The measurements were made in an applied
magnetic field of 36kG. The transitions to the $2p_{-1}$ state are
shown in Figure 6. The dashed curve for sample A shows the
spin conserving transitions for the Si donor in this Si doped

14. REYNOLDS *Characterization of GaAs*

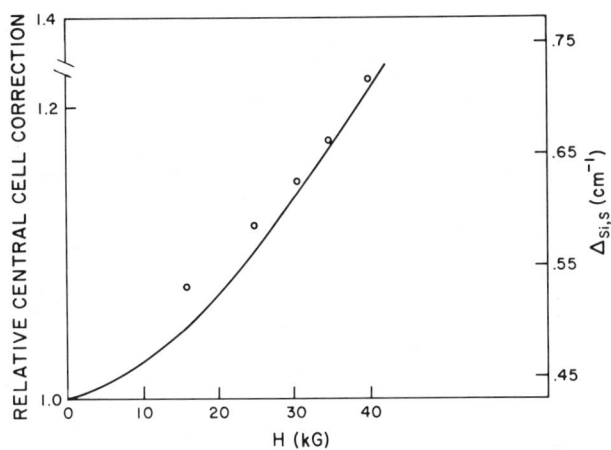

Figure 3. Separation of the Si and S donors as a function of magnetic field is shown by the closed circles. The solid line is the theoretical curve of $|\chi_{1s}(0)|^2$ as a function of magnetic field. The zero-field value of $\Delta_{Si,S}$ is 0.43 cm^{-1}. (Reproduced with permission from Ref. 24. Copyright 1983 American Physical Society.)

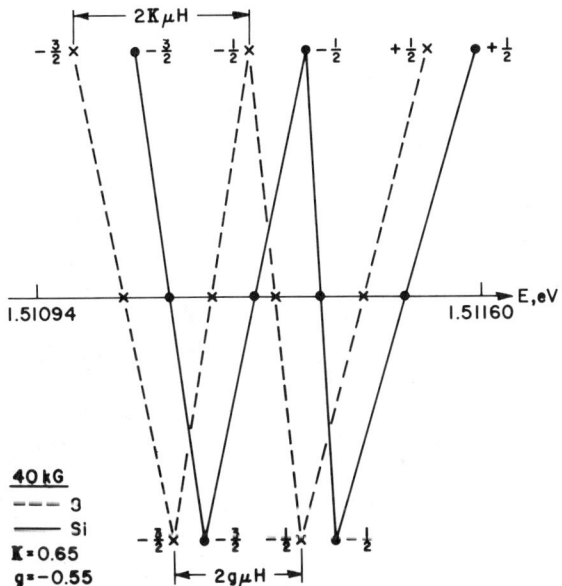

Figure 4. Nomogram for interpretation of the 2p$_{-1}$ magnetic field components of the Si and S donors. The free-electron g value is assumed, a hole k value of k = 0.65 is then measured. (Reproduced with permission from Ref. 24. Copyright 1983 American Physical Society.)

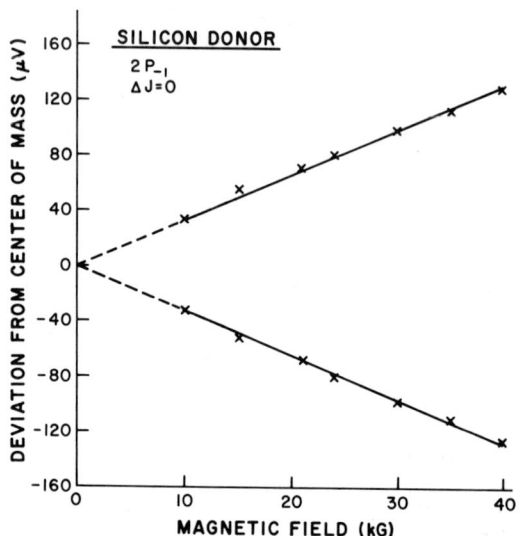

Figure 5. Deviation from the center of mass is plotted as a function of magnetic field for the Si donor. Reproduced with permission from Ref. 24. Copyright 1983 American Physical Society.)

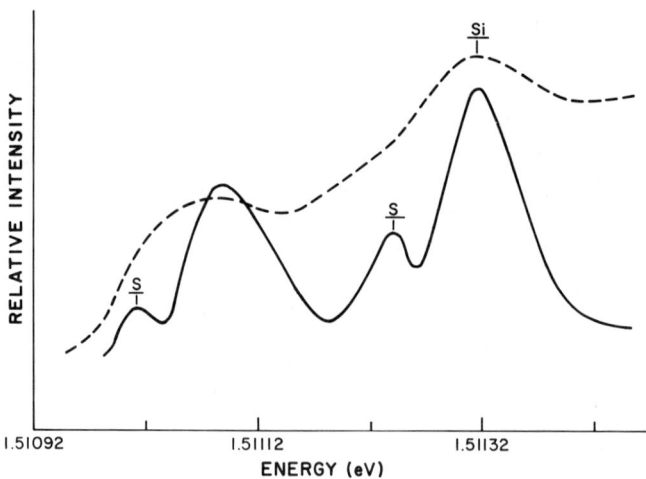

Figure 6. Shallow donors identified in two different GaAs samples from the $2p_{-1}$ transition. The dashed curve shows the Si donor in sample A; the solid curve shows the Si donor as well as the residual S donor in sample B.

sample. The solid curve for sample B, also Si doped shows the Si donor with the well resolved S residual donor also present. This demonstrates that the technique is capable of identifying donors at relatively high concentrations (5×10^{15} cm^{-3}). These results could not have been achieved in zero magnetic field.

The magneto-optical spectroscopy technique has certain advantages over FTIR: (1) It can penetrate layers such as active layers as well as layers associated with some heterostructures, (2) it does not require contacts, (3) donors and acceptors can be identified simultaneously permitting an estimate of sample compensation, and (4) since it samples a very thin layer of material it can be used in profiling layers. These two techniques provide methods for identifying very shallow levels in semiconductors.

Conclusions

In order to make judgement about the quality of an epitaxial layer, or any semiconductor material, it is necessary to characterize that material. The characterization may take many forms, depending on the intended end product of the material. In this paper optical characterization has been emphasized, focusing on the magneto-optical photoluminescent spectroscopy technique for identifying impurities. The early experiments from which different chemical donors were identified in materials like GaAs employed the FTIR technique. Both of these techniques are very useful for identifying impurities however, in this paper, it is pertinent to point out some of the advantages of the magneto-optical spectroscopy technique: (1) It can penetrate layers such as active layers as well as layers associated with some heterostructures, (2) It does not require contacts, (3) donors and acceptors can be identified simultaneously permitting an estimate of sample compensation, and (4) since it samples a very thin layer of material it can be used in profiling layers.

Literature Cited

1. Rashba, E. I., Opt. Spektrosk. 1957, 2,508.
2. Rashba, E. I. and Gurgenishvili, G. E., Fiz. Tverd. Tela 1962, 4,1029 (English trans.: Sov. Phys. Solid State 1962, 4 759.
3. Stillman, G. E.; Wolfe, C. M. and Korn, D. M., Proc. Int. Conf. Physics of Semicond. (Warsaw), 1972, p. 863.
4. Fetterman, H. R.; Larsen, D. M.; Stillman, G. E.; Tannenwalk, P. E. and Waldman, J., Phys. Rev. Lett. 1971, 26,975.
5. Almassy, R. J.; Reynolds, D. C.; Litton, C. W.; Bajaj, K. K. and G. L. McCoy, Solid State Commun. 1981, 38,1053.
6. Reynolds, D. C.; Litton, C. W.; Smith, E. B.; Yu, P. W. and Bajaj, K. K., Solid State Commun. 1982, 42,827.
7. Thomas, D. G. and Hopfield, J. J., Phys. Rev. 1962, 128,2135.
8. Dean, P. J.; Cuthbert, J. D.; Thomas, D. G., and Lynch, R. T., Phys. Rev. Lett. 1967, 18,122.
9. Reynolds. D. C.; Litton, C. W. and Collins, T. C., Phys. Rev. 156,3 (1967); 156,881 (1967).
10. Reynolds, D. C.; Litton, C. W. and Collins, T. C., Phys. Rev. 1969, 177,1161.
11. Reynolds, D. C.; Litton, C. W. and Collins, T. C., Phys. Rev. 1968, 174,845.

12. Reynolds, D. C. and Collins, T. C., Phys. Rev. 1969, 185,1099.
13. Merz, J. L.; Kukimoto, H.; Nassau, K. and Shiever, J. W., Phys. Rev. 1972, B6,545.
14. Hiesinger, P.; Suga, S; Willmann, F. and Dreybrodt, W., Phys. Status Solidi 1975, B67,64.
15. White, A. M.; Dean, P. J.; Taylor, L. L.; Clarke, P. C.; Ashen, P. J. and Mullin, J. B. J. Phys. 1972, C5,1727.
16. White, A. M.; Dean, P. J. and Day, B. J. Phys. 1974, C7,1400.
17. Henry, C. H. and Nassau, K., Phys. Rev. 1970, B2,977.
18. Benoit a' la Guillaume, C. and Lavallard, P., Phys. Status Solidi 1975, B70,K143.
19. Ruhle, W. and Klingenstein, W., Phys. Rev. 1978, B18,7011.
20. Herzberg, G. and Spinks, J. W. T., Proc. R. Soc. London Ser. A, 1934, Ser. a 147,434.
21. Reynolds. D. C.; Langer, D. W.; Litton, C. W.; McCoy, G. L. and Bajaj, K. K., Solid State Commun. 1983, 46,473.
22. Dean, P. J., Herbert, D. C. and Lahee, A. M., J. Phys. 1980, C13,5071.
23. Cabib, D.; Fabri, E. and Fiorio, G., Nuovo Cimento 10B, 185(1972). 24.
24. Reynolds, D. C.; Bajaj, K. K.; Litton, C. W. and Smith, E. B., Phys. Rev. 1983, 28B,3380.

RECEIVED November 8, 1985

Thermal-Wave Imaging in a Scanning Electron Microscope

Allan Rosencwaig

Therma-Wave, Inc., Fremont, CA 94539

>Thermal-wave imaging is a new technique that permits the detection and imaging of surface and subsurface thermal features in a sample through the interaction of thermal waves with these features. A scanning electron microscope, modified to perform high-resolution thermal-wave imaging, has been used to detect subsurface mechanical defects such as microcracks, to image grains, grain boundaries and dislocation networks, and to detect and image dopant regions and lattice variations in crystals.

There has been considerable interest lately in imaging techniques that employ thermal waves. (1-4) In thermal-wave imaging, a beam of energy, usually a laser or electron beam, is focused and scanned across the surface of a sample. This beam is generally intensity-modulated at a frequency in the range of 10kHz to 10MHz. As the beam scans across the sample it is absorbed at or near the surface, and periodic surface heating results at the beam modulation frequency. This periodic heating is the source of thermal waves, which propagate from the heated region. The thermal waves are diffusive waves similar to eddy current waves, evanescent waves, and other critically damped phenomena that travel only to two wavelengths before their intensity becomes negligibly small. Nevertheless, within their range, the thermal waves interact with thermal features in a manner that is mathematically similar to the scattering and reflection processes of conventional propagating waves. (5) Thus any features on or beneath the surface of the sample that are within the range of these thermal waves and that have thermal characteristics different from their surroundings will reflect and scatter the waves and thus become visible. These thermal features can be defined as those regions of an otherwise homogeneous material that exhibit variations, relative to their surroundings, in either the thermal conductivity k, the volume specific heat ρC,

0097-6156/86/0295-0253$06.00/0
© 1986 American Chemical Society

and in some instances the thermal expansion coefficient α_t. Variations in these thermal parameters arise, most commonly, from variations in the local lattice structure of the material, and, for metals and other good electrical conductors, from variations in the local carrier concentration as well.

Imaging of these thermal features requires detection of the scattered and reflected thermal waves. This detection is currently accomplished by several different techniques. However, thermal-wave imaging at the high spatial resolution needed for microelectronics investigations, where micron-sized features must be resolved, requires, as seen in Table I, the detection of high frequency (MHz range) thermal waves. To date only the thermoacoustic detection method has been used routinely for detecting such high frequency thermal waves. (1) (6)

Table I. Thermal-Wave Spatial Resolution for Thermal Conductors (Silicon, Metals) and Thermal Insulators (Oxides, Ceramics, Biological Materials)

Modulation Frequency	Thermal Conductor	Thermal Insulator
100 Hz	200 to 300 µm	20 to 30 µm
10 kHz	20 to 30 µm	2 to 3 µm
1 MHz	2 to 3 µm	2000 to 3000 Å

Thermoacoustic signals occur in any sample with a nonzero thermal expansion coefficient because of the periodic stress-strain conditions in the heated volume defined by thermal waves. thus, as depicted in Figure 1, the 1-MHz thermal waves give rise to 1-MHz acoustic waves. These thermoacoustic waves are propagating waves of much longer wavelengths, typically a few millimeters at 1MHz, compared to a few microns for thermal waves at the same frequency. The magnitude and phase of the thermoacoustic waves are directly related to the temperature profiles in the heated volume and thus are directly affected by the presence of scattered and reflected thermal waves.

Thermoacoustically generated ultrasonic waves were predicted by White (7) and first used for imaging by von Gutfeld and Melcher. (8) These and subsequent experiments were concerned, however, with the use of thermoacoustic waves for ultrasonic imaging, where the interactions of the acoustic waves with the elastic features in the sample produce the principal contrast mechanisms. The possibility of using the thermal waves, which precede the thermoacoustic waves for the imaging of thermal features was not realized at the time. Although acoustic waves are detected, thermal-wave imaging with the thermoacoustic probe is not acoustic imaging, since the acoustic waves always have wavelengths many orders of magnitude greater than the thermal waves and thus are not able to image the same small features.

15. ROSENCWAIG   *Thermal-Wave Imaging in an SEM*   255

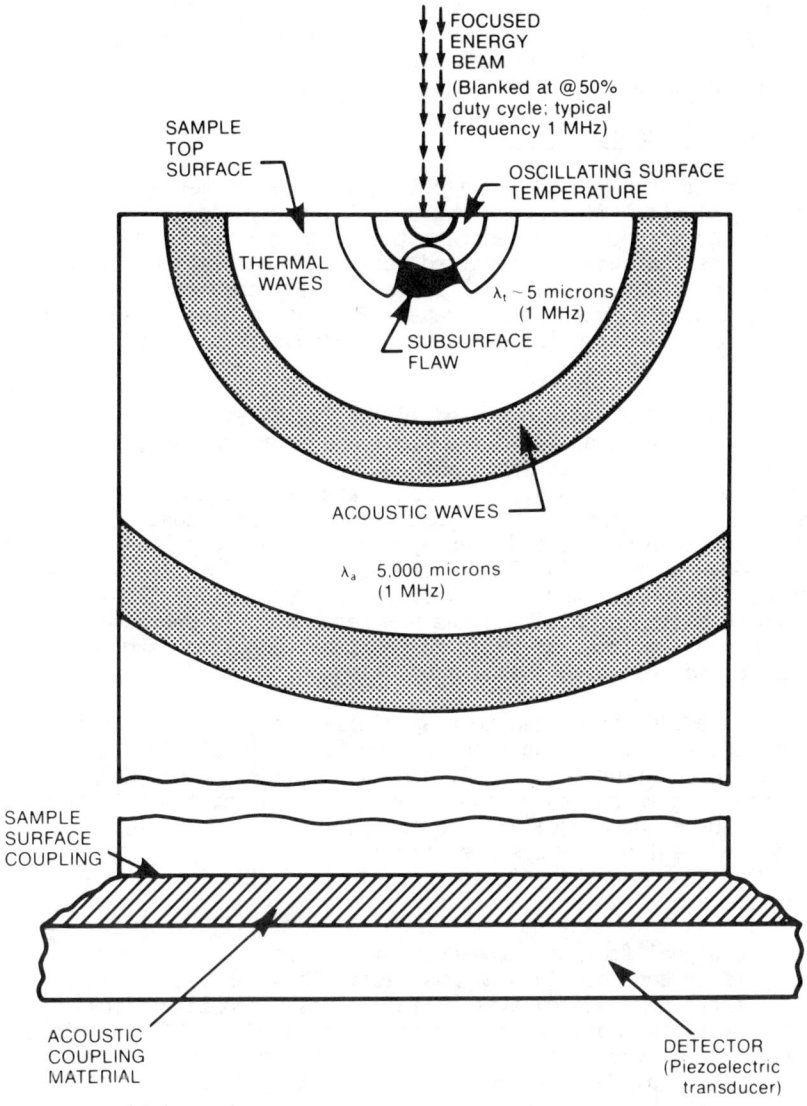

Figure 1: A schematic representation of the physical processes that occur during thermal-wave imaging.

That is, the acoustic waves simply carry or amplify the
waves and thus are not able to image the same small features.
That is, the acoustic waves simply carry or amplify the
information describing the thermal-wave event. In thermal-wave
imaging, the thermoacoustic waves are thus used as a monitor to
detect the presence of thermal waves scattered or reflected from
thermal features.

Experimental Configuration

Most high-frequency thermal-wave imaging is performed in a
scanning electron microscope where an intensity-modulated
electron beam is used to generate the thermal waves. The
experiments that will be reported here all have been performed
with a Therma-Wave, Inc. Model 101 system attached to a Hitachi
S-520 scanning electron microscope. The system block diagram for
the scanning electron/thermal-wave microscope is shown in Figure
2.

The SEM is usually operated in the 20-30 keV range, with the
final condenser relaxed and the final aperature removed, to
provide a beam of up to 2µamp with a spot size of 0.5µm. the
beam is intensity-modulated at frequencies from 100kHz-2MHz using
electrostatic plates inserted just below the gun anode. The
electrostatic plates are driven by a driver that supplies a 100V
square-wave pulse train with a 50% duty cycle. The rise and fall
times of the square waves are < 35 nsec. We have found that
during this type of blanking the SEM resolution is degraded by
less than 0.1µm, thus allowing conventional electron imaging to
be performed even with the beam blanking on.

The sample is mounted to a specially designed piezoelectric
transducer that takes the place of the conventional SEM stub.
The sample mounting is commonly done with silver paint, although
wax, vacuum grease, cement, and even mechanical attachments have
all been used successfully. The transducer employs a piezo-
ceramic element and is designed for high sensitivity and good
shielding from the stray fields and electron flux in the SEM
chamber.

The signal from the transducer is fed into a low-noise
preamplifier and then into an analyzer. This analyzer
incorporates a frequency synthesizer, a phase-sensitive lock-in
amplifier system capable of operating from 64kHz to 2MHz, a video
amplifier, and various image control functions. The frequency
synthesizer provides the input to the blanking driver and the
reference to the phase-sensitive lock-in amplifier system. The
video output from the analyzer is fed directly into the imaging
CRT of the SEM, thereby generating a thermal-wave image as the
SEM electron beam rasters over the sample. Alternatively, the
video output is fed into an image buffer which can then provide a
time-averaged steady-state thermal-waver image on a video
monitor.

Applications. We have performed, with the system described above, high-resolution, thermal-wave imaging on many different materials. We have detected and imaged subsurface mechanical defects such as microcracks and voids, grain boundaries, grains, and dislocations, and dopant regions and lattice variations in crystals.

Subsurface Mechanical Defects. Subsurface mechanical defects such as voids, cracks and delaminations represent substantial thermal features and thus are readily detected with a thermal-wave microscope. (1) (9) (10) One illustration of this application is shown in Figure 3. Figure 3a is the electron image of a Si device. The thermal-wave image in Figure 3b shows a large subsurface crack network in the Si chip and a delamination or subsurface chip-out region at the top in the center. Neither of these mechanical defects is visible with either optical or electron microscopy. Other examples of thermal-wave imaging of subsurface mechanical defects have been presented elsewhere. (9) (10).

Metallography. Thermal waves can also be used to image variations in single and multi-phase crystals. This ability is of considerable utility in metallography, since different metallic phases or grains can be readily imaged with no special sample preparation. We illustrate this in Figure 4 where the columnar or dendritic grains and the various transistion and precipitation zones are clearly visible in the thermal-wave image of a weld region in a Co-Cr alloy. In addition, we can see in the thermal-wave image, a crack forming along the sharp precipitation zone in the center of the dendritic grains. This image dramatically reveals that we are dealing with an inherently weak weld due to the presence of the large dendritic grains and the sharp transistion or precipitation zone. In addition, the thermal-wave image shows that this weld already has an incipient crack forming in it along the precipitation zone. Such images cannot be obtained by any other nondestructive means.

Another example is shown in Figure 5 which shows the electron and thermal-wave images of an Al-Zn alloy. The electron image (a) shows only topographical fetures, while the thermal-wave image (b) clearly shows both the grain structures and the presence of Fe or Sn precipitates.

As seen from the above Figures and from those obtained by other investigators (2) (11) grain structures in single and multi-phase metals and alloys show up very clearly in thermal-wave images. It should be kept in mind that conventional SEM images will usually show grain contrast only for multi-phase materials with substantial atomic number differences between grains, and for single-phase materials only if they have been previously etched or if one can perform electron channeling experiments. There are several possible mechanisms for the exceptionally strong contrast seen in the thermal-wave images of both grains and grain boundaries. Certainly grain boundaries

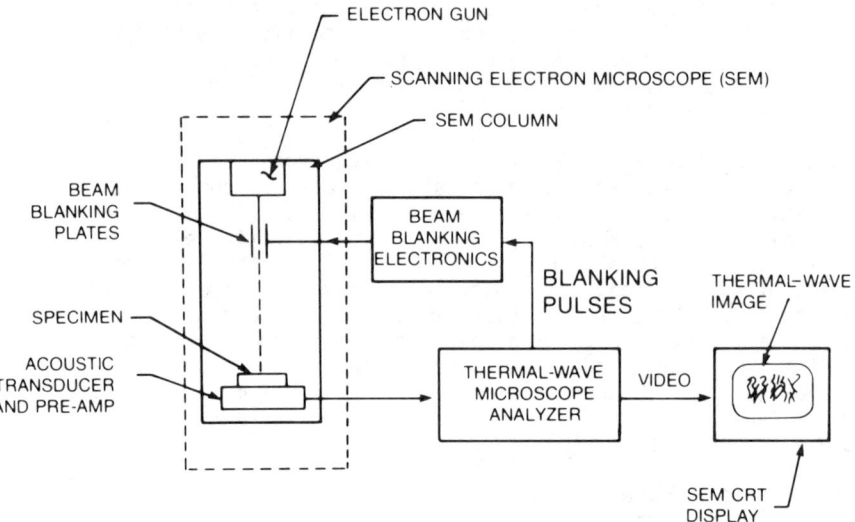

Figure 2: Block diagram of the thermal-wave microscope system as part of a scanning electron microscope.

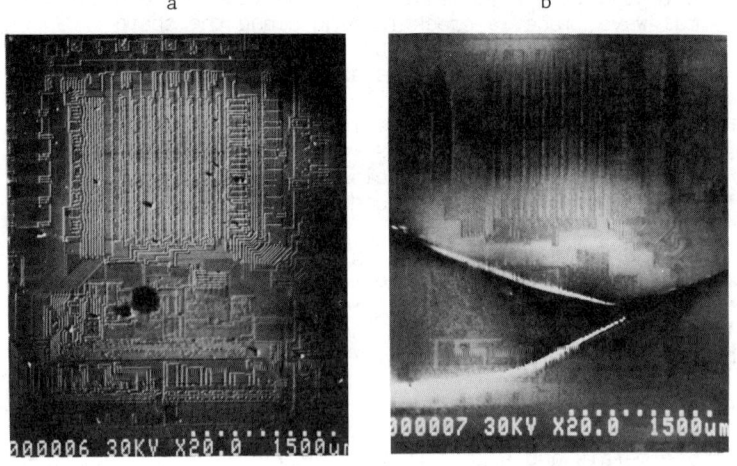

Figure 3: Examples of subsurface mechanical defects in a Si device. The electron micrograph (a) shows no defect features. The thermal-wave image (b) shows a subsurface network of microcracks in the lower half of the device and a subsurface delamination or chip-out in the top center.

 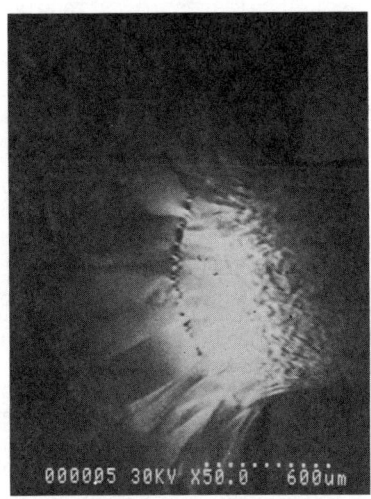

Figure 4: a.) Electron micrograph of a polished weld region in a Co-Cr alloy. b.) Thermal-wave micrograph of same region showing all of the grain microstructure including the dendritic grains and transition zones in the weld region. Note the presence of a subsurface microcrack forming along the sharp precipitation zone near the center of the weld.

Figure 5: Electron (a) and thermal-wave (b) micrographs at 50X of an Al-Zn alloy. The thermal-wave micrograph shows the Al-Zn grains, and the presence of Fe or Sn precipitates.

represent local areas of strongly perturbed lattice structure, and thus of considerable variations in the local thermal parameters. Likewise, different grains in a multi-phase material may have different chemical compositions, and thus quite different thermal parameters as well. The contrast mechanism for grains in a single-phase material is somewhat more difficult to explain. Other investigators (2) (11) have suggested that this contrast is due primarily to anisotropy in the elastic properties with the result that differently oriented grains respond differently to the localized thermal deformation. However, local variations in the elastic properties affect the thermal-wave signal only through Poisson's ratio, and the antisotropy in Poisson's ratio will be much smaller than that in the elastic constants themselves. Thus this does not appear to be the major source of contrast as it would be in acoustic imaging. Instead we should look to local variations in the thermal parameters, ρC, α_t and most of all k. Note that even in a cubic material, significant anisotropies may arise in the thermal conductivity k, and, to a lesser extent, the expansion coefficient α_t for grains at the surface of the sample because of cold work and other surface-related conditions.

Crystalline Distruptions and Variations. When a crystal lattice is highly ordered, minor changes in lattice structure can produce measurable changes in the local thermal conductivity of the material and thus can be imaged with a thermal-wave microscope. (10) (12) This capability is illustrated in Figure 6 which shows a GaAs device. The optical and electron micrographs image the visible features of the gate structures in the devices. The thermal-wave image shows, in addition, the Si-doped regions of the GaAs, since these regions have a different thermal conductivity than the undoped regions. Such images permit a rapid and nondestructive analysis of the effects of lateral diffusion of dopant in semiconducting crystals.

Correlations with EBIC and XRT. Thermal features that arise either from mechanical defects or from metallic grains and grain boundaries are usually easy to recognize. However, thermal features arising from more subtle crystalline disruptions and variations, such as those described above are more difficult to identify. Before thermal-wave microscopy can be accepted as a routine, standard analytical technique, one needs to establish a direct correlation of some of these less obvious thermal-wave images with those obtained with other more widely accepted techniques. Below, we discuss two such correlations, one with electron beam induced current (EBIC) and the other with x-ray topography (XRT).

EBIC analysis is an important tool for characterizing electrically active features in semiconducting materials (13). EBIC images are generated by rastering the electron beam of a SEM over the sample and collecting the current generated in electron-hole production. A p-n junction or Schottkey barrier must be established on the sample to collect the charge.

Contrast is produced in the image when either the probing beam scans across a region in which the electron-hole production changes measurably, or when the probe beam comes close to a strong recombination center. EBIC image resolution and EBIC sampling depth are both controlled by the size of the region within which the primary electrons lose energy as they enter the sample. The limits of this volume are defined by the surfaces at which the electrons no longer have sufficient energy for ionization. A typical EBIC resolution limit is 0.5μm, and a typical sampling depth is approximately 5μm.

Thermal-wave microscopy has similarities in resolution and sampling depth. While the contrast mechanisms are very different, we would expect to find many of the same features appearing in both EBIC and thermal-wave images since both electrical conductivity and thermal conductivity are transport properties of the material, and thus will respond to changes in material properties in similar ways.

The correlation study between EBIC and thermal-wave imaging was performed with Si solar cell material. This is large-grain ribbon material which has been polished to produce a smooth surface for the EBIC studies. Figure 7 is a thermal-wave micrograph of the Si solar cell material showing grain boundaries and other thermal features. Figure 8a is an electron image of a smaller area of the sample, in which the only features seen are the contact metallization lines from the solar cell. Figure 8b is the EBIC picture of the same area showing detail due to subgrain boundaries. The thermal-wave image in Figure 8c shows all the same subgrain boundary features as the EBIC micrograph with similar resolution. This suggests that thermal-wave microscopy can be used to study the same features that EBIC would show, but without the need to make the p-n junctions.

X-ray topography is a technique which uses x-ray diffraction to image features which produce diffraction contrast, (14) (i.e., produce changes in diffracted beam angles). Thus it is applied primarily to single crystal or, at least, highly oriented polycrystalline samples. Contrast is produced at crystalline defects such as dislocations, twins, or stacking faults which are regions of high strain and often the location of precipitate particles. In general, the XRT images local variations in crystal strain. The diffracted beam which is used to produce the image can be selected to be either a beam that is transmitted through the sample (in which case the image is the integral of all contrast producing features within the sample), or a beam that is reflected in the sample surface region (in which case the sampled volume is only a layer approximately 20 μm thick at the surface). The x-ray topograph is usually recorded by exposing a photographic emulsion, thus producing a unity magnification image of the sample. Further magnification is then obtained photographically by enlarging the image during printing. The magnification and resolution limits are imposed by the grain size in the emulsion and this typically limits the resolution to approximately 1-5 μm.

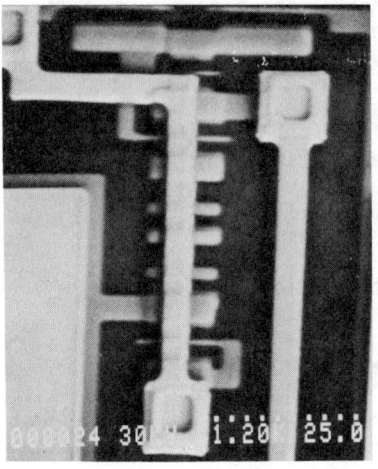 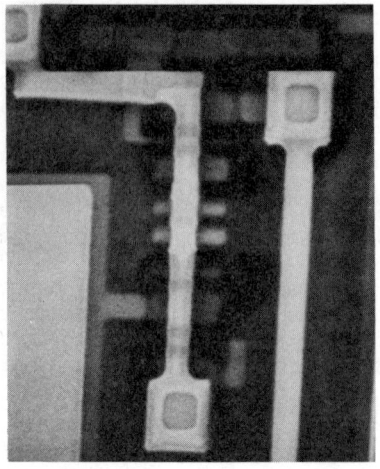

a b

Figure 6: Images of a GaAs device. The electron micrograph (a) shows the surface circuit features. The thermal-wave image (b) shows, in addition, the Si-doped regions around and underneath some of the circuit structures.

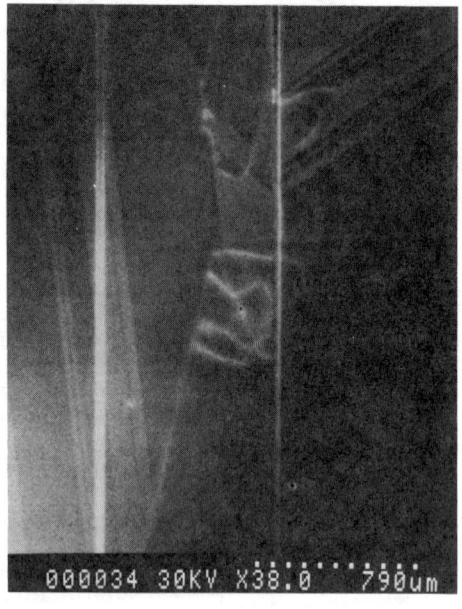

Figure 7: Thermal-wave image of Si solar cell material showing grain structure.

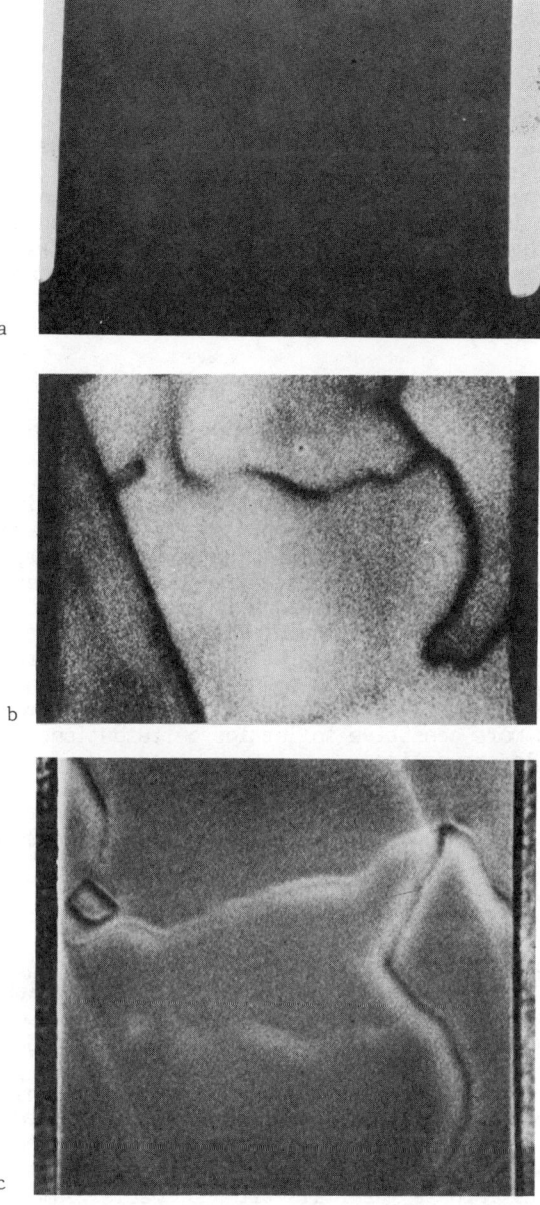

Figure 8: a.) Electron micrograph of a single grain region of Si solar all material. b.) EBIC image of same region showing the sub-grain boundaries that affect electron-hole current in the p-n junction. c.) Thermal-wave image of same region showing the same features as the EBIC image.

Figure 9a is a part of a reflection topograph taken on an LEC-grown GaAs crystal. A 30-hour exposure was required to obtain this image. The patterns seen in this topograph are primarily very dense dislocation networks.

For comparison, Figure 9b is a thermal-wave micrograph of the same area of the crystal. Clearly, the same features are visible in the two imaging techniques and, in fact the dislocation networks appear to be somewhat better resolved in the thermal-wave image. However, whereas the x-ray topograph required 30 hours to obtain the thermal-wave image was obtained in two minutes. We have also seen individual dislocations at higher magnification with thermal-wave imaging, as shown in Figure 10, where the individual dislocations appear as small points.

The contrast mechanism in the XRT micrograph is clearly the variations in the local elastic strain. That the elastic strain variations are not the direct contrast mechanism for the thermal-wave micrograph can be deduced both from theoretical arguments and from recent work related to acoustic microscopy. (15) (16) This work has shown, both experimentally and theoretically, that dislocations in semiconducting crystals are not detectable by even the micron-sized ultrasonic waves generated in a high-frequency acoustic microscope, because the dislocation strain fields produce negligible variations in the local elastic constants. Thus the thermal-wave contrast mechanism must be attributable primarily to variations in the local thermal conductivity, which, being a dynamic transport property, is more sensitive to lattice perturbations than are the static elastic properties.

Thermal-wave imaging thus appears to be as powerful a technique as x-ray topograhy for imaging dislocations in GaAs materials, but is much faster and simpler.

Conclusion

Thermal-wave imaging is a means of detecting and imaging previously invisible thermal features. Its applications in materials analysis are many and diverse. These include the detection and imaging of subsurface defects, including interfacial flaws and microcracks; the detection and characterization of areas of a crystalline lattice that have been modified through the introduction of foreign ions or defects; and the imaging of dislocations both singly and in networks. In addition, the ability to image crystalline phases and grains in single and multi-phase materials with no special sample preparation makes possible more convenient, nondestructive in-situ, and perhaps even dynamic studies of metals, composites, and other materials. In biology, the method may be used in the future to analyze and map the microscopic structure of membranes and cells in terms of their local thermal parameters, thereby providing potentially valuable data.

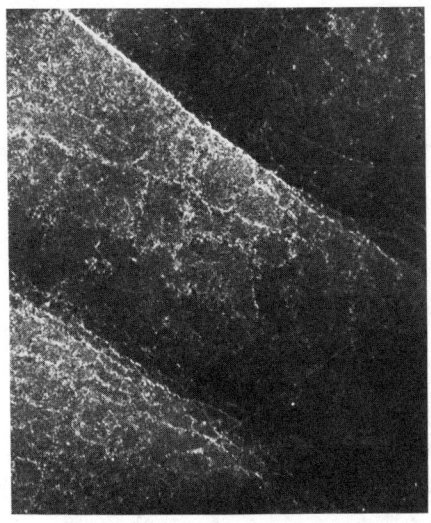 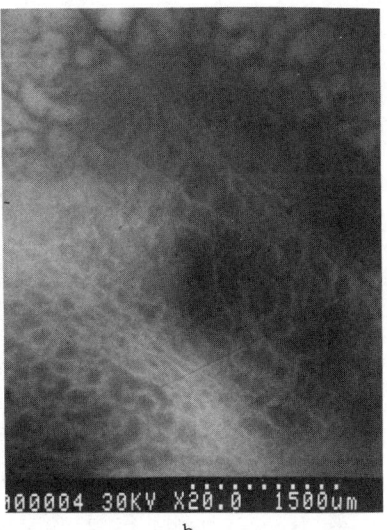

Figure 9: a.) X-ray topograph (XRT) of a section of LEC-grown GaAs crystal at 20X magnification, showing a dense network of dislocations. b.) Thermal-wave image of same region showing the same features as the XRT image. Note the crack that appears in both images in the lower right-hand corner.

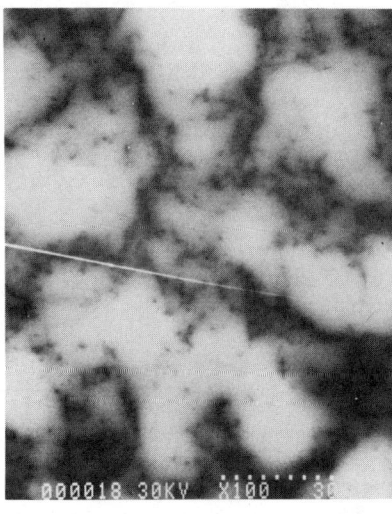

Figure 10: Higher magnification (100X) thermal-wave image of the LEC-grown GaAs crystal in the regions of the crack, showing both dislocation networks and features (individual dark points) related to individual dislocations.

Literature Cited

1. Rosencwaig, A. *Science* 1982, 218, 223.
2. Cargill, G. S. *Physics Today* 1981, 234, 27.
3. Busse, G. In "Scanned Image Microscopy", Ash, E.A., Ed.; Academic: London, 1980; p. 341.
4. Thomas, R.L. Favro, L.D.; Grice, K.R.; Inglehart, L.J.; Kuo, P.K.; Lhota, J.; Busse, G. *Proc. 1982 Ultrasonics Symp.*, 1982, (B.R. McAvoy, ed.) IEEE: New York; p. 586
5. Opsal, J.; Rosencwaig, A. *J. Appl. Phys.* 1982, 53, 4240.
6. Rosencwaig, A.; Busse, G. *Appl. Phys. Lett.* 1980, 36, 725.
7. White, R.M. *J. Appl. Phys.* 1963, 34, 559.
8. von Gutfeld, R.J.; Melcher, R.L. *Appl. Phys. Lett.* 1977, 33, 257.
9. Brandis, E.; Rosencwaig, A. *Appl. Phys. Lett.* 1980, 37, 98.
10. Rosencwaig, A. *Solid State Technol.* 1982, 25, 91.
11. Davies, D.G. *Scanning Electron Microsc.*, III 1983, 1163.
12. Rosencwaig, A.; White, R.M. *Appl. Phys. Lett.* 1981, 38, 165.
13. Ravi, K.V.; Varker, C.J.; Volk, C.E. *J. Electrochem. Soc.* 1973, 120, 533.
14. Tanner, B.K.; Bowen, D.K. "Characterization of Crystal Growth Defects by X-ray Methods"; Plenum Press: New York, 1980.
15. Sulewski, P.E.; Dynes, R.D.; Mahajan, S.; Bishop, J.; *Appl. Phys.* 1983, 54, 5711.
16. Sulewski, P.E.; Bishop, D.J. *Appl. Phys.* 1983, 54, 5715

RECEIVED October 31, 1985

16

Fourier Transform Mass Spectrometry in the Microelectronics Service Laboratory

W. H. Penzel

Defense Systems Division, Honeywell Inc., Hopkins, MN 55343

> Mass spectrometry has become more useful in the support of electronic development and manufacturing processes. Fourier transform mass spectrometry, the latest advance in this analytical method, is another step forward in versatility, sensitivity and reproducibility in analytical characterization, qualification and quantification of raw materials and contaminants as used in electronic devices. A review will be provided of basic instrument hardware and interfacing, significant operating parameters and limitations, and special inlet systems. Emphasis will be placed on material evaluation, process control and failure analysis. Data handling will be reviewed using appropriate examples encountered in material and failure analysis.

Over the last 50 years, mass spectrometry has seen the introduction and development of several types of mass spectrometers. These fall into the general categories of time-of-flight, quadrupole and electrostatic/magnetic sector instruments. The application and utility of these increased significantly as data handling technology began to catch up with the needs of the analyst.

Ion cyclotron resonance (ICR) experiments were performed in the early '70's, which applied the theory that ion spin frequency was directly related to ion mass, and which, by application of Fourier transforms, permitted the development of the theoretical base and practical application for Fourier transform mass spectrometry (FTMS) (1-7). While initial experiments utilized conventional field sources, i.e. electromagnets, the desire for greater field strength and stability soon led to the use of super conducting magnets. Experimentation confirmed advantages of this approach to mass spectrometry. Since mass determination is based purely on ion spin frequency, it is possible to form and to detect ions in the same region of the instrument, only separating these events by time (1-7). Additionally, all ions are detected simultaneously (not sequentially as in the other types of instruments) thus minimizing time related distortion of the mass spectra.

In FTMS, several analytical techniques are available simply by varying cell parameters. The techniques are electron ionization (EI), chemical ionization (CI), self-CI, and MS/MS (Mass Spectrometry/Mass Spectrometry, providing re-analysis of specific previously obtained mass fragments). Wide mass range, high sensitivity, ultra high resolution, high scan rates, and the elimination of the need for internal standards are possible, due to the high field strength and stability of the super conducting magnet (6-10).

Commercial instrumentation using the superconducting magnet was first introduced by Nicolet, Inc., in 1982. Our laboratory is equipped with one of the first instruments in industrial service use, and it has demonstrated many of the outstanding capabilities mentioned previously. An example of the high stability of the instrument in terms of mass is given in Table I, which lists various mass fragments as calibrated on "Day 1," and reanalyzed on three consecutive days. The test sample was perfluorotributylamine (PFTBA) and utilized an instrument mass range of 10-3,000 and a field strength of 1.9 Tesla. The data shows a maximum deviation of 3.3 ppm at m/z 18, 124 ppm at m/z 100, and 35 ppm at m/z 502, which is noteworthy particularly in view of the fact that operating parameters were changed frequently between the analyses cited, and that the instrument was placed in the stand-by mode overnight between operating days. Although this data would not qualify for high resolution research work, it more than suffices for service laboratory application.

Table I. Stability of Selected Ions Over Four Days In The Absence of Instrument Recalibration

m/z	Day I	Day II	Delta PPM	Day III	Delta PPM	Day IV	Delta PPM
18	18.01018	18.01022	+2.2	18.01024	+3.3	18.01023	+2.8
100	99.99742	99.98869	-87	99.98498	-124	99.98674	-107
131	130.98685	130.98430	-20	130.98635	-3.8	130.98410	-21
219	218.98919	218.99589	+30	218.99600	+32	218.99513	+27
502	501.95394	501.94941	-9.0	501.94529	-17	501.97144	+35

However, if the same fragments are repeatedly examined over a short period of time, they yield more reproducible data. The results obtained over a 90 minute period are shown in Table II; representing reanalysis without intervening calibration. Here the maximum deviation at m/z 18 is 2.2 ppm, at m/z 100 is 23 ppm and at m/z 502 is 95 ppm.

Table II. Short-Term Stability of Selected Ions Over 90 Minutes
In the Absence of Instrument Calibration

m/z	Initial	30 Min.	Delta PPM	60 Min.	Delta PPM	90 Min.	Delta PPM
18	18.010313	18.010284	-1.6	18.010291	-1.2	18.010272	-2.2
100	99.997673	99.995857	-18	99.996349	-13	99.995405	-23
131	130.998072	130.996902	-8.9	130.996897	-9.6	130.996452	-12
219	218.985515	218.968526	-78	218.973151	-56	218.970313	-69
502	502.040229	502.015054	-50	502.008728	-63	501.992678	-95

Similarly, stability comparisons were made over short periods of time using low field strength at 0.7 Tesla. This setting is necessary for analyses of ions below m/z 10, such as when backfill gases are being analyzed. Table III gives typical results for this condition. It should be noted that the total mass range of the instrument is limited to a maximum of approximately m/z 250 at this field strength.

Table III. Stability of Selected Ions Over One Hour In The Absence of Instrument Recalibration

m/z	Initial	30 Minutes	Delta PPM	One Hour	Delta PPM
18	18.00172	18.00194	+12	18.00179	+3.8
131	130.35528	130.34772	-58	130.33418	-162
219	220.37685	220.37341	-16	220.40116	+110

While this variation is significantly greater than in the case of the higher field strength data, it is not considered significant in our application because individual peaks of interest are still readily identifiable and do not represent quantification difficulties. In addition, it should be noted that accuracy can be enhanced greatly by using concurrent calibrants during any analysis. This approach would be very helpful if absolute mass determinations must be performed, but is a process not usually necessary in our application.

Equipment

The FTMS consists of several major sections, which include the analyzer assembly, the inlet systems and the electronic control system. The analyzer assembly consists of a four-inch diameter vacuum chamber which extends through the super conducting magnet (Figure 1). The magnet is liquid helium cooled, with a liquid nitrogen jacket providing external heat protection. The vacuum chamber is equipped with a turbomolecular pump providing pressures in the 10^{-9} Torr range, an ionization gauge, an inlet system flange, and a cell connector flange. The one cubic inch analytical cell resides in the center of the magnetic field (10, 13), within the vacuum chamber. Both the cell and the chamber are heatable to above $250°C$. The sample inlets include a

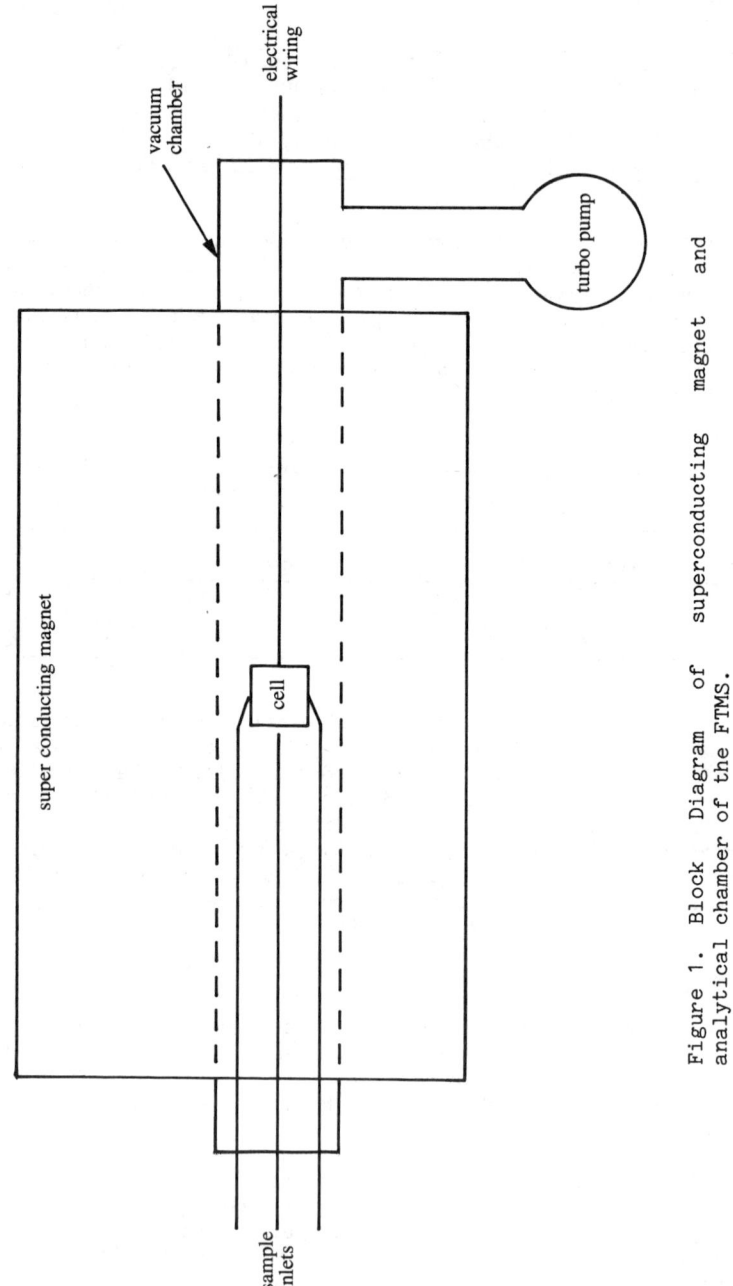

Figure 1. Block Diagram of superconducting magnet and analytical chamber of the FTMS.

gas chromatograph interface, a gas inlet system, a solids insertion probe, and a custom residual gas analysis (RGA) and moisture analysis fixture (Figure 2).

The electronic control system includes cell operating controls; an RF generator; vacuum, temperature and liquid helium/liquid nitrogen controls; and a 196 Kbyte (20 bit) data system. The data system includes, in addition, 10 Mbyte and 160 Mbyte disk drives, as well as appropriate printers/plotters. (Figure 3). This control system not only permits adjustment of operating conditions affecting resolution, sensitivity, scan rates and cell parameters, but also permits experiment selection, data manipulation, peak measurement, background subtraction and library searches.

The basic operation of the instrument (3) consists of the ionization of the sample in the cell by a timed electron beam which is followed, after a short interval, by an RF pulse applied to the plates of the cell. This pulse coherently excites all ions in the cell into cyclotron motion. The motion continues after cessation of the pulse, and the resonance is detected by the plates of the cell, amplified and the data stored in the computer. The excitation/detection cycle is repeated numerous times and the collected data summed. The data is then subjected to fourier transformation and the frequency spectrum resulting converted into a mass spectrum. The spectrum is normalized to the major peak. For quantitative work, the calibration can be based either on peak height or peak area. Here, major considerations will include the resolution chosen and the relative concentrations of the constituents under investigation.

Only a few specific operating parameters have a major impact on the spectra obtained. These include the excitation level (the power of the RF pulse), the filament emission current, the electron beam voltage, the trapping voltage, (which is applied to the cell trapping plates), and the signal attenuation. Each of these has to be tuned specifically to maximize response for any peak(s) of interest, and this is particularly important for the lower field strength analyses.

Applications

Freedom from contamination is increasingly important as solid state devices increase in complexity and decrease in size. Since corrosion control is a major problem in such devices, tests have been devised to assure the cleanliness level of manufactured parts.

Moisture. Control of moisture has become a major challenge as part of the contamination control picture. Mil-Std-883C, Method 1018.2, recognizes this and has established mass spectrometry as one of the methods found suitable for moisture analysis in packaged devices with internal volumes of 10 to 800 microliters.

Just as other types of mass spectrometers require special inlet fixturing to accomplish this analysis, so does the FTMS (Figure 2). The fixturing developed in our laboratory consists of a temperature controlled specimen holder (Figure 4) which also provides for back up (protective) vacuum. The specified "transfer

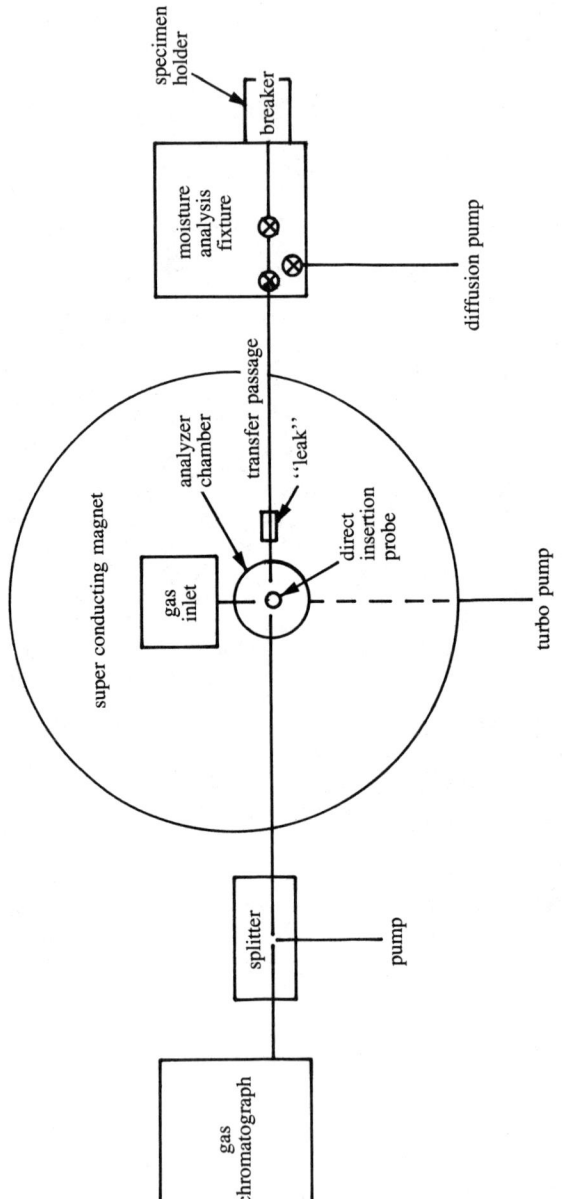

Figure 2. Block Diagram of inlet systems for the FTMS including custom moisture analysis fixture.

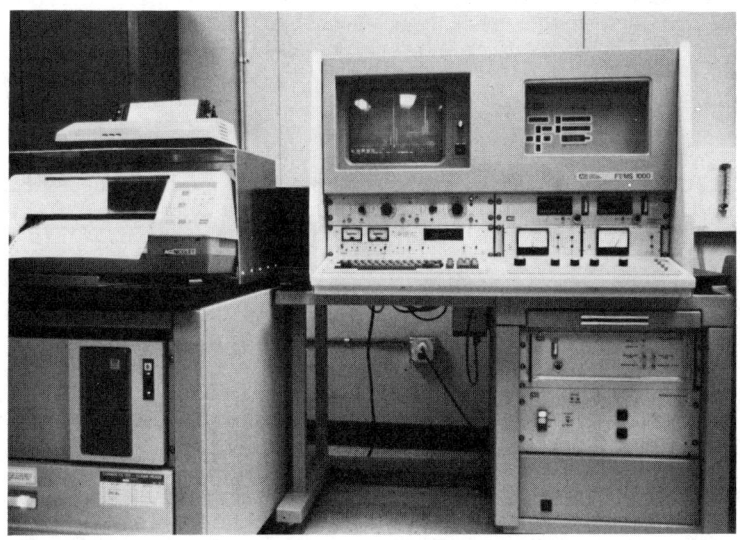

Figure 3. Data system of Nicolet, Inc. FTMS-1000.

Figure 4. Component holder for internal moisture analysis of sealed devices.

passage" consists of a piercing block with a high vacuum linear-motion feedthrough, appropriate valving, a 0-10 Torr differential capacitance pressure measuring head, a replaceable add-on dead volume, and a connecting line with fixed "leak." This complete assembly is maintained at the specified 125°C and is prepumped by a four-inch diffusion pump.

The analysis performed utilizes instrument background spectra and calibration factors based on nitrogen with known moisture concentration and room air. These standards permit calculation of response factors for the masses of interest, including m/z 2, 4, 18, 28, 32, 40, 44, and any other as necessary for a specific analysis. Specific standardization is accomplished using a three volume generator (Figure 5) which covers the sample volume range and simulates, as much as possible, the actual piercing of a test specimen.

Prior to analysis, each test specimen is conditioned at 100°C for at least 1 hour. The specimen is then placed into the 100°C sample holder and pressed against the piercing block to form a vacuum seal. The "transfer passage," which includes the piercing block, is evacuated to 10^{-6} Torr, minimum, and background spectra are collected. The specimen is then pierced, the contained atmosphere allowed to expand into the transfer passage and the mass spectrometer, and sample spectra are collected. This sequence requires that an appropriately sized add-on volume has been chosen for the transfer passage to assure that the final sample pressure in the mass spectrometer is in the optimum range. This particular condition is necessitated by one of the few drawbacks of the FTMS. Care must be taken to maintain total analytical chamber pressure below 2×10^{-7} Torr. Should the sample pressure rise in the cell to the point where the mean free path of individual ions is impacted by adjoining ions, then any additional sample will actually cause signal attenuation, which results in erroneous test data. Experimentally, the threshold for this phenomenon has been established at 2×10^{-7} Torr in our instrument, thus establishing a limit above which experiments should not be performed. By providing for a total expansion volume of the transfer passage sized to keep the cell pressure below this threshold, one can readily avoid the attenuation problem and produce reliable data.

Data evaluation can be accomplished using either peak height or peak area. Peak height has been found to produce more consistent results due to the difficulty of consistently applying baselines to small peaks in the mass spectra in the process of making area determinations. Peak height data avoids this problem to the greatest degree. The raw data is then calculated into component concentrations, in the process of which provision has to be made for instrument background and specific peak response factors. A typical result of analyses of five metal-can integrated circuits is shown in Table IV.

Table IV. A Typical Contained Atmosphere Composition In TO-5 Can
Operational Amplifier, Nominally Back Filled
with Dry Nitrogen

Sample Number	1	2	3	4	5
Water, ppm	3460	2930	2230	2230	2250
Nitrogen, %	99.6	99.6	99.7	99.7	99.7
Oxygen, %	N.D.	N.D.	N.D.	N.D.	N.D.
Argon, %	N.D.	N.D.	N.D.	N.D.	N.D.
Carbon Dioxide, %	0.1	0.1	0.1	0.1	0.1

N.D.: Not detected.

Materials. Another aspect of semiconductor manufacture is the control of materials used in the fabrication process. A typical example is a spin-on fluid used to deposit inorganic films by hydrolysis of an organometallic. Capillary or packed column GC/MS can readily be performed on the FTMS (11, 12), utilizing appropriate splitters and glass jet separators between the gas chromatograph and the mass spectrometer. Figures 6 and 7 illustrate packed column GC/MS results by showing the gas chromatogram, the total ion chromatogram, and a mass spectrum of the 3.1 minute GC peak. A complete analysis is listed in Table V. Note that the concentration data is not response factor corrected.

Table V. Typical Spin-On Fluid Solvent Composition

Peak	Retention Time Minutes	Concentration %	Identification
1	2.42	7.0	Methanol
2	3.09	51.6	Ethanol
3	4.66	5.4	Methyl Acetate
4	5.40	35.4	Ethyl Acetate

Residues. During failure analysis or as a process control, it is frequently of interest to determine whether residual solvents or potential offgassing/outgassing products exist in coatings applied to components or assemblies. Application of a modified commercial purge and trap apparatus (Hewlett-Packard, Model 7675A) has proven effective for this type of investigation.

Known size specimens are placed in the tube and are flooded with a purge gas, typically helium. The tube has been modified with a heater and controls, and is thus able to subject the specimen to a controlled elevated temperature environment. Should highly volatile compounds be suspected, pre-chilling of the sample and tube prior to purging is possible. The time of purge necessary to obtain an optimum sample has to be established experimentally and will vary with the nature of the specimen under test.

Once the purge time has elapsed, a normal desorption/GC injection cycle follows. Capillary GC/MS has been found to be the most effective approach in completing the analysis. An

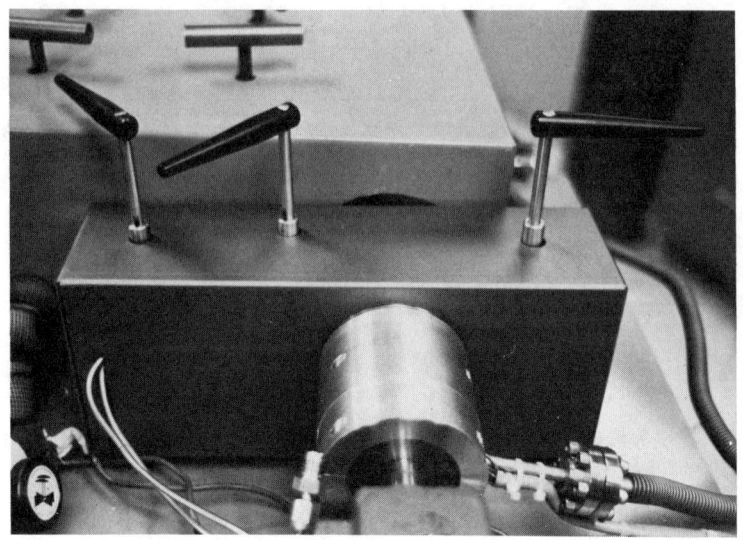

Figure 5. Three volume generator for moisture analysis calibration.

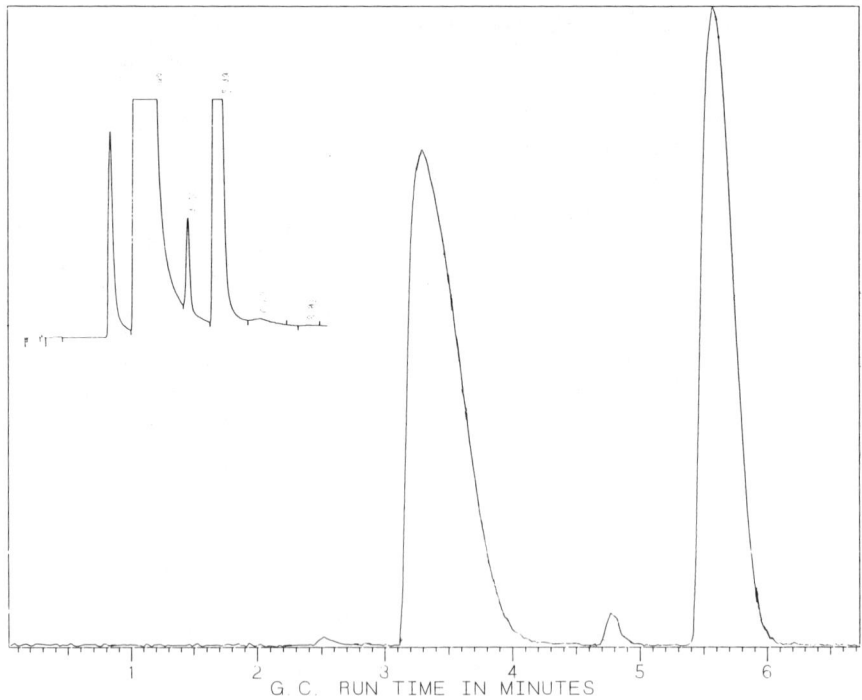

Figure 6. Gas chromatogram (inset) and total ion chromatogram of typical spin-on fluid solvent package.

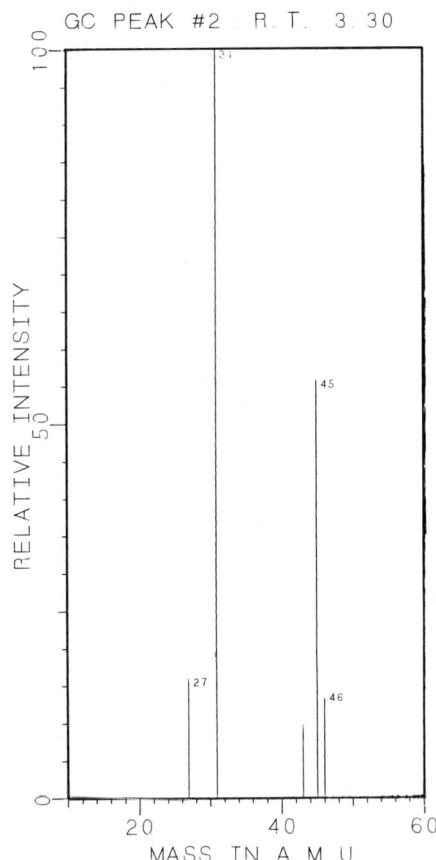

Figure 7. Mass spectrum of 3.1 minute GC peak, identified as ethanol, of Figure 6.

illustrative example is shown in Table VI, which demonstrates the sensitivity of the technique as applied to residual solvents in a conformal coated printed wiring board, which had been subjected to fluorocarbon solvent cleaning several hours earlier. The gas chromatogram/total ion chromatogram (Figure 8) and mass spectrum (Figure 9) illustrate typical results obtained.

Table VI. Composition of Printed Wiring Board Offgassing/Outgassing Products After Fluorocarbon Solvent Degreasing

GC Peak No.	GC R-Time	Identification	Probable Origin
1	4.03	1,1,2-Trichloro-2,2,1-Trifluoroethane	Degreasing solvent
2	5.13	2-Methyl-1,3-Butanediol	Flux residue
3	7.04	Unidentified	--
4	8.55	3-Bromopentane	Fire retardant

Process. Gases are occasionally suspect when contamination or side reactions are observed in the manufacturing process of solid state devices. For this type of sample, the gas inlet system permits direct connection to the gas supply or transfer vessel and controlled injection into the mass spectrometer via precision bleed valves. Usually, separation of constituents is not necessary, since most gases yield characteristic peaks in the mass spectrum. Typical examples examined have included chlorine in nitrogen and methane in argon. Caution must be observed in this approach, however, since the dynamic range of the instrument limits relative sensitivity to 0.1%. This clearly establishes that trace contamination analysis is not viable through this method, but that the method can be used to screen process gases.

Contamination. Another approach to contamination identification for solid specimens is temperature programmed direct insertion solids probe analysis (13). In this instance, a solid specimen, say a surface scraping of an IC die, is placed into a capillary of a direct insertion probe and is then heated from a -50 to 450°C under programmed conditions while continuous scans are being acquired. Evaluation of such spectra indicates a quasi-separation of components by boiling point and does permit cleaner identification than in the case of pyrolysis. Figure 10 is a typical example of a spectrum obtained from a discolored IC die which was analyzed in this manner. The spectrum was obtained at 400°C and shows remnants of the die bond epoxy at the lower mass end, while the peaks at m/z 207, 208 and 209 strongly suggest diffusion pump oil, as identified in previous analyses. A comparison of this spectrum with a spectrum obtained from an undiscolored die showed that they were essentially alike, except for the peaks at m/z 207-209. Similarly, comparison with previously obtained spectra of diffusion pump oil confirmed the initial interpretation of the contaminant in the suspect die spectrum.

16. PENZEL *Microelectronics Service Laboratory* 279

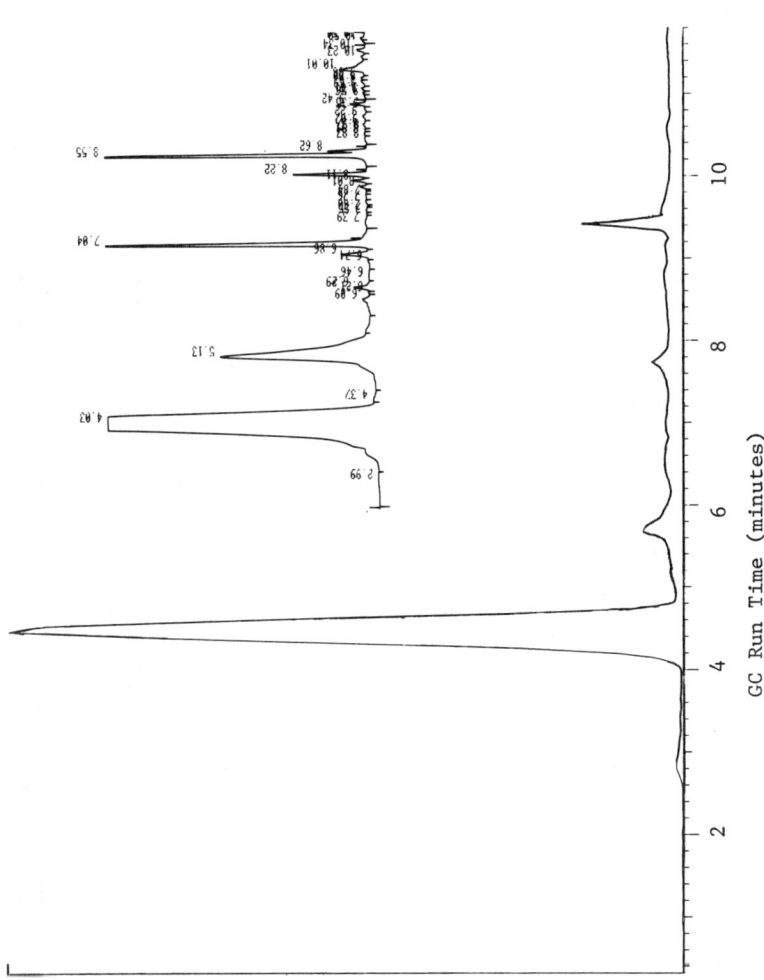

Figure 8. Typical gas chromatogram (inset) and total ion chromatogram obtained in purge-and-trap type analysis of conformal coating on printed wiring board.

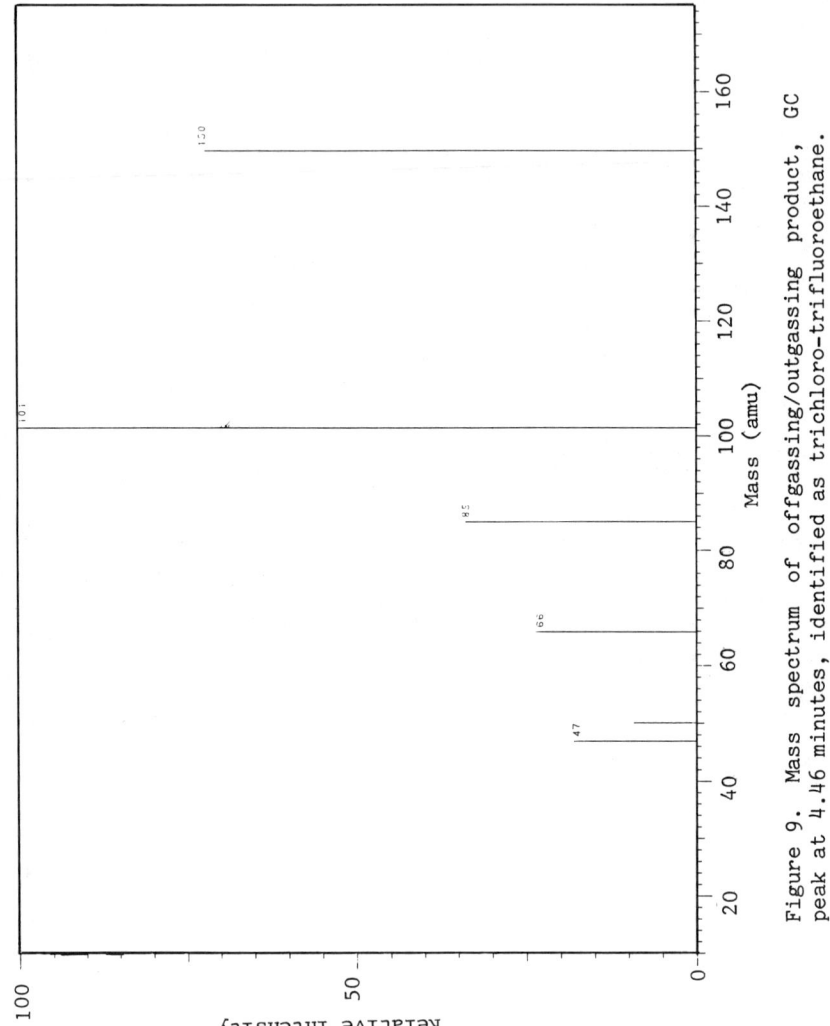

Figure 9. Mass spectrum of offgassing/outgassing product, GC peak at 4.46 minutes, identified as trichloro-trifluoroethane.

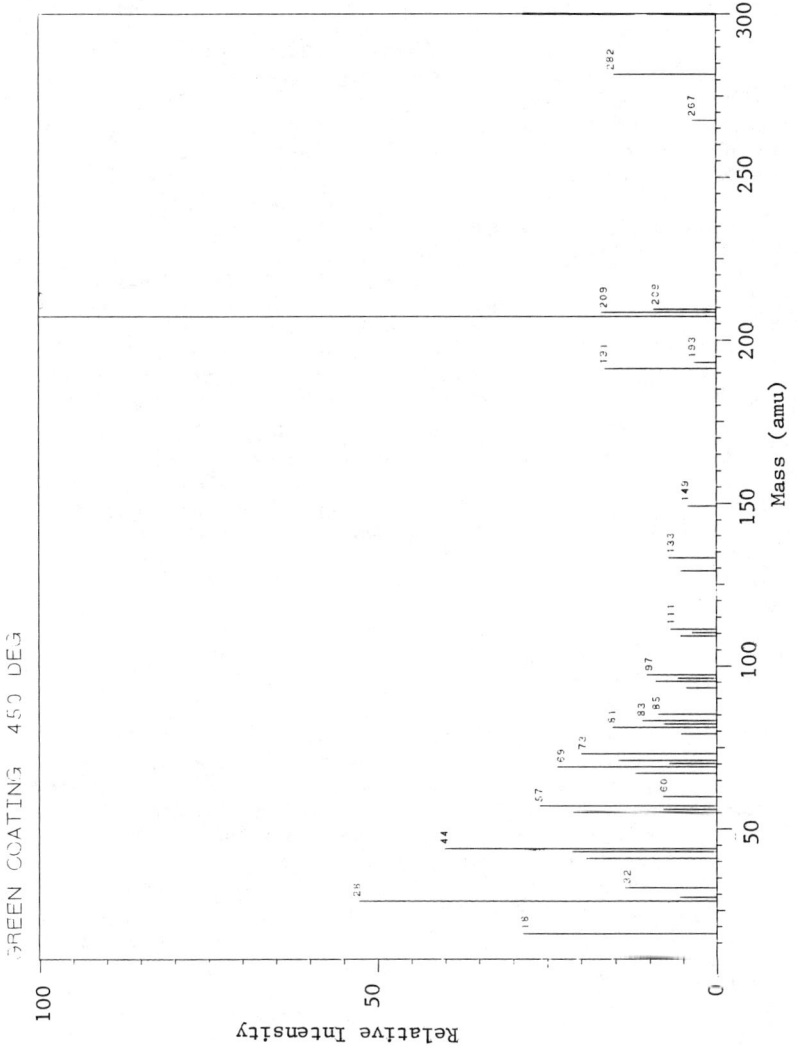

Figure 10. Mass spectrum of an epoxy-bonded die contaminant suggesting diffusion pump oil.

Data Handling. Consideration of data handling must be given on initial experiment set-up, since the number of data points selected will directly affect the resolution of the spectra obtained. It is possible to obtain as many as 64K data points, or as few as 1K per spectrum, thus providing for a wide latitude in resolution. The data processing speed is, unfortunately, directly affected, inasmuch as high data point counts result in high processing time.

On the other hand, the selection of spectrum mass range width is not affected by the previous choice, but may range from less than 1 unit to over 3000. In addition, while a spectrum may have been collected over a wide mass range, narrow ranges can be examined and magnified during the evaluation process.

The data system also provides the opportunity of ejecting a mass of choice from the cell just prior to spectrum acquisition, thus permitting the acquisition of spectra from which specific peak related attenuation or interference has been removed. This type of peak removal may remove whole interference spectra, if the parent mass of a compound is chosen for ejection initially.

Part of most mass spectral examination involves the subtraction of background spectra or of known interferences. The data system permits subtractions in two ways. The first is a subtraction in display only, while a given data set is being examined, while the second also alters the data stored in RAM. As a valuable assist in identifying spectra, we have an NIH/EPA library and attendant search routines on-line. This library is expandable and can provide as many as twenty likely matches for an unknown spectrum.

Data system outputs consist of graphic or tabular information, both on-screen or by printer/multi-color plotter.

Summary

Application of the FTMS to microelectronic laboratory service has proven to be very successful. Instrument stability and resolution are unmatched by conventional instruments. In moisture analysis, comparison of quadrupole and magnetic sector instruments with FTMS, has, in addition, shown higher sample through-put and data consistency for both known atmospheres and device lot samples.

More typical applications of mass spectrometry, such as GC/FTMS, direct insertion probe sampling and gas/vapor analysis have also been demonstrated as routine using this instrument.

Used in conjunction with a state-of-the-art data handling system and reference library, the FTMS is capable of meeting all microelectronic service laboratory mass spectrometry needs.

Acknowledgments

The author wishes to thank J. E. Alberg and T. L. Harrington for their invaluable assistance with the experimental work performed in connection with this manuscript.

Literature Cited

1. Comisarow, M. B.; Marshall, A. G. Chem. Phys. Lett. 1974, 25, 282-283.
2. Comisarow, M. B.; Marshall, A. G. J. Chem. Phys. 1975, 62, 293-295.
3. Wilkins, C. L. An. Chem. 1978, 50, 493-497.
4. Comisarow, M. B. Adv. Mass Spectrometry 1980, 8B, 1698-1706.
5. Marshall, A. G.; Roe, D. C. J. Chem. Phys. 1980, 1581-1590.
6. Allemann, M.; Kellerhals, Hp Chem. Phys. Lett. 1980, 75, 328-331.
7. Wilkins, C. L.; Gross M. L. An. Chem. 1981, 53, 1661-1676.
8. Ghaderi, S.; Kulkarni, P. S.; Ledford, E. B., Jr.; Wilkins, C. L.; Gross, M. L. An. Chem. 1981, 53, 428-436.
9. Marshall, A. G. An. Chem. 1979, 51, 1710-1713.
10. Ledford, E. B., Jr.; Ghaderi, S.; White, R. L.; Spencer, R. B.; Kulkarni, P. S.; Wilkins, C. L.; Gross, M. L. An. Chem. 1980, 52, 463-468.
11. Ledford, E. B., Jr.; White, R. L.; Ghadari, S.; Wilkins, C. L.; Gross, M. L. An. Chem. 1980, 52, 2450-2451.
12. White, R. L.; Wilkins, C. L. An. Chem. 1982, 54, 2443-2447.
13. Ledford, E. B., Jr.; Ghadari, S.; Wilkins, C. L.; Gross, M. L. Adv. Mass Spect. 1980, 8B, 1707-1724.

RECEIVED August 5, 1985

17

Materials Characterization Using Elemental and Isotopic Analyses by Inductively Coupled Plasma Mass Spectrometry

B. Shushan, E. S. K. Quan, A. Boorn, D. J. Douglas, and G. Rosenblatt

SCIEX, Thornhill, Ontario, Canada L3T 1P2

> A new technique for elemental and isotope analysis of materials used in the semiconductor industry is presented. The technique involves the use of an inductively coupled plasma to convert trace elements to their gaseous ions followed by analysis of these ions by mass spectrometry. Examples include the quantitative analysis of trace copper by isotope dilution and the analysis of trace contaminants in boron, indium phosphide and reagent acids.

Inductively Coupled Plasma Mass Spectrometry (ICP/MS) is a powerful technique for elemental and isotope analysis which has undergone extensive development over the last decade by a number of research groups (1,2,3). These research efforts have led to the recent development of commercially available ICP-MS systems which, by virtue of their novelty, are now just beginning to demonstrate capabilities well suited to the characterization of materials used in the semiconductor industry. This paper will outline the principles of ICP-MS and provide some examples of the instrumentation's analytical capabilities.

Instrumentation

The ICP-MS used in the present work was the ELAN 250 manufactured by SCIEX. A schematic of this instrument is shown in Figure 1. As well as being a good source for optical emission, the ICP is an excellent ion source; in fact, most elements inside the 7000°K plasma exist, to a major extent, as singly positively-charged ions (2). As in optical emission ICP systems, solutions to be analyzed are nebulized into the high temperature argon plasma. The ions produced therein are sampled through a differentially pumped interface linked to the mass spectrometer. Ions are then separated electronically by means of a quadrupole mass filter and detected using a high sensitivity pulse counting sytem. All operations exclusive of the plasma system are controlled through the instrument's computer which also provides for data manipulation and presentation.

0097-6156/86/0295-0284$06.00/0
© 1986 American Chemical Society

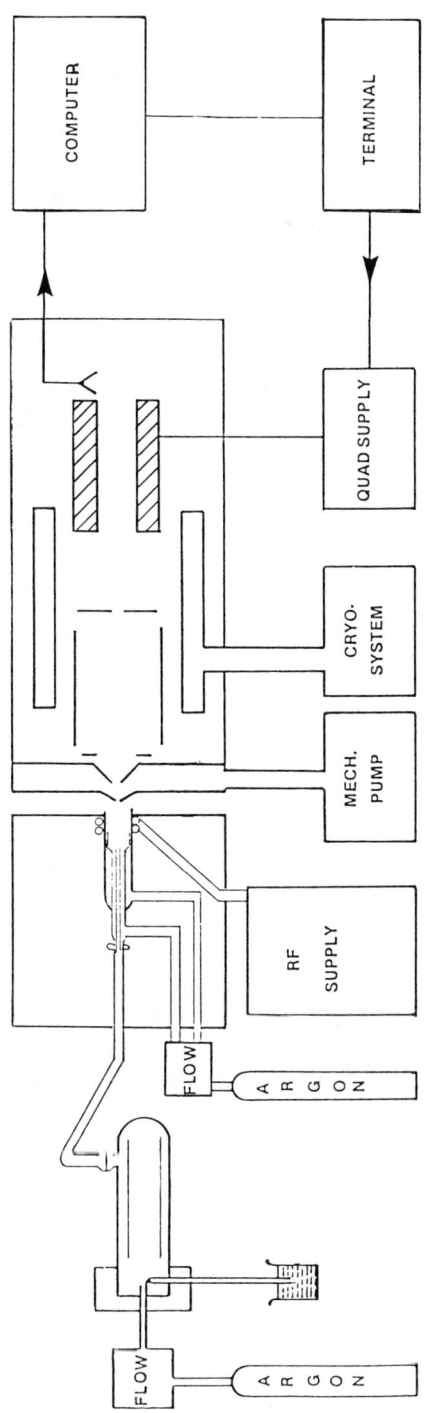

Figure 1. Schematic Representation of the ELAN 250 ICP/MS System.

Results & Discussion

The ICP has proliferated as a method of converting chemical compounds into their elemental constituents which subsequently emit light of characteristic wavelengths. Accordingly, ICP has been used extensively as an emission source for optical detection systems in order to perform elemental analysis. Since each element can emit hundreds of optical lines, the use of ICP/AES for multiple element analysis, or for the detection of elements in unknown or concentrated matrices, can suffer from interferences due to spectral overlap. By contrast, ICP-MS provides inherently simpler spectral information. An example of such a spectrum is demonstrated in Figure 2 showing a typical ICP-MS scan for a 10 ug ml^{-1} solution of mixed transition metals. The demonstrated sensitivity here is 10^4 to 10^5 counts s^{-1} per ug ml^{-1} and, coupled with the nearly universal ionization efficiency of the ICP ion source, provides typical detection limits in a narrow range between 0.1 to 10 ng.ml^{-1} for most elements. In fact over 90% of the elements in the periodic table are accessible for such analytical determinations.

The inherent simplicity of ICP-MS spectra implies that spectral interferences from major matrix elements can be minimal. This is evident in Figure 3 which shows a portion of ICP-MS scan of an indium phosphide sample. A tin impurity is most visible showing peak signals at the characteristic isotopes (masses 116,117,118, 119,120,122 and 124). Since indium and tin represent a worst case scenario (In^{115} and Sn^{116} differ by only one mass unit, ie., the smallest increment) it is of great practical importance that the signals for these two elements do not interfere with each other. The degree of separation between elements of neighbouring mass is referred to as "abundance sensitivity" and is an indication of the system's mass spectral resolving power. Abundance sensitivities for the ELAN™ instrument are typically 1 in 10^5 on the low mass side of a peak, and better than 1 in 10^7 on the high mass side. In the present example, the In^{115} peak intensity is estimated to be more than 10^5 times that of Sn^{116}, and yet there is no mass spectral overlap between those two signals.

High abundance sensitivities help in the accurate determination of isotope ratios of elements, permitting the quantitation of these dissolved elements by isotope dilution methods. This quantitative technique entails adding to the sample a known amount of a stable isotope of the element one wishes to quantitate. The corresponding increase in signal at that isotope's mass is directly proportional to the amount of material added. Comparison of this increase in signal with the signal intensity for one of the undiluted native isotopes provides a method of direct quantitation <u>within</u> the matrix of the sample and requires just one or two measurements for each quantitative result. For example, Figure 4 shows the quantitative determination of copper in an orchard leaf digest using the technique of isotope dilution (4). This figure constitutes the hard-copy output from the ELAN instrument for this analytical procedure and consists of three mass spectra over the copper region

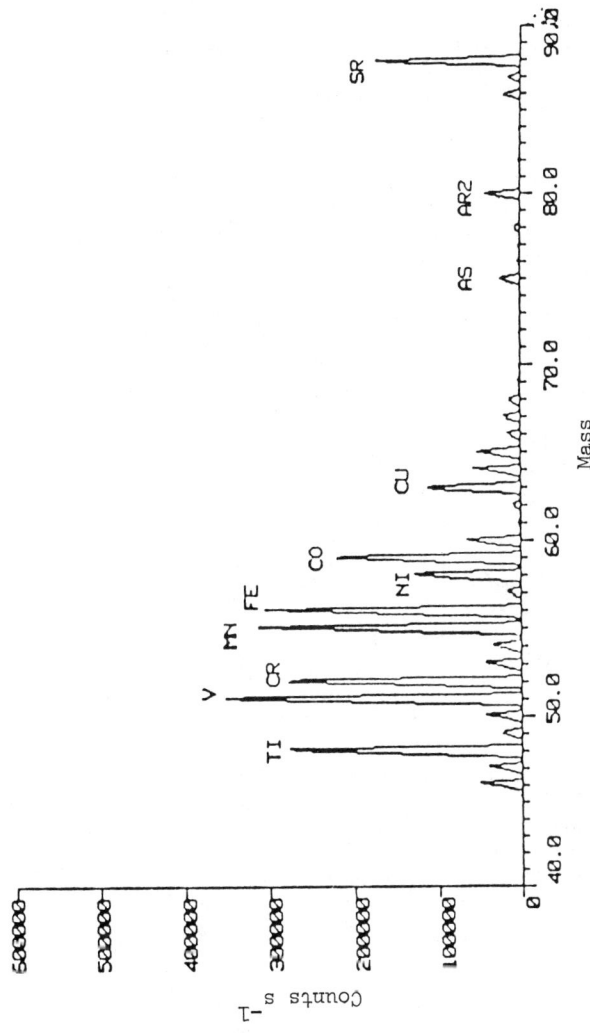

Figure 2. Mass spectrum of a solution of elements each present at 10 ug ml^{-1}.

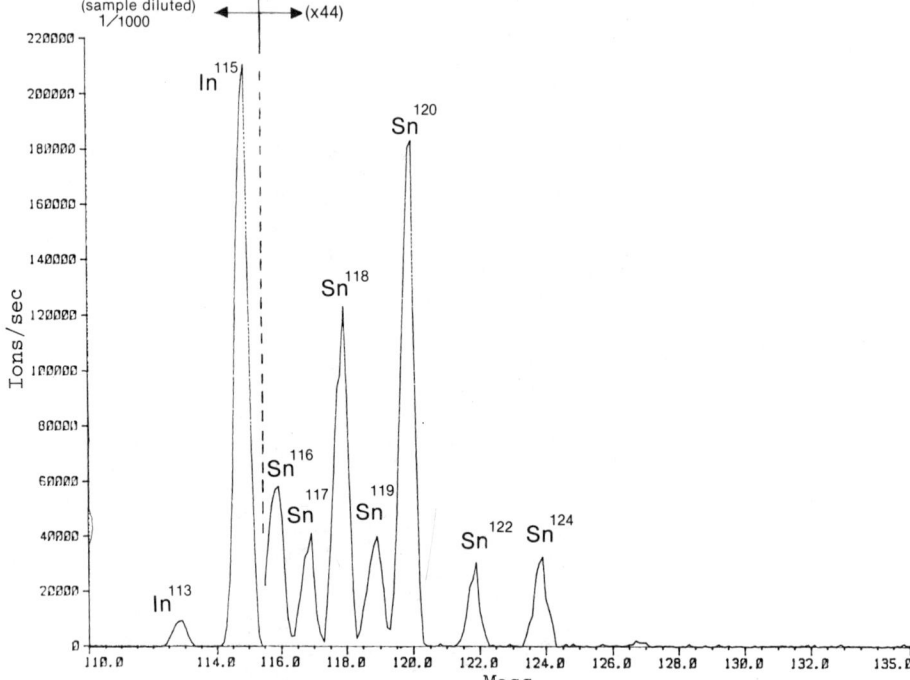

Figure 3. Mass Spectrum of an Indium Phosphide Sample (0.77% in Aqua Regia showing tin impurity at ca. 2 ppm) In order to bring indium peaks on scale the solution was diluted 1000 times before scanning over the indium region (masses 110-115). The tin portion of the spectrum (masses 116-135) was run undiluted.

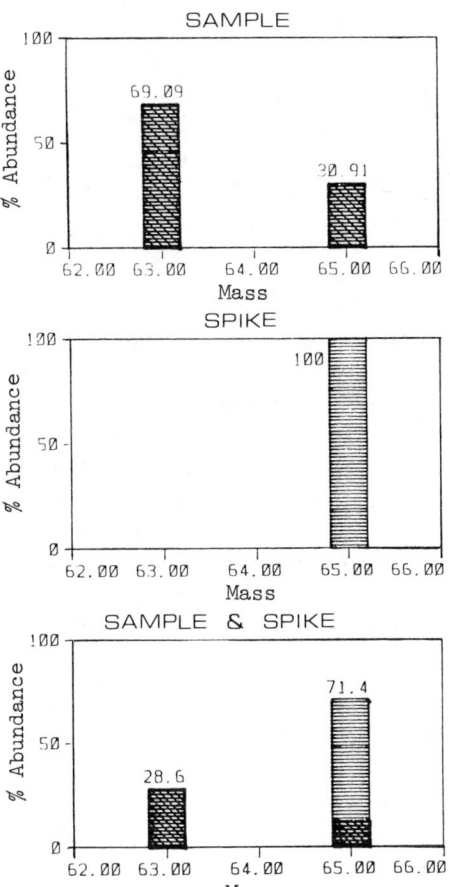

Figure 4. Results for the determination of copper in an organic matric (Orchard Leaf) by isotope dilution.
(TOP) Mass spectrum between 62 and 66 mass units for the unspiked sample containing 1% dissolved organic matter. The mass spectral peaks for copper were normalized and displayed in histogram format.
(MIDDLE) Same as above obtained on the isotope dilution standard which contained only Cu^{65}.
(BOTTOM) Same as above obtained on the sample solution; 25 ml volume spiked with 4 ug Cu^{65}.

(mass 62 to 66) where the signals are normalized and displayed in histogram format. Figure 4 (top) is the natural isotopic distribution of copper in the sample showing the percent-abundance of each isotope (ie., 69.09% and 30.91% for Cu^{63} and Cu^{65} respectively). The middle spectrum is that of the isotope-dilution standard which was, as expected, 100% Cu^{65}. The bottom spectrum is that of 25 ml of the orchard leaf digest containing 0.25gm orchard leaf material spiked with 4ug Cu^{65}. The ratio R of Cu^{65} to Cu^{63} was measured as (71.4/28.6) = 2.50. Since the contribution of Cu^{63} in the bottom spectrum is purely native, the contribution of native Cu^{65} to the mass 65 intensity can be easily accounted for, and is shown as the slightly darker portion of the mass 65 histogram. The concentration of copper in the digest may now be determined directly:

$$[Cu] = \frac{(A_{sp} - R \times B_{sp}) \times W_{sp}}{(R \times B_s - A_s) \times W_s}$$

Where: [Cu] is in ppm
R is the ratio Cu(65)/Cu(63)
A_{sp} is the spike Cu(65) abund.
B_{sp} is the spike Cu(63) abund.
W_{sp} is the spike weight in ug
B_s is the natural Cu(63) abund.
A_s is the natural Cu(65) abund.
W_s is the sample weight in grams

Thus [Cu] was thus calculated to be 11.3 ppm for this orchard leaf sample which was NBS certified as 12 ppm. The sampling procedure required only a few seconds for each of the measurements, and only three scans or less are needed for this type of procedure. Precision is typically of the order of $\pm 0.5\%$ and combined with the system's speed the technique is obviously suitable for multielement quantitative determinations, and those applications requiring high sample throughput.

The spectral simplicity and sensitivity of ICP-MS make it ideally suited for multi-element screening applications. Figure 5, for example, is a scan between 50 and 90 atomic mass units performed on a boron sample dissolved in "pure" HNO_3. With this single scan it was possible to qualitatively (and semi-quantitatively) determine the presence of chromium, germanium and iron impurities in the boron matrix (0.71% in HNO_3). A similar scan of the HNO_3 solvent (not shown) revealed that those elements in parentheses were solvent impurities including vanadium, copper and gallium. Since absolute sensitivities of the technique to the various elements is roughly similar (if their first ionization potentials (IP) are lower than 9.0 eV) semi-quantitative information can be easily obtained by the addition of suitable internal standards ie., elements whose IP's are similar to that of the analyte elements one wishes to semi-quantitate.

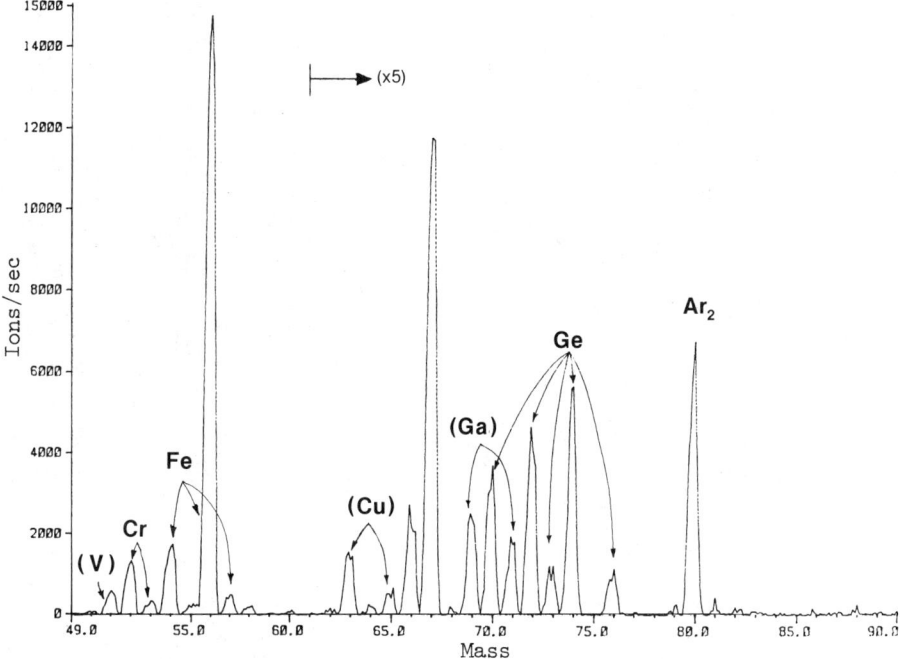

Figure 5. The ICP/MS analysis of a boron solution (0.71% in HNO_3) showing chromium, germanium, and iron impurities (ca. 180, 20, and 700 ppb, respectively) between 49 and 90 mass units.

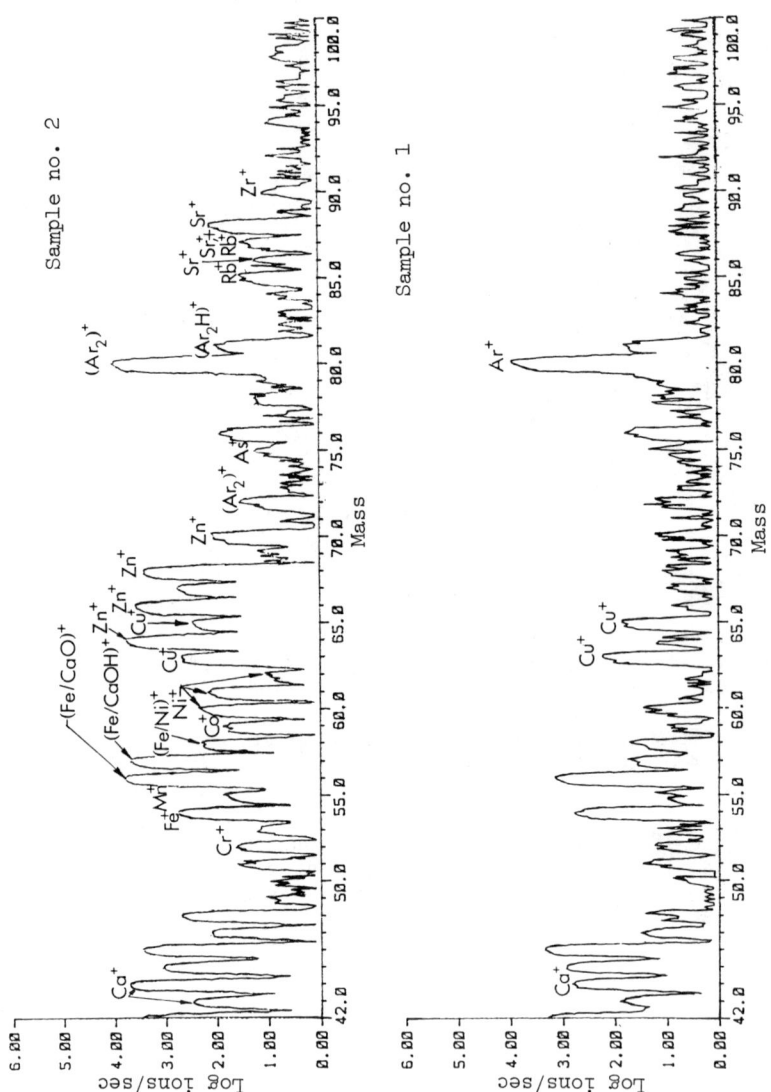

Figure 6. The comparison of ICP/MS spectra obtained on lead samples (ca. 0.5% is HNO₃) of varying purities.

Many quality control applications require the rapid multi-element analysis of feed-stock samples to compare their impurity levels to those of suitably pure standards. Figure 6 demonstrates such a procedure showing the ICP-MS analysis of two lead samples used in battery production. Many impurities are observed in the sample No. 1 spectrum at substantial levels, whereas sample No. 2 exhibited fewer impurity signals and at lower concentrations. Even though the lead is at a level of concentration which is many orders of magnitude more than any elemental impurity, there is no spectral interference from lead since its signals are located well above this mass range. The levels and types of impurities in samples of "pure" elements such as those evident in the present spectra also serve as a fingerprint alluding to sample origin or method of production.

Conclusion

The ICP has demonstrated itself to be an efficient ion source capable of ionizing most elements in the periodic table with similar efficiencies. The combination of the ICP with the speed, flexibility and spectralsimplicity of mass spectrometry provides a powerful technique for the characterization and quality control of bulk materials used in the semiconductor industry.

Literature Cited

1. D.J. Douglas, E.S.K. Quan and R.G. Smith, Spectrochimica Acta. 38B (1983) 39

2. R.S. Houk, V.A. Fassel and H.J. Svec, Dynamic Mass Spectrometry6 (1981) 234, Heydon, London, ed-D/Price and J.F.J. Todd
3. A.L. Gray and A.R. Date, Ibid P.252

4. SCIEX Application Note

RECEIVED August 12, 1985

18

Activation Analysis of Electronics Materials

Richard M. Lindstrom

Center for Analytical Chemistry, National Bureau of Standards, Gaithersburg, MD 20899

> Neutron activation analysis is one of a small number of methods capable of multi-elemental analysis of subnanogram quantities of contaminants in semiconductors and other materials. Milligram to gram-sized samples of silicon, quartz, graphite, or organic materials are nearly ideal for the method. The physics of the processes involved is simple, and qualitative identification of components is an integral part of the quantitative analysis. Except for the need for access to a nuclear reactor, the equipment required is readily available commercially, and is comparable in cost and complexity to that used in other advanced analytical techniques.

The need for elemental analysis of materials important to the electronics industry need not be belabored, since the correlation between trace element contaminants and device performance has been amply demonstrated in the literature. A systematic study of these effects requires that analytical methods be available for accurate measurement of sub-part-per-million concentrations. As technology progresses from large-scale to very-large-scale integration, the characterization of materials at sub-part-per-billion levels will be required (1). What was true in 1970 is equally true today: "While the future is bright, the technological problems are formidable, and materials characterization must contribute in large part to their solution (2)".
Matters to be considered in the selection of a method of chemical analysis include the suite of elements for which the method is useful, and the lower limit of concentration at which each element can be detected and measured, as well as the accuracy of the analysis and the cost in money and time. Few analytical methods are applicable at the concentration levels required, and none is optimum by all these criteria. This paper describes one method in common use, neutron activation analysis (NAA). The essential characteristics of NAA will

This chapter not subject to U.S. copyright.
Published 1986, American Chemical Society

be given in sufficient detail for the solid-state physicist or engineer to judge whether this method is the one of choice in solving a particular problem.

Keenan and Larrabee (1) have recently written a comprehensive summary of the characterization of VLSI silicon, in which the available analytical methods are compared and contrasted; in a sense the present paper can be considered an expansion of their treatment of NAA. An earlier book, also by the group at Texas Instruments (2) discusses NAA in detail as the state of the art existed in 1970. Since that time the development of large, high-resolution gamma-ray detectors has made NAA less dependent on radiochemistry and transformed it into a high-precision spectroscopic technique. There are many references in the literature to the characterization of electronics materials by NAA and other nuclear methods (3,4,5,6,7). Several excellent textbooks on activation analysis are available (8,9,10), to which the reader is referred for more complete discussion of the matters outlined in the present summary.

Neutron activation is a major contributer to modern elemental analysis. In a recent compilation of 11,000 published analyses of 75 environmental and biological Standard Reference Materials issued by the National Bureau of Standards (NBS), over half the analyses reported were performed by NAA (11). A considerable part of contemporary trace-element geochemistry is reliant on INAA, largely because of the method's high sensitivity, broad elemental coverage, and the ease of analysis of large numbers of small samples with a modest investment of time (12).

Basic Features of the Neutron Activation Method

The foremost characteristic of neutron activation analysis is that **nuclear** reactions are employed for **chemical** measurements. Reactor irradiation produces neutron-rich isotopes which usually decay by emission of beta particles, often accompanied by gamma rays as the daughter nucleus deexcites. Gamma-ray spectra are composed of discrete narrow lines which are usually well resolved by modern detectors. As a familiar example, NAA is used to determine cobalt through the capture of slow neutrons by the Co-59 isotope. The radioactive Co-60 product is detected by measurement of the delayed gamma rays emitted at discrete energies of 1173.238 and 1332.502 keV when it decays to stable Ni-60. In a recently developed extension of NAA, the gamma rays emitted instantaneously in the neutron absorption step are measured; this method is most often called neutron-capture prompt-gamma activation analysis, or PGAA.

The elemental scope of neutron activation is wide, with as many as 51 elements having been determined instrumentally in one material by a combination of NAA and PGAA (13). Several tables of detection limits have been published (14,15,16,1), and a table of experimentally measured detection limits in Si and SiO_2 is included here (Table I). A striking characteristic of NAA is the wide range of detection limits across the periodic table. Many elements are measurable in the sub-nanogram range, but on the other hand some elements cannot be detected at all. A histogram of interference-free detection limits is given in Figure 1.

Table I. Detection Limits Measured at NBS for Trace Element Analysis of Si and SiO_2 by Instrumental Neutron Activation

Element	Nuclide measured	Half-life	Detection Limit (ng/g)
Na	Na-24	15.0 h	0.1
Mg	Mg-27	9.46 m	3000
Al	Al-28	2.24 m	500
Cl	Cl-38	37.2 m	5
K	K-42	12.4 h	100
Ca	Ca-49	8.70 m	2000
Sc	Sc-46	83.8 d	0.003
Ti	Ti-51	5.75 m	1000
V	V-52	3.75 m	10
Cr	Cr-51	27.7 d	1
Mn	Mn-56	2.58 h	0.1
Fe	Fe-59	44.5 d	50
Co	Co-60	5.27 y	0.1
Ni	Ni-65	2.54 h	2000
Cu	Cu-64	12.7 h	0.5
Zn	Zn-65	244 d	1
As	As-76	26.3 h	0.02
Br	Br-82	35.3 h	0.02
Zr	Zr-95	65.5 d	500
Ag	Ag-110m	250 d	0.01
Sb	Sb-122	2.70 d	0.01
Cs	Cs-134	2.06 y	0.05
Ba	Ba-131	11.8 d	10
La	La-140	40.3 h	0.01
Ce	Ce-141	32.5 d	0.2
Sm	Sm-153	46.7 h	0.05
Eu	Eu-152m	9.32 h	0.01
Hf	Hf-181	42.4 d	20
Ta	Ta-182	115 d	0.05
W	W-187	23.9 h	0.02
Au	Au-198	2.70 d	0.0005
Th	Pa-233	27.4 d	0.1
U	Np-239	2.36 d	0.3

Samples of 100 mg were irradiated for up to 6 hr at 5×10^{13} $n/cm^2 \cdot s$ and counted at the maximum possible efficiency. The detection limit is defined according to Currie (17).

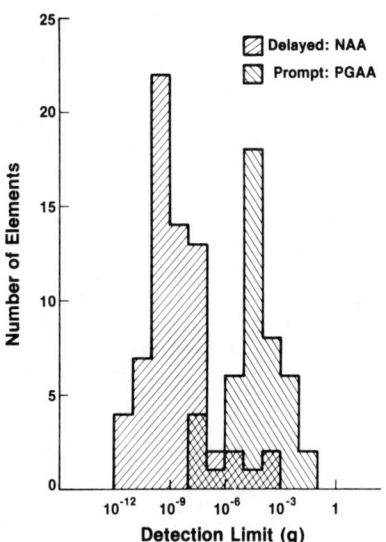

Figure 1. Histogram of detection limits for neutron activation analysis (NAA) and neutron-capture prompt gamma-ray activation analysis (PGAA). Data from Reference 15.

Nuclear Consequences. The use of nuclear rather than chemical reactions has several non-chemical consequences:
1. Both the activating and indicating particles are highly penetrating in gram-sized or smaller samples of common materials, so that errors are not generally incurred by analytical complications due to the nature of the matrix.
2. There is little correlation between detectability of an element and its position in the periodic table, except for the useful fact that the major elements of the earth's surface and atmosphere -- H, C, N, O, and Si -- generally have low cross sections for neutron capture. Consequently, plastics, graphite, quartz, and silicon are nearly ideal materials to irradiate because little or no radioactivity is produced from the major constituents. When the matrix consists of any of a number of geologically rare elements -- Ga, Ge, As, Cd, In, Sb, and Te in the present context -- this selectivity for trace elements is lost because of the large quantity of radioactivity produced when samples of these elements are irradiated (Table II).
3. Because chemical manipulations are ordinarily performed only after the irradiation or are omitted entirely, activation analysis is free of reagent blank.
4. The number of interfering nuclear reactions that can possibly occur is limited by the conservation of energy, and all likely interferences can be enumerated by inspection of

the Table of Nuclides (18) or by consulting published tables (19). Correction for those few interferences that can occur can be very accurate (20).
5. Qualitative analysis is an integral part of the quantitative analysis, since the gamma-ray spectrum and half-life of a radionuclide are highly diagnostic fingerprints of the species. The set of gamma-ray lines emitted in radioactive decay is well characterized, in the sense that few important lines are unknown. Although a recently published table contains 48,000 entries (21), the density in energy space is low compared to the detector resolution, and most radionuclides emit several lines. Hence the number of candidate species for each peak in a spectrum is small.
6. Since radioactive decay follows Poisson statistics, a lower limit to the precision of an analysis can be obtained by a single measurement. In practice, counting statistics generally is the limiting uncertainty, since chi-squared tests often show that the single-measurement precision is an excellent predictor of sample-to-sample repeatability.

<u>Kinetics of Activation Analysis</u>. The time scale of the measurement is determined by the half-life of the radionuclides produced: it is often nearly optimal to irradiate for one half-life, wait one half-life, and count for one half-life.

The amount of radioactivity A(0) of a given radionuclide present at the end of an irradiation interval IT is

$$A(0) = N \sigma \phi (1-e^{-\lambda \cdot IT}) \quad (1)$$

where N = number of target atoms irradiated
 σ = cross section or reaction probability, cm^2
 ϕ = neutron flux, $cm^{-2}s^{-1}$
 λ = ln 2 / t(1/2), s^{-1}
 t(1/2) = halflife of the radionuclide, s
 IT = irradiation time, s.

The activity decays exponentially after the irradiation ends, the amount remaining at a time WT after irradiation being

$$A(WT) = A(0) e^{-\lambda \cdot WT} \quad (2)$$

If this activity is measured during the time from WT to WT + CT, with an efficiency of ε counts per disintegration, then the number of counts observed C is the integral of (2), or

$$C = \varepsilon A(WT) (1-e^{-\lambda \cdot CT}) \quad (3)$$

From (2) and (3) we have

$$A(0) = \frac{C \lambda e^{\lambda \cdot WT}}{\varepsilon (1-e^{-\lambda \cdot CT})} \quad (4)$$

Table II. Activities of Semiconductors After Neutron Irradiation*

Material	Radio-nuclide	Half-life	Activity (mCi/g)**	Radiations
Si	Si-31	2.62 h	5.0	beta
GaP	P-32	14.28 d	0.59	beta
	Ga-72	14.10 h	170	beta, gamma
GaAs	Ga-72	14.10 h	120	beta, gamma
	As-76	26.3 h	160	beta, gamma
Ge	Ge-71	11.2 d	4.2	beta
	Ge-75	82.8 m	130	beta, gamma
	Ge-77	11.30 h	0.99	beta, gamma
CdTe	Cd-115	44.6 d	0.92	beta, gamma
	Cd-117m	3.4 h	0.26	gamma
	Cd-117	2.6 h	0.61	beta, gamma
	Te-127	9.4 h	9.6	beta, gamma
	Te-129	70 m	22	beta
InSb	In-114m	49.51 d	0.25	gamma, beta
	Sb-122	2.72 d	37	beta, gamma
	Sb-124	60.3 d	1.3	beta, gamma

* Irradiation for 1 hour at a thermal neutron flux of 10^{13} n/cm^2·s is assumed. Nuclides with half-lives less than one hour have been arbitrarily omitted. These include isotopes of Ga, Ge, Cd, In, and Te. Minor nuclides and daughters are also omitted.

** Local regulations generally require radiation shielding and other safeguards for work with 1 mCi or more of radioactivity to assure that the dose to personnel is acceptably low.

If an unknown sample (x) and a standard (s) are irradiated together and their decay-corrected activities A(0) are measured with the same efficiency, then by applying equations (1) and (4) to both sample and standard we have

$$\frac{A(0,x)}{A(0,s)} = \frac{N(0,x)}{N(0,s)} \qquad (5)$$

where σ, ϕ, IT, and ϵ cancel in the ratio. The result is that the ratio of the number of atoms of the target element in the sample to that in the standard is the same as the ratio of the corresponding activities. In more chemical terms, if [Y] is the concentration of element Y in mass m of material, then

$$[Y](x) = [Y](s) \cdot \frac{m(s)}{m(x)} \cdot \frac{A(0,x)}{A(0,s)} \tag{6}$$

In this demonstration a host of assumptions and simplifications have been made, which in practice can be corrected for accurately, shown to be unimportant, or made to cancel in the design of the measurements. These include neutron and gamma-ray attenuation in the sample, spatial gradients and temporal changes in neutron flux, differences in counting efficiency between sample and standard, multistep nuclear reactions, and instrumental nonlinearities at high counting rates.

Making an Activation-Analysis Measurement. The most prominent technique in nuclear analytical chemistry is instrumental neutron activation (INAA), in which thermal neutrons from a nuclear reactor are used to irradiate the sample and the induced radionuclides are measured nondestructively with a germanium gamma-ray spectrometer. Sensitivity may be enhanced by chemically separating the elements of interest before radionuclide assay.

High-sensitivity NAA, of course, requires access to a reactor. In the United States there are at present 61 research and testing reactors with a power of 100 KW or greater, operated by universities, state and national laboratories, and industries in 25 states, most of which can be used for activation analysis (22). Many analysts work only with long-lived activation products and thus need not be present at the reactor to perform irradiations. The equipment required for gamma-ray assay's a detector and pulse-height analyzer's is comparable in cost and complexity with that needed for other modern analytical methods.

Suitable sample sizes for irradiation are usually a gram or less. There is no inherent lower limit, and the upper limit is set by the finite transparency of thick samples to neutrons and by the practical problems of inserting a sample into a neutron field. Very large samples, such as entire silicon ingots for neutron-transmutation doping (23) and even oil paintings (24,25), have been studied by NAA with special facilities.

Step 1: Pretreatment of Samples

Despite all possible care in handling before reaching the analytical laboratory, any sample received should be considered to have a contaminated surface. Since the recoil atoms produced in the irradiation may have energies of several hundred eV, any activated impurity may be driven into the surface of the sample. A judgment must be made about the risk of adding more contaminant by imperfect precleaning; generally a light etch after irradiation is preferable if the nature of the experiment permits. If the matrix will become highly radioactive after neutron irradiation (Table II), it may be desirable to separate the elements of interest beforehand, for which clean-room chemistry and high-purity reagents may be required to keep blanks within tolerable bounds (26,27).

Step 2: Packaging

Prior to an activation analysis, the samples and standards are packaged for irradiation. Double containment is usual, the outer container being designed for safety and mechanically suited to the particular reactor facility used. Common container materials are polyethylene, polypropylene, polystyrene, quartz, aluminum, or graphite. The inner container must be reasonably radiation resistant and non-contaminating to the sample. Specially cleaned plastics, high-purity quartz, reactor- or semiconductor-grade graphite, pure metal foil, or pure silicon may be used.

Step 3: Irradiation

The samples are irradiated in the reactor for a time between seconds and weeks, depending on the half-life of the radionuclide to be measured or the tolerable waiting time, whichever is shorter. Care must be taken to assure that the samples and standards receive equal exposure to neutrons, or that unequal doses be measured quantitatively by irradiation of monitors. The flux in a reactor which operates at constant power can be stable to better than 1% from day to day, while that in a training reactor may vary greatly on a time scale of minutes. The very fact that the reactor core is a neutron source guarantees that there will be a spatial gradient in neutron intensity, which may amount to tens of percent or more across a large sample container (28).

Step 4: Transfer and Wait

The irradiation container is unloaded and the samples and standards prepared for radionuclide assay. After irradiation no care whatever need be taken to avoid contamination of the sample by non-radioactive impurities. Time performs the same function as a skilled chemist if the radionuclide to be measured has a longer half-life than the interferences. For example, trace Cr is easily measured in the presence of Ti even though the neutron-capture products of both elements emit gamma rays of exactly the same energy. Gram for gram, Ti emits 1000 times more photons per second than does Cr immediately after a 1-minute irradiation, but Cr is 2000 times more radioactive than Ti two hours later.

Step 5: Chemistry

If interfering activities are present which are longer-lived than the nuclide of interest, it may be necessary to remove them by post-irradiation chemistry. A mild etch is sufficient to remove many surface contaminants if those constitutents are washed away with a flowing etch solution or held in solution by complexing agents (?) to prevent redeposition on the sample. For example, silicon is a sufficiently powerful reducing agent to reduce nitrogen eight oxidation numbers to ammonium ion during dissolution in $HF + HNO_3$.

The advent of efficient high-resolution gamma-ray detectors during the past decade has nearly eliminated the tedious radiochemical separation of each trace element from all others which was once necessary, although group separations are often a powerful aid to specificity and sensitivity. If the matrix itself is the major contributor to the activity (as with Ge or most III-V compounds), safety precautions may be necessary to protect the chemist against radiation. Many procedures have been published for matrix removal and for the separation of trace constituents (29,3,2,30). Ga, Ge, As, and Sb may be separated from other elements by distillation, a procedure which is well suited to hands-off operation. Solvent extraction and ion exchange are often used to remove both large and small amounts of undesirable components. Radiochemistry remains, however, a discipline requiring substantial amounts of skill and time.

Step 6: Radioactivity Assay

Gamma-ray spectroscopy owes a debt to semiconductor technology for making high-resolution spectroscopy possible (31). A detector consists of a large, high-quality germanium crystal, 100 cm^3 or more of near-intrinsic resistivity, held at liquid-nitrogen temperature in vacuum during use. The crystal, reverse biased to operate as a p-i-n junction diode, acts as an ionization chamber, converting the energy of a 1-MeV gamma-ray photon into 340,000 electron-hole pairs. The energy resolution in practice can be as good as 0.16 percent (full width at half the maximum peak height) at 1 MeV.

Charge pulses from the detector are amplified and sorted according to size into a histogram of intensity as a function of gamma-ray energy. Several firms produce multichannel pulse-height analyzers which are capable of sorting pulses into 8000 or even 16000 channels at a rate of 10^5 events per second with a proportionality between channel number and energy of better than 0.1 percent. The large number of data points in a spectrum makes computer-aided analysis practically mandatory whenever more than one or two peaks are to be measured. Although a busy spectrum may contain upward of 200 peaks, each is locally well represented by a narrow near-Gaussian peak on a smooth baseline. As a result, the spectrum can be analyzed piecewise. Laboratory mini- and microcomputers are adequate for these calculations, especially since any disadvantage in computing speed compared with a mainframe computer is offset by the immediacy of response with a dedicated machine. Programs have been written by numerous researchers and equipment manufacturers for the analysis of gamma-ray spectra (32).

The counting time of the activated sample, like the irradiation time, is usually comparable to the half-life of the radionuclide measured, or as much as a day for small quantities of long-lived nuclides. Standards are counted under identical conditions of counting efficiency.

Step 7: Calculation of Elemental Concentration

Corrections are made to the net areas of the gamma-ray peaks in the spectra of sample and standard for contributions from unresolved components, radioactive growth and decay, dead time and other rate-related counting effects, and nonuniform irradiation conditions. Once these corrections are made, the concentration is obtained from Equation 6.

This procedure, as described, requires that a known quantity of each element to be measured be irradiated under the same conditions as the unknown sample. If the relative production rates of the radionuclides used are measured accurately in a particular irradiation facility, subsequent measurements can be done with less labor by irradiating only a few monitor elements to characterize the neutron spectrum. Published values of cross sections have been used as a measure of these relative reaction rates, but the inaccuracy of tabulated data and the poorly known neutron spectrum of most reactors makes this approach unsatisfactory when good accuracy is needed.

Neutron-capture prompt-gamma ray activation analysis (PGAA) is a recent addition to the nuclear analytical arsenal. In this technique the instantaneous gamma ray emission from a sample is measured as it is irradiated in a flux of reactor neutrons (33,34,35). Because the sample must be several meters from either the core of the reactor or (less commonly) from the detector, the sensitivity of this technique is generally poorer than in conventional NAA. However, it is possible to measure small quantities of many elements which do not give radioactive neutron-capture products, notably 0.01 mg of H, 50 ng B, and 1 mg P in an electronics context.

Some Applications of NAA in Electronic Materials Studies

Trace-Element Characterization of Silicon. In order to define "solar grade" silicon with sufficient precision to make the optimum economic choice among possible production processes, Davis et al. doped a series of silicon ingots with single transition metal impurities and produced curves relating normalized photovoltaic efficiency to the concentration of the contaminant (36). Establishing the x-axis of these curves was not entirely straightforward. As a contribution to this work, NBS measured concentrations of a number of dopants in these samples, with the detection limits found in Table I. An illustration of the difficulty faced in this work is that forty percent of the concentrations determined at NBS differed from the nominal concentration by a factor of two or more.

Spatial Location of Constituents. Despite the fact that NAA is a method for bulk analysis, its sensitivity has been used to advantage in diffusion studies, by successive etching or mechanical lapping of surface layers (5,2). Especially if the radioactivity of the layers removed is measured directly rather than as the difference in activity of the substrate, the technique can be sensitive and accurate.

Contamination Studies. Schmidt and Pearce (37), in a classic study, using NAA evaluated sources of contamination of silicon during device processing. Their work demonstrated clearly that high-temperature oxidation steps are particularly likely to add transition-metal contaminants. Mechanical polishing compounds, boats and handling tools, and components of ion-beam equipment were also identified as being potentially troublesome.

When platinum resistance thermometers at NBS were found to behave irregularly, contamination of the surface of quartz insulator tubing was suspected of being a contributor. Another cause was found after an NAA survey of the material used showed that the trace element content was within the manufacturer's specifications.

Boron in Silicon and Glasses. As a complement to earlier comparisons of the relation between B content and resistivity of silicon ($\underline{38}$), a set of B-doped Si wafers was analyzed by PGAA. The concentration range from 3×10^{17} to 1×10^{20} atoms/cm^3 was measured with irradiation times of 10 minutes to 3 hours. Bulk boron has also been determined in a series of borosilicate glasses as part of a study of B concentration near the glass surface ($\underline{39}$). The reproducibility among replicates was 1%, within the counting statistics obtained after an irradiation of 15 to 30 minutes. Agreement with ICP analysis of the same samples was within 2% ($\underline{40}$).

Film-Thickness Measurements. Although NAA is usually regarded (and most often used) as a technique for measuring constituents at trace levels, it has proved valuable for the accurate measurement of major elements, such as the determination of the thickness of metallic films. For example, a set of gold films evaporated on silicon for use as interlaboratory standards for Rutherford backscattering has been calibrated by INAA ($\underline{41}$). After the analysis was completed, comparison with gravimetric measurements of the gold content showed agreement to within 0.3 percent. Other unpublished work at NBS ($\underline{42}$) has shown NAA to have the sensitivity to detect a 2-nm Fe film on graphite, the selectivity to determine the thickness of a 10-nm film of Pd under 5 μm of Ag, and the accuracy to determine the quantity of Ni and Cr in a multilayer film of 260 nm total thickness to within 0.5 percent.

Related Nuclear Analytical Techniques

Thermal neutrons are capable of inducing a small number of important nuclear reactions which produce charged particles. The outstanding example is of course U-235, whose fission fragments can be registered by insulating solids and etched into optically visible tracks ($\underline{43}$) to provide a two-dimensional image of the uranium distribution on a surface. Similarly, neutron-induced alpha particles from B-10, Li-6, and O-17, tritons from Li-6, and protons from N-14 have been used to visualize and quantify the distribution of elements on surfaces or in cut sections. Alternatively, the energy of the light charged particles can be measured during irradiation. The shape of the spectrum gives graphically the depth distribution of the atoms which absorbed the neutrons ($\underline{44}$); this neutron depth profiling is the subject of another paper at this symposium ($\underline{45}$).

Radioactive tracers are ideal tools for studying problems of diffusion, phase distribution equilibria, and processing contamination. Kane and Larrabee state "without reservation, that all reliable values for segregation coefficients for impurities in semiconductors have been determined by using radiotracer methods" ($\underline{2}$). With high-efficiency radiation detectors the quantity of radioactivity that must be handled need not be large. The method is deserving of wider use.

The distribution of radioactive atoms over the face of a crystal can be made visible by exposing x-ray film to the surface. Autoradiography with neutron-activated specimens or those doped with radiotracers has been used for localization of defects, for studying the uniformity of doping in the crystal, and for measuring the uniformity of surface films (_3_,_4_,_2_,_46_,_47_).

Summary

This paper has given in outline form the major considerations on which a successful application of neutron activation analysis should be based. The most effective applications of this method have involved close and continual interaction between the physicist or engineer and the analyst from the beginning of the project, in order to ensure measurements that are meaningful and lead to the solution of the problem being studied. Activation analysis is only one of many analytical techniques available for materials characterization; its sensitivity, accuracy, simple physics, and nondestructive nature should make it an important contributor for years to come.

Literature Cited

1. Keenan, J. A.; Larrabee, G. B. In "VLSI Electronics Microstructure Science, Vol.6: Materials and Process Characterization"; Einspruch, N. G.; Larrabee, G. B., Eds.; Academic Press: New York, 1983; p. 1.
2. Kane, P. F.; Larrabee, G. B. "Characterization of Semiconductor Materials" McGraw-Hill: New York, 1970.
3. Heinen, K. G.; Larrabee, G. Anal. Chem., 1966, 38, 1853.
4. Martin, J. A. In "Semiconductor Silicon"; Haberecht, R. R., Ed.; Electrochemical Society: New York, 1969; p. 547.
5. Kudo, K. J. Japan Soc. Appl. Phys., 1975, 44, 319.
6. Riotte, H. G.; Herpers, U.; Weber, W. Radiochem. Radioanal. Lett., 1977, 30, 311.
7. Bouldin, D. J. Radioanal Chem., 1982, 72, 35.
8. Kruger, P. "Principles of Activation Analysis" Wiley-Interscience: New York, 1971.
9. DeSoete, D.; Gijbels, R.; Hoste, J. "Neutron Activation Analysis" Wiley-Interscience: London, 1972.
10. Amiel, S. "Nondestructive Activation Analysis" Elsevier: Amsterdam, 1981.
11. Gladney, E. S.; Burns, C. E.; Perrin, D. R.; Roelandts, I.; Gills, T. E. "1982 Compilation of Elemental Concentration Data for NBS Biological, Geological, and Environmental Standard Reference Materials (NBS Spec. Pub. 260-88)" National Bureau of Standards; Washington, D.C., 1984.
12. Laul, J. C. Atomic Energy Rev., 1979, 17, 603.
13. Germani, M. S.; Gokmen, I.; Sigleo, A. C.; Kowalczyk, G. S.; Olmez, I.; Small, A. M.; Anderson, D. L.; Failey, M. P.; Gulovali, M. C.; Choquette, C. E.; Lepel, E. A.; Gordon, G. E.; Zoller, W. H. Anal. Chem., 1980, 52, 240.
14. Meinke, W. W. Science, 1955, 121, 177.
15. Vogt, J. R. (Univ. of Mo., Columbia), unpublished, 1981.
16. Koch, O. G.; LaFleur, P. D.; Morrison, G. H. Pure and Appl. Chem., 1982, 54, 1565.

17. Currie, L. A. Anal. Chem. 1968, 40, 586.
18. Walker, F. W.; Kirouac, G. J.; Rourke, F. M. "Chart of the Nuclides" General Electric Company: San Jose, 1977.
19. Erdtmann, G. "Neutron Activation Tables" Verlag Chemie: Weinheim, 1976.
20. Lindstrom, R. M.; Fleming, R. F. In Proc. 3rd Int. Conf. Nucl. Methods in Envir. and Energy Res. (CONF-771072); Vogt, J. R., Ed.; U of Mo.: Columbia, 1977; pp. 90.
21. Erdtmann, G.; Soyka, W. "The Gamma Rays of the Radionuclides" Verlag Chemie: Weinheim & New York, 1979.
22. Burn, R. R.; Bilof, R. S. "Research, Training, Test, and Production Reactor Directory" American Nuclear Society: LaGrange Park, Ill, 1983.
23. Guldberg, J., Ed. "Neutron-Transmutation-Doped Silicon" Plenum: New York, 1981.
24. Ainsworth, M. W., Ed. "Art and Autoradiography: Insights into the Genesis of Paintings by Rembrandt, Van Dyck, and Vermeer" Metropolitan Museum of Art: New York, 1982.
25. Carter, R. S.; Ganozcy, M.; Olin, C. H.; Schroder, I. G; Cheng, Y-T.; Olin, J. S. In "Proc. Conf. Use and Develop. of Low and Medium-Flux Research Reactors"; Harling, O., Ed.; MIT Press: Cambridge, Mass, 1982.
26. Das, H. A. Pure & Appl. Chem., 1982, 54, 755.
27. Murphy, T. E. in "Accuracy in Trace Analysis: Sampling, Sample Handling, Analysis" (NBS Spec. Pub. 422); LaFleur, P. D., Ed; U. S. Government Printing Office, Washington, 1976, p. 509.
28. LaFleur, P. D.; Becker, D. A. J. Radioanal. Chem., 1974, 19, 149.
29. Thompson, B. A.; Strause, B. M.; Leboeuf, M. B. Anal. Chem., 1958, 30, 1023.
30. Cerofolini, G. F.; Polignano, M. L.; Picco, P.; Finetti, M.; Solmi, S.; Gallorini, M. Thin Solid Films, 1982, 87, 373.
31. Hansen, W. L.; Haller, E. E. In "Nuclear Radiation Detector Materials", Haller, E. E.; Kraner, H. W.; Higinbotham, W. A., Eds.; North-Holland: New York 1983; p. 1.
32. Carpenter, B. S.; D'Agostino, M. D.; Yule, H. P., Eds. "Computers in Activation Analysis and Gamma-Ray Spectroscopy"(CONF-780421), U. S. Department of Energy: Washington, DC, 1979.
33. Failey, M. P.; Anderson, D. L.; Zoller, W. H.; Gordon, G. E.; Lindstrom, R. M. Anal. Chem., 1979, 51, 2209.
34. Glascock, M. D. In "Neutron-Capture Gamma-Ray Spectroscopy and Related Topics"; von Egidy, T., Gonnenwein, F., Maier, B., Eds.; Institute of Physics: London, 1982; p. 641.
35. Anderson, D. L.; Zoller, W. H.; Gordon, G. E.; Walters, W. B.; Lindstrom, R. M. In "Neutron-Capture Gamma-Ray Spectroscopy and Related Topics"; von Egidy, T.; Gonnenwein, F.; Maier, B., Eds.; Institute of Physics: London, 1982; p. 655.
36. Davis, J. R.; Rohatgi, A.; Hopkins, R. H.; Blais, P. D.; Rai-Choudhury, P.; McCormick, J. R.; Mollenkopf, H. C. IEEE Trans. Elect. Dev., 1980, ED-27, 677.
37. Schmidt, P. F.; Pearce, C. W. J. Electrochem. Soc., 1981, 128, 630.
38. Thurber, W. R.; Carpenter, B. S. J. Electrochem. Soc., 1978, 125, 654.

39. Riley, J. E. Jr.; Downing, R. G.; Fleming, R. F.; Lindstrom, R. M.; Vincent, D. H. <u>J. Solid State Chem.</u>, 1984, to be published.
40. Riley, J. E. Jr., Lindstrom, R. M. In prep, 1985.
41. Lindstrom, R. M.; Harrison, S. H.; Harris, J. M. <u>J. Appl. Phys.</u>, 1978, 49, 5903.
42. Fleming, R. F. Private communication, 1984.
43. Fleischer, R. L.; Price, P. B.; Walker, R. M. "Nuclear Tracks in Solids: Principles and Applications " University of California Press: Berkeley, 1975.
44. Downing, R. G.; Fleming, R. F.; Langland, J. K.; Vincent, D. H. <u>Nucl. Inst. Meth. Phys. Res.</u>, 1983, 218, 47.
45. Downing, R. G.; Fleming, R. F.; Maki, J. T. This symposium.
46. Szabo, E. <u>J. Radioanal. Chem.</u>, 1974, 19, 23.
47. Rausch, H. <u>J. Radioanal. Chem.</u>, 1978, 44, 119.

RECEIVED June 14, 1985

19

Trace Element Survey Analyses by Spark Source Mass Spectrography

Fredric D. Leipziger and Richard J. Guidoboni

Northern Analytical Laboratory, Inc., Amherst, NH 03031

>The role of Spark Source Mass Spectrography (SSMS) as a high sensitivity trace element analytical method is discussed. The unparalleled combination of sensitivity and complete element coverage makes SSMS especially suitable for the analysis of liquid and solid materials involved in semiconductor processing. Sample requirements are discussed. The application of SSMS to semiconductor materials, process reagents, dopants, and metals, is illustrated. Advantages and disadvantages of the technique as well as sensitivity and accuracy are discussed.

Exactly twenty-three years ago this week a conference was held in Boston on ultrapurification of semiconductor materials. One third of the papers at that conference were devoted to impurity analyses (1). Spark source mass spectrography was the newest and most promising analytical technique available at the time, and I would like to compare the status of SSMS at that time to its present status.

The first SSMS instrument was reported by Dempster (2) in 1946. Hannay of Bell Laboratories was responsible for the first applications to semiconductor materials (3,4) in the mid 50's. The technique was so promising that commercial instrumentation became available in 1960. The attractive features of the technique were complete element coverage (all elements on the periodic table could be detected) and excellent sensitivity (to 1 part per billion atomic). The two major applications of SSMS at that time were semiconductor and nuclear reactor materials -- both new technologies and both extremely impurity sensitive. The biggest disadvantage of the technique, although not clearly realized at the time, was lack of quantitation.

The need for information regarding trace level bulk impurities in semiconductors and nuclear reactor materials was urgent and could not be satisfied by other techniques existing at the time. In the next decade well over 200 instruments were operating in industrial and government laboratories here and in Europe and Japan. During this period the lack of quantitation became apparent and a major effort was made in many laboratories to overcome this fault.

At the present time there are probably fewer than 80 instruments operating in the free world, and 20 or fewer in the United States. The lack of quantitation for routine and diverse samples still exists except in two special cases -- isotope dilution and electrical detection. These two cases are generally not applicable to semiconductor materials. However, the combination of complete element coverage and excellent sensitivity are still unparalleled by modern analytical techniques devoted to bulk impurity analyses. Thus, despite its inherent semi-quantitative character, we have a most useful method which efficiently fills a need in the semiconductor and high purity material fields. Figure 1 compares sensitivity and element coverage of SSMS with other techniques. Many European laboratories using sophisticated densitometry and computer data handling are producing data with accuracies claimed to be ±5%. However, many of these analyses require two to five days per sample (5).

The three key factors governing the successful use of SSMS today are knowledge of the limitations of the technique, operator experience, and cleanliness. The customer must be made aware of the semi-quantitative nature of the data and the fact that certain elements such as sodium, potassium, carbon, and nitrogen require special interpretations. The operator must be experienced in the interpretation of the data -- he must be aware of interferences from compound ions, multiply charged species and other sources.

The interpretation of SSMS data falls into two distinct areas -- element identification and estimates of quantity. The criteria used in our laboratory for positive elemental identification are the presence of the doubly ionized species and the identification of the isotopic pattern when possible. Quantitation will be discussed later. Last, the instrument source must be cleaned regularly to avoid memory problems. Our approach to the memory problem is to have a complete set of source parts for each matrix. A set of parts are dedicated for silicon analyses, another for gallium arsenide, etc. These parts and the source itself are cleaned on a regular and frequent basis. When these factors are under control, SSMS has proved to be a reliable, reproducible technique for the bulk analysis of trace impurities.

An examination of the recent literature shows that over 90% of the papers published are of European or Japanese origin. Several excellent reviews (5,6) indicate that fewer than 10% of the articles are concerned with semiconductors. The greatest emphasis appears to be in the fields of geology, biology and metallurgy. These workers uniformly are attempting to produce quantitative data and their approach is to evaluate a relative sensitivity factor (RSF) for each element for a particular matrix. Then the RSF can be applied to future determinations in the same matrix. Cooperative studies show, beyond any question, that the RSF cannot be transferred from one laboratory or instrument to another.

Instrument Design

The ability to ionize all the elements of the periodic table is the major strength of SSMS but it imposes some severe restrictions on the design of the mass spectrograph. The high energy required to produce these ions results in an ion beam with a large energy spread which adversely effects resolution. In order to reduce the energy spread

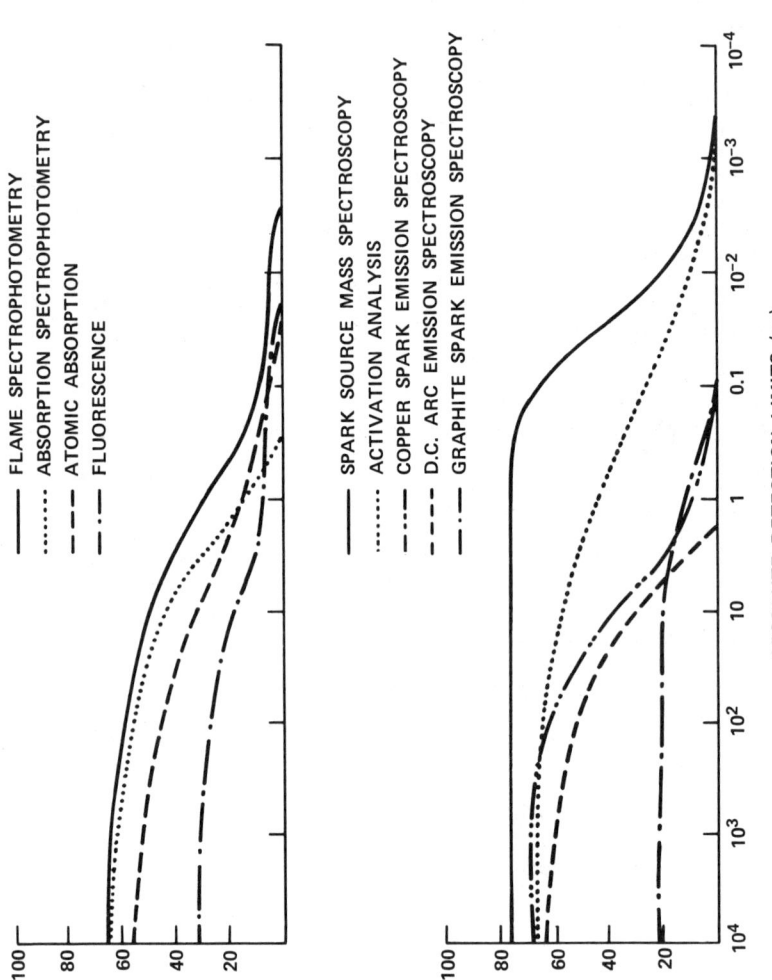

Figure 1. Scope of Trace Methods of Analysis. Reproduced with permission from Ref. 11 Copyright 1965 Interscience Publishers.

of the beam, a double focussing analyzer is required. A schematic diagram of a typical SSMS is shown in Figure 2. The energy spread present in an rf spark ion beam may be several thousand volts compared to less than ten volts for electron impact ion beams.

The ion source is a radio frequency vacuum discharge source and is illustrated in Figure 3. The ionization efficiency of this source is uniform within a factor of three, due to the fact that the ionizing energy available is so much greater than the ionization potential of any element. This is one of the most attractive features of the technique, since it allows us to analyze every element on the periodic table without large variations in sensitivity.

The detector for most SSMS analyses is the Ilford Q-II photoplate. This detector integrates the signal and also provides a permanent record of the analysis. The integrating properties of the photoplate and its wide dynamic range make it an excellent detector for SSMS. These properties also allow us to detect inhomogeneities in a sample when the spectra do not increase in a regular fashion with increasing charge collection. It is also possible, by using a gold or graphite counter electrode, to do some probe type analyses. If there are small areas of particular interest those areas can be analyzed using the counter electrode technique. The need to place a gelatin emulsion in the high vacuum of the analyzer section necessitates a prepumping chamber for the photoplates. Details of instrument design and source configurations are discussed by Herzog and Franzen (7,8).

Sample Requirements

Samples for SSMS analysis must be conducting self supporting solids capable of withstanding vacuum. Two electrodes are required, ideally ½" long by .050". Conductors and semiconductors need only to be cut to size and cleaned before mounting. Non-conducting powders are mixed with high purity graphite powder and pressed into the appropriate electrode shape. Non-conducting solids must first be ground before mixing with graphite and compacting. Liquids are pipetted onto high purity graphite powder, mixed and pressed into shape after drying at low heat. An increased sensitivity is possible with liquids since the impurities are concentrated on the graphite.

Sample preparation requires extreme care to avoid the introduction of contamination in the cutting and grinding operations. This is an area where operator experience is of prime importance. Extensive cleaning and etching may be required to provide an analysis representative of the sample composition. A further cleaning step is provided in the instrument itself by sparking the sample surface energetically to clean away the first few atomic layers. This prespark can be recorded on the photoplate to provide qualitative information about the chemistry of the surface.

This brief description of sample preparation points to one of the reasons why the sensitivity of SSMS is so high. We are working with the solid sample itself -- there is no dilution factor due to the necessity to dissolve the sample. Techniques such as AAS and ICP will lose a factor of 10 to 100 in ultimate sensitivity due to dilution occurring when a sample is dissolved. Also, when high purity materials are subjected to dissolution, the chances for contamination are high, and reagent blanks must be evaluated. In many cases the

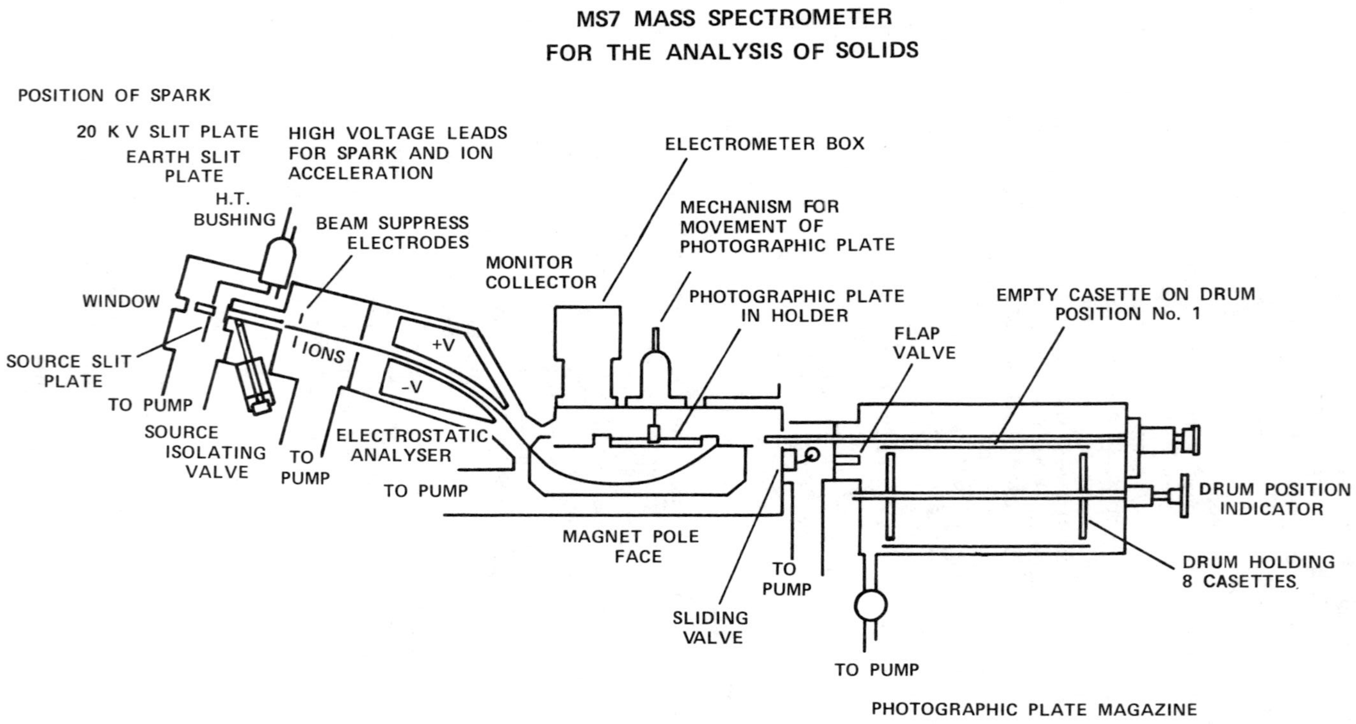

Figure 2. Typical Spark Source Mass Spectrograph.

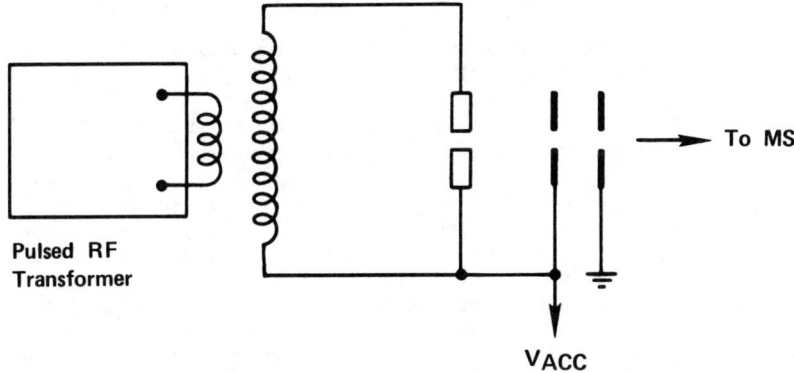

Figure 3. Radio Frequency Spark Source. Reproduced with permission from Ref. 11 Copyright 1965 Interscience Publishers.

reagent blanks will be significantly higher than the elemental concentration.

Mass Spectra

The spectra produced by the rf spark are quite simple. Ions produced by the spark are accelerated into the analyser region, are mass separated, and fall on the photoplate where a latent image is generated and later developed. Figure 4 shows a series of mass spectra generated from a silicon sample and is chosen to illustrate most of the common ion types. Singly and multiply charged species can be seen as well as compound ions and molecular ions. The strongest series of lines of masses 28, 29, and 30 are due to the silcon isotope ions $^{28}Si^+$, $^{29}Si^+$, and $^{30}Si^+$. The lightest exposures give information concerning the isotopic abundance. The next most abundant series of lines are at masses 14, 14.5, and 15. These are due to the ions $^{28}Si^{2+}$, $^{29}Si^{2+}$ & $^{30}Si^{2+}$. A triply ionized series is also seen at masses 9.33, 9.67, and 10.00. These groupings of multiply ionized species of the parent element may be seen as low as 6^+ or even lower.

Another feature of silicon spectra is molecular ions. Ions due to the silicon dimers (Si_2^+) are visible at masses 56, 57, 58, 59 and 60. Since ^{28}Si is the most abundant isotope the dimer ions containing ^{28}Si are also most abundant. Starting at mass 84 ($^{28,28,28}Si_3^+$) another series of molecular ions may be seen.

The ion responsible for the line at mass 31 is, in part, due to $^{30}Si^1H^+$. This hydride ion is of great interest since the most common element producing a line at mass 31 is phosphorus. However, in this case, the absence of a line at mass 15.5, due to $^{31}P^{2+}$ indicates that $^{30}Si^1H^+$ is the cause of the line.

This figure shows a graded series of exposures ranging from 0.001 nanocoulombs to 50 nanocoulombs. At the longest exposure a nominal sensitivity of 0.02 parts per million atomic is obtained for a monoistopic element.

Table I lists a few of the most common spectral interferences. The determination of iron in silicon is difficult because of the $^{28}Si_2$ dimer which interferes with the major iron isotope at mass 56. In this case it is possible to use one of the less abundant iron isotopes as the analytical line.

Table I. Some Common Interferences in SSMS

$^{28}Si_2^+$	$-$	$^{56}Fe^+$	
$^{93}Nb^{3+}$	$-$	$^{63}Cu^{16}O^+$	$- \;\; ^{79}Br^+$
		$^{31}P^+$	
$^{30}Si^1H^+$		$^{65}Cu^{16}O^+$	$- \;\; ^{81}Br^+$
		$^{31}P^+$	
		$^{69}Ga^{3+}$	$^{23}Na^+$

The niobium - phosphorus case is one where multiply charged

matrix ions obscure an impurity line. The only solution in this case where both matrix and impurity are monoisotopic would be if the instrument had sufficient resolving power to separate the two lines. The phosphorus line is at m/e 30.9738 while the niobium line is at m/e 30.9686. A resolving power of 6000 would therefore be required to analyze phosphorus in niobium.

The analysis of phosphorus in silicon is complicated by the presence of a small contribution from the $^{30}Si^1H^+$ ion. This ion is probably the result of the combination of silicon with hydrogen from water vapor in the source. Since phosphorus is monoisotopic no other lines are available and the analyst must resort to the doubly ionized phosphorus line at m/e 15.5. The use of doubly ionized lines for analysis in this case is possible since compound ions such as $^{30}Si^1H^+$ normally do not exist as doubly charged species.

The resolving power of these spectrographs can be as high as 10000. However, use of the slits necessary to attain this resolution results in a marked increase in exposure time due to lower ion transmission.

Data Interpretation

Semiquantitative estimates of impurity concentrations are made by comparing the exposure necessary to produce a "just detectable" line for an isotope or element of known concentration with the exposure necessary for the "just detectable" line of the unknown impurity. This procedure produces satisfactory results within a factor of three.

Attempts to obtain quantitative results entail emulsion calibration, microphotometry of line densities, corrections for ion mass and ion energy, and computer handling of the data. Data interpretation is discussed in detail by Kennicott [9] and Owens [10]. Results obtained by this procedure vary from accuracies of ± 5% to deviations of orders of magnitude from true values.

Application to Semiconductor Technology

SSMS is used successfully for the analysis of materials involved in all phases of semiconductor manufacturing. These include semiconductors such as silicon and gallium arsenide, high purity metals such as gold and aluminum, acids and etchants such as nitric and hydrofluoric acids, dopants such as phosphorus trichloride and boron tribromide, and solvents such as trichloroethane. Starting crystal growth materials such as gallium and arsenic as well as crucible materials are also suitable candidates for SSMS analyses. This list includes both conducting and insulating solids as well as liquids.

The analysis of semiconductor materials is illustrated in Tables II and III. Table II shows comparison of the impurity content of a series of silicon samples. These materials are received as chunks (in the case of starting materials) or slices from a crystal. They are cut to the appropriate size and etched for cleaning with high purity acid. The figure illustrates high sensitivity analysis of three polycrystalline silicon samples with the -20 material showing much higher aluminum content than the others. SSMS performs well in this type of comparative analysis.

Table II. Comparison of Three Silicon Samples' Impurities in ppm Weight

	-08	-90	-20
B	<0.001	<0.001	0.003
C	<1	<1	<1
Mg	<0.003	<0.003	0.01
Al	0.005	0.005	0.1
P	0.005	<0.003	0.005
Cl	<0.003	0.01	0.006
Ca	0.01	0.01	0.04
Cr	<0.006	*0.01	<0.006
Mn	<0.006	<0.006	<0.006
Fe	<0.1	*0.2	<0.1
Cu	<0.007	*0.02	0.006
As	<0.008	<0.008	<0.008

63 elements <0.003 ppma

*inhomogeneous

Table III is the analysis of a high purity gallium sample used as starting material for gallium arsenide growth. Figure 5 shows the change necessary in sample mounting due to the low melting point of gallium. The sample is melted while still in its original container and a drop is poured into a high purity graphite cup electrode. A counter electrode of high purity graphite is used to complete the electrode pair. The graphite used here as well as the graphite powder needed to mix with insulating powders has previously been analyzed by SSMS for impurity content.

Table III. Analysis of Gallium Metal Impurities in ppm Weight

B	0.03	Cr	<0.7
Mg	<1	Mn	0.04
Al	0.2	Fe	0.4
Si	0.1	Ni	0.2
S	<2	Cu	0.1
Cl	0.2	Zn	0.2
K	0.2	As	0.1
Ca	0.2	In	1

57 elements <0.2 ppma

Table IV shows the usefulness of SSMS for the analysis of crucible materials used in crystal growth. Samples of fused silica and graphite were analyzed and the impurities found are listed. The ability to examine all phases of production is of great help in pinpointing contamination sources.

Figure 4. Mass Spectra of Silicon.

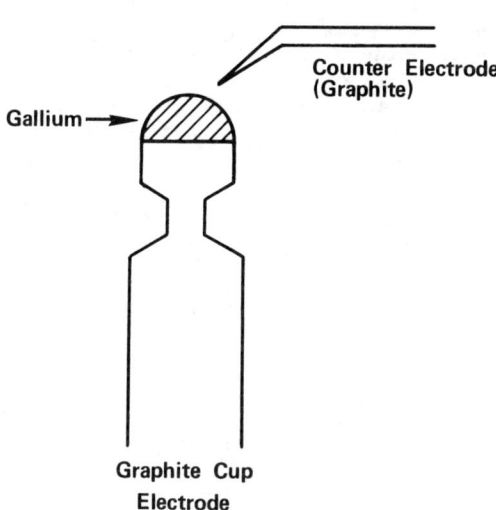

Figure 5. Analysis of Gallium Metal.

Table IV. Analysis of Crucible Materials

	Fused Quartz	Graphite impurities in ppm weight
B	0.04	0.02
Al	7	0.07
Si	–	2
P	0.03	0.03
Cl	1	0.3
K	1	–
Ca	1	0.1
Cr	–	0.2
Mn	0.1	–
Fe	<10	0.2
Cu	0.1	–
64 elements <0.02 ppma		65 elements <0.03 ppma

The analysis of most dopants is complicated by their physical state and extreme reactivity. Table V shows the impurities found in samples of arsenic trichloride and boron tribromide by SSMS. For these materials a known volume is pipetted onto a weighed portion of high purity graphite powder. The mixture is warmed with a heat lamp, mixed, and pressed into electrodes. It is possible to obtain an increased sensitivity for these and other liquids by evaporating up to 50 ml on a gram of graphite and concentrating the non-volatile impurities.

Table V. Analysis of Volatile Dopants

	$AsCl_3$	BBr_3 impurities in ppm weight
Be	<0.006	–
B	0.03	–
Na	0.1	<0.02
Mg	<0.6	<0.1
Al	0.2	0.03
Si	–	0.2
P	<0.08	0.01
S	<0.8	<0.1
Ca	1	–
V	–	<0.002
Cr	0.02	0.005
Mn	0.02	0.002
Fe	1	0.02
Cu	0.04	0.006
Zn	0.07	0.004
As	–	0.01
57 elements <0.1 ppma		56 elements <0.1 ppma

Process chemicals include such diverse materials as inorganic acids, hydrocarbon solvents, halogenated solvents, alcohols, and organic acids. The analysis of a trichloroethane sample is shown in Table VI as illustrative of the utility of SSMS to analyze this entire range of materials. As in the previous example, a known volume of the solvent was pipetted onto high purity graphite, warmed to evaporate the matrix, and pressed into electrodes.

Table VI. Impurities in CH_3CCl_3, ppm Weight

Be	0.001	Ca	0.02
B	0.01	Cr	0.003
Mg	<0.2	Mn	0.04
Al	0.01	Fe	0.04
Si	0.04	Ni	<0.02
P	0.008	Cu	0.04
S	0.4	Sn	<0.007
K	0.08	As	0.007

57 elements <0.05 ppma

Summary

Spark source mass spectrography is the most useful tool available to the microelectronics industry for bulk trace level impurity analyses of a wide variety of materials. The technique routinely examines crystal growth start materials, crucibles, finished crystals, dopants, solvents, metals and all substances used in microelectronics manufacture.

The inability of the technique to routinely provide anything better than semiquantitative data is more than offset by its wide range of applicability and unusual properties. The combination of high sensitivities - to one part per billion - coupled with the ability to detect any element on the periodic table make SSMS a tool with tremendous usefulness to the microelectronics industry.

Literature Cited

1. Morrison, G. H. in "Ultrapurification of Semiconductor Materials"; Brooks, M. S., Kennedy J. K., Eds: MacMillan: New York, 1961; p. 267.
2. Dempster, A. J; The Mass Spectrograph in Chemical Analysis: USAEC Report MDDC 370, 1946.
3. Hannay, N. B. Rev. Sci. Instr. 1954, 25, 644.
4. Hannay, N. B; Ahearn, A. J. Anal. Chem. 1954, 26, 1056.
5. Beske, H. E; Gijbels, R; Hurrle, A; Jochum, K. P; Fresenius Z. Anal. Chem. 1981, 309, 329.
6. Beske, H. E: Int. J. Mass. Spec. Ion Phys. 1982, 45, 173.
7. Herzog, R. I; in "Trace Analysis by Mass Spectrometry"; Ahearn, A. J., Ed; Academic: New York, 1972; p. 57.
8. Franzen, J., Ibid., p. 11.
9. Kennicott, P. R., Ibid., p. 179.
10. Owens, E. B; Giordano, N. A. Anal. Chem. 1963, 35, 1172.
11. Morrison, G. H "Trace Analysis, Physical Methods" Interscience; 1965.

RECEIVED June 27, 1985

20

Characterization of Components in Plasma Phosphorus-Doped Oxides

Jana Houskova, Kim-Khanh N. Ho, and Marjorie K. Balazs

Balazs Analytical Laboratory, Inc., Mountain View, CA 94043

> Three compounds of phosphorus have been verified in plasma deposited phosphosilicate glass and borophosphosilicate glass films. They are phosphorus pentoxide, phosphorus trioxide, and phosphine. Evidence is given for the presence of a fourth phosphorus-containing compound. Data are given from analysis by colorimetry, ion chromatography, and P-31 NMR on standards, plasma, plasma enhanced, and atmospheric CVD films. Some information on LTO/LPO deposited films is included.

In past years, the analysis of doped dielectric MOS films has always led to the conclusion that the dopant materials existed in their fully oxidized state. Thus, phosphorus existed as phosphorus pentoxide and as such could easily be quantitatively measured in aqueous solutions as the phosphate ion. Although many suspected the presence of phosphorus trioxide, there was no evidence of its existence.

In the spring of 1983, however, during a routine analysis of a set of PSG silicate films, a distinct foul odor was detected coming off the wafers during etching. The suspicion that this odor was phosphine led to a renewed study of all glass films. Because the set of wafers that yielded the foul smelling gas was plasma deposited doped glass, our first study was on plasma and plasma enhanced CVD PSG films.

The immediate concern about the presence of phosphine in doped oxide films was safety in handling these materials. The permissible exposure limit (PEL) set by OSHA is 0.3 ppm, and the concentration immediately dangerous to life and health (IDLH) is 200 ppm. All precautions must be taken when etching plasma PSG of BPSG wafers. A basket of twenty 3 inch wafers, 5000 angstroms thick, containing 5 wt. % phosphorus could release as much as 2500 ppm of phosphine or 12.5 times the IDLH.

To identify the foul smelling gas, a closed system was built in which the film was dissolved. The liberated gas was driven to a reaction chamber and oxidized. If the gas were phosphine, it would be converted to phosphate ion and as such could be analyzed. This test was positive for every plasma and plasma enhanced phosphorus doped oxide that was analyzed. Deductively it was concluded that the gas was phosphine. These results and the suspicion of suboxides of phosphorus were immediately publicized through the trade journals in September 1983 (1).

An attempt was made to positively identify the gas by mass spectrometry. The film was dissolved with HF in a closed chamber and the resultant gas was injected into a GC/MS. No signal was obtained, and it was concluded that the gas did not reach the mass analyzer or that it was lost during the process of injecting it into the MS.

Meanwhile, the study to determine suboxides of phosphorus proceeded. It is well known that the phosphate ion reacts with molybdate ion to form a yellow phosphomolybdate salt and that this salt produces a blue solution in the presence of a reducing agent. Samples of known compounds containing phosphate ion readily and quickly form this blue solution, whereas samples containing phosphite ion do not react positively unless oxidized first to phosphate ion.

Tests were run on plasma oxides with and without oxidizing the samples. In every case, the results yielded higher weight percents of phosphorus from the oxidized sample than from an unoxidized sample from the same wafer. These results confirmed the existence of suboxides of phosphorus in the plasma phosphosilicate glasses. It was believed that this compound was phosphorus trioxide, which would not react with the molybdenum ion and form a moly blue complex unless it was first oxidized according to the following reaction:

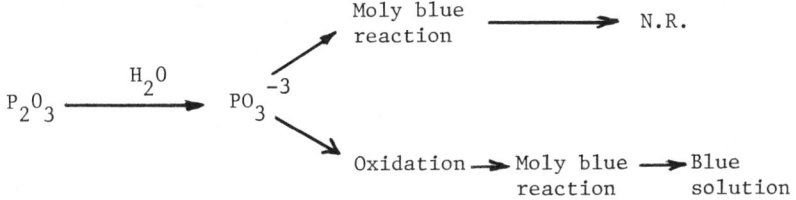

It is also well known (2) that not all oxoacids of phosphorus react with molybdenum ion to form a positive blue complex at the same rate. Time studies were done on plasma PSG wafers. Figure 1 shows distinct differences between the phosphate ion and doped oxide film from a wafer when analyzed with and without oxidation.

A time study was also conducted on various oxo phosphorus acids (Figure 2) and on several kinds of PSG films (Figure 3). The time studies were conducted by dissolving and quantitatively removing the films from the wafers and then reacting them with a molybdate ion simultaneously. The shape of the curves indicated the presence of a slow reacting phosphorus compound in plasma, plasma enhanced, and low temperature/low pressure oxides (LTO/LPO) produced PSG films. Only atmospheric CVD films produced a typical phosphate ion curve.

Many possible compounds of phosphorus exist. The simplest are listed in Table I. Even more complicated oxoacids of phosphorus were studied and reported in the paper by Ohashi and Yoza (3). Their study of oxoacids of phosphorus with molybdenum ion was done in neutral and acid solutions in the presence and absence of a reducing agent. Their results showed that acids containing a -P-P-P straight or cyclic chain gave color development similar to those we found in

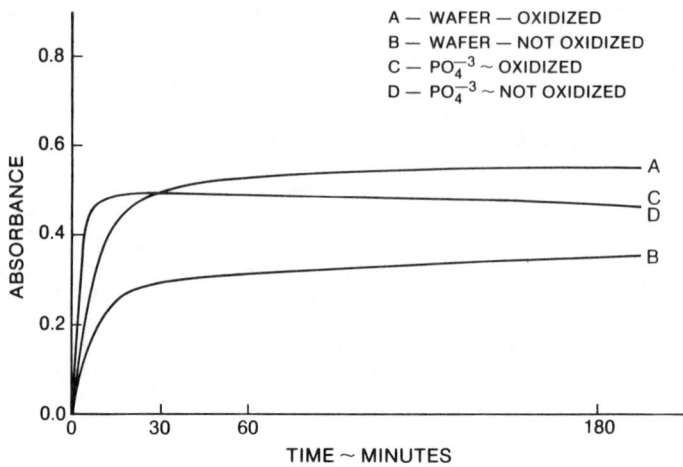

Figure 1. Color development vs. time.

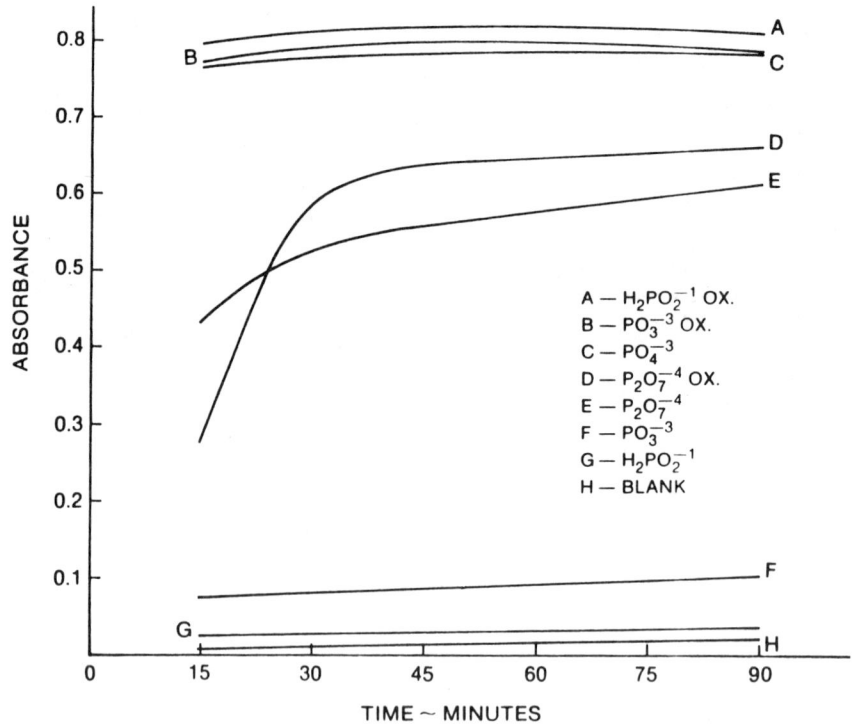

Figure 2. Time study on oxoacids of phosphorus molybdate blue reaction.

Table I. Phosphorus Oxides and Acids

Basic Oxide Forms	Combination Oxide Forms	Acid Forms*		
		H_3PO_2	hypo -	
	P_4O_6	HPO_2	meta -	phosphorous acids
P_4O_6	P_4O_7	$H_4P_2O_5$	pyro -	
$(PO_2)_n$	P_4O_8	H_3PO_3	ortho-	
	P_4O_9	$H_4P_2O_6$	hypo -	
P_4O_{10}	P_4O_{10}	HPO_3	meta -	phosphoric acids
		$H_4P_2O_7$	pyro -	
		H_3PO_4	ortho-	

*These acids form various ionic species depending on the pH of the solution.

plasma time studies (Figure 4). These acids, plus $H_4P_2O_5$, gave a blue green color in a molybdate solution when no reducing agent was added. Tests on plasma PSG samples gave a positive blue green color when no reducing agent was added to a molybdate solution.

The second method of our choice capable of identifying and quantifying the acids of phosphorus was ion chromatography (IC), (4, 5). IC is an excellent analytical tool which can chromatographically separate a mixture of components and make it possible to identify and quantify them.

The system used for this study is illustrated in Figure 5. Briefly, the eluent is introduced into the system using the pulseless pump. It then passes through the injection valve where the sample is injected. The sample is pushed into the separator column where the special ion exchange resin retains each ionic species with different strength. The separated sample together with the eluent, passes into the suppressor column where the background conductivity of the eluent is reduced. Finally, the sample passes the conductivity detector where a response to the changes in ionic concentration are visualized by the chart recorder or an integrator.

With the anion detection system on the ion chromatograph and sodium carbonate solutions in various pH ranges as an eluent, an effort was made to distinguish the oxo phosphorus acids that existed in a hydrofluoric acid solution of PSG films. The phosphate and phosphite ions were readily detected and quantitated when compared with known standards of these ions, as shown in Figure 6 (a) - (i). The standard solution gave a single peak for the phosphite and the phosphate ions, with retention times (RT) of 3.05 and 18.07, respectively. The chromatographic spectra of the samples contained as many as seven peaks. Those that were identified were hydrofluoric acid

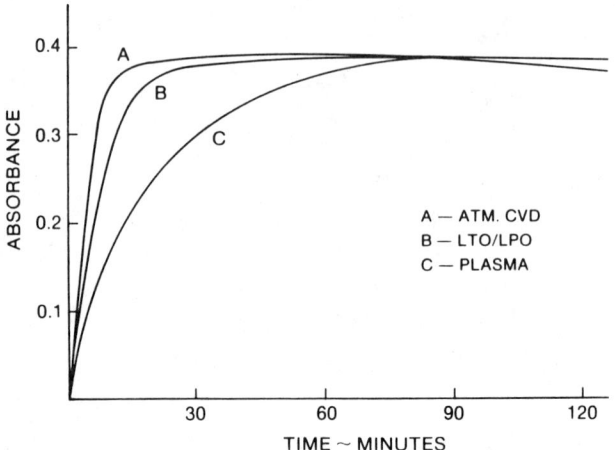

Figure 3. Time study on PSG films molybdate blue reaction.

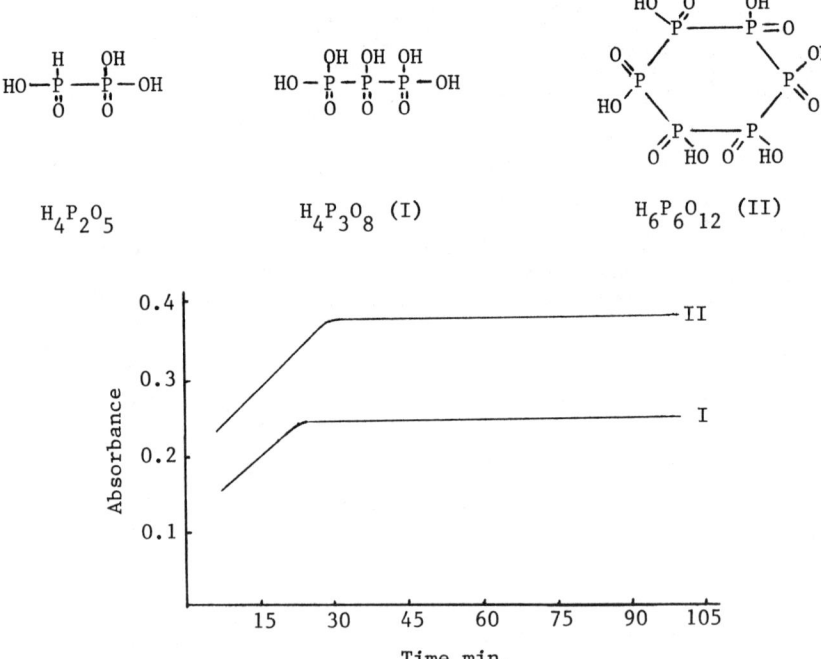

Figure 4. Phosphorus compounds that give positive Mo ion tests without a reducing agent.

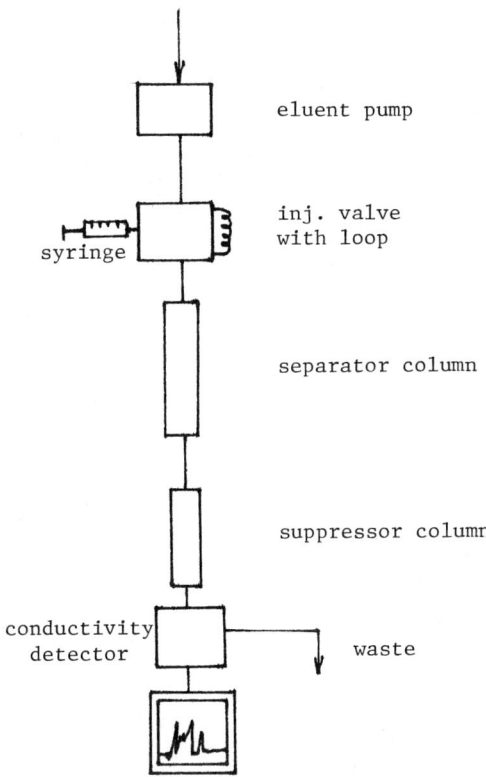

Figure 5. Configuration of the ion chromatograph.

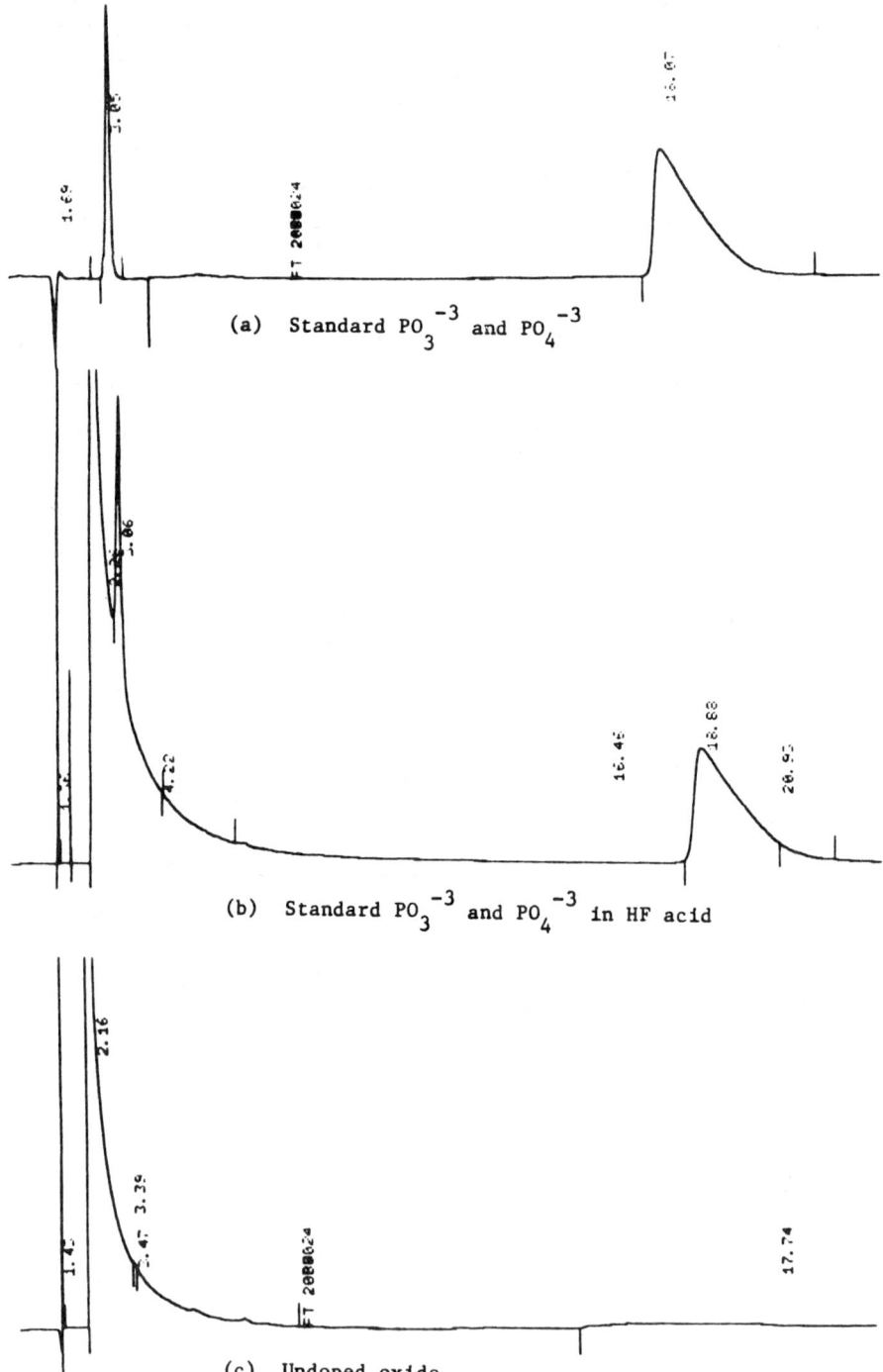

Figures 6a-c. Ion chromatography study.

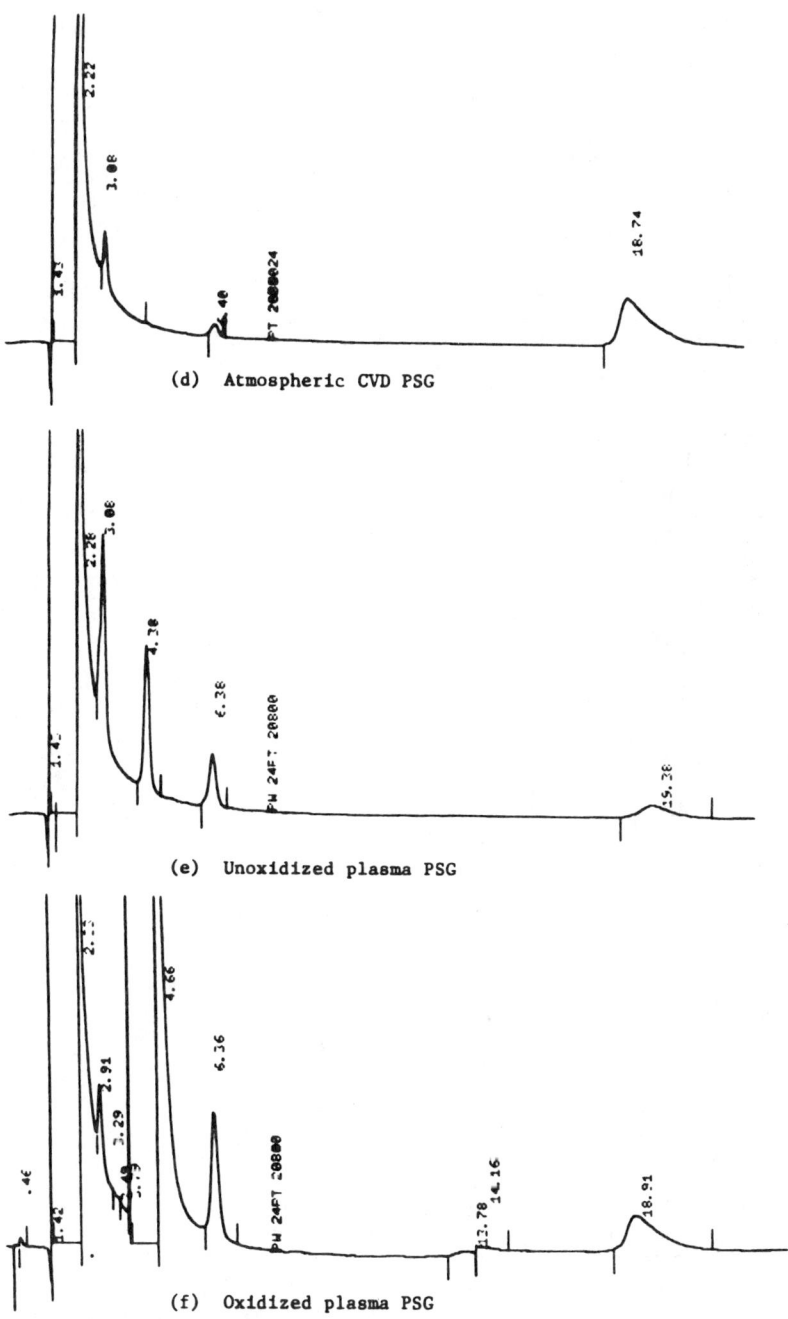

Figures 6d-f. Ion chromatography study.

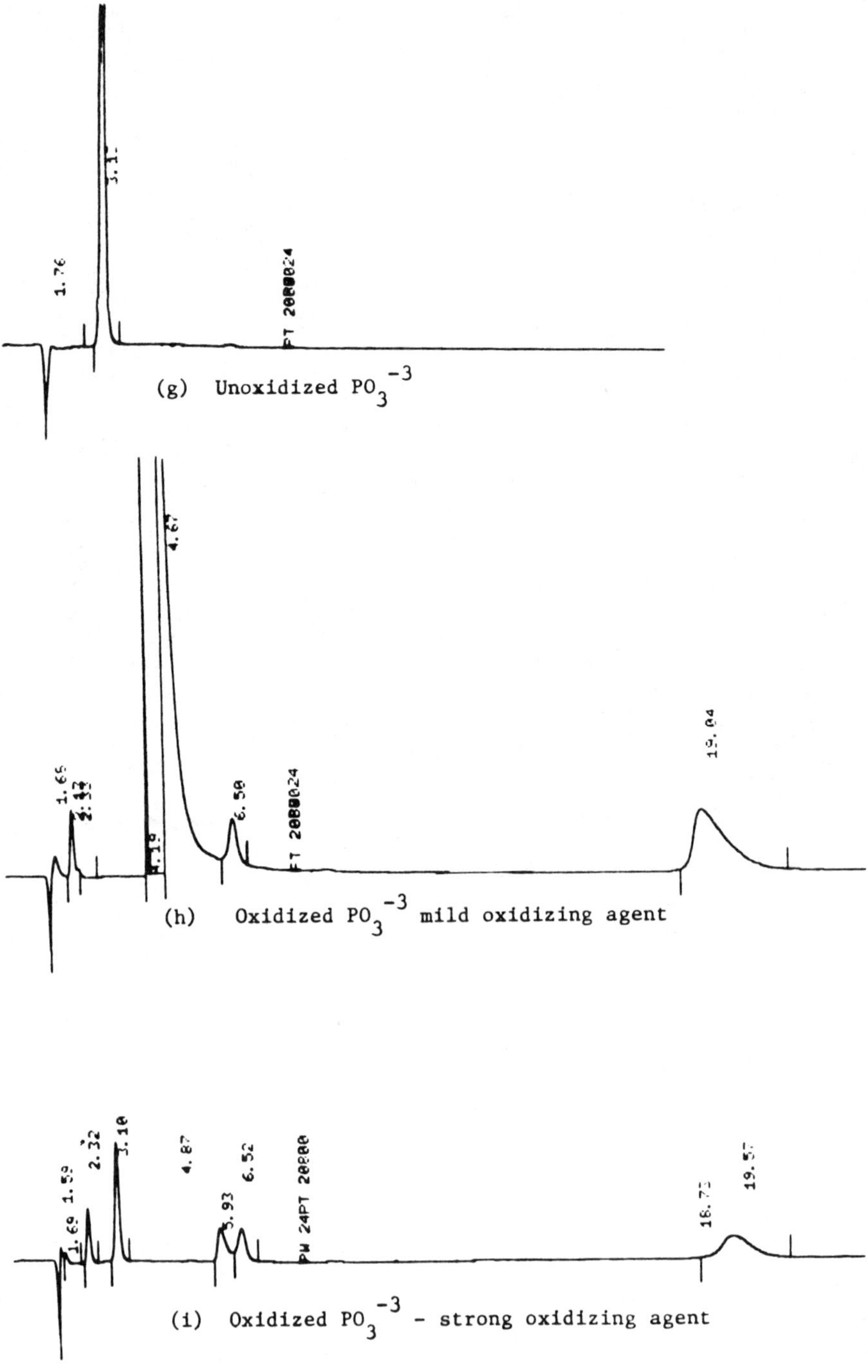

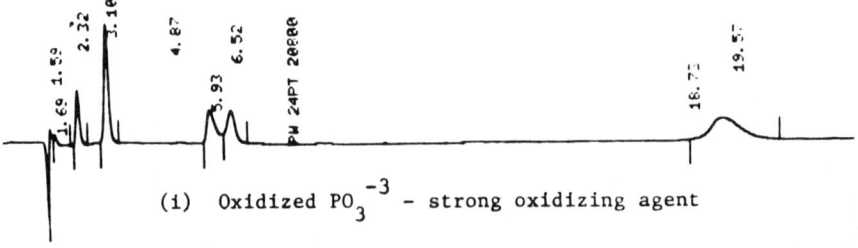

Figures 6g-i. Ion chromatography study.

peaks (RT = 1.43, 2.16) and the known phosphite and phosphate peaks, figure 6 (a) - (c).

Two peaks were found in the plasma films that were not identified and later proved to be important. They were those with RT of 4.38 and 6.38, see figure 6 (e). Atmospheric CVD samples, Figure 6 (d), do not have a peak at 4.38 and the peak at 6.38 is much smaller. Also, the ratio of phosphite to phosphate ion is noticeably higher on plasma wafers.

The disadvantage of chromatographic systems is that compounds can remain on the columns or move down the column so slowly that they are missed. For example, under the conditions for phosphate and phosphite ions, no peaks were found for pyrophosphate or hypophosphite ions. (Other oxophosphorus acids, including those mentioned in Ohashi and Yoza's paper, were not available for this study.) Different chromatographic conditions may reveal these and other phosphorus compounds in a PSG film.

Since only two compounds were found by ion chromatography, whereas the wet chemical data indicated the presence of three compounds, the total quantity of phosphorus found by colorimetry using the molybdenum blue complex procedure with and without oxidizing the samples was compared with the sum of phosphite and phosphate by IC. The results from this study are found in Table II.

Table II. Colorimetry Vs. Ion Chromatography for Weight % Phosphorus

Sample Type	Colorimetry			Ion Chromatography		
	P_2O_5	P_2O_3	Total	P_2O_5	P_2O_3	Total
Atm CVD	4.8	0.0	4.8	4.01	0.78	4.79
LTO/LPO	10.0	0.8	10.8	8.05	0.87	8.92
Plasma Enhanced	ND	ND	5.0	1.34	0.45	1.79
Plasma	3.8	1.1	4.9	1.89	1.35	3.24

ND = Not determined

The same weight percent of phosphorus from atmospheric CVD films was found by both methods. For all plasma, plasma enhanced, and LTO/LPO samples, however, the total percent phosphorus found by IC was significantly less than that found by colorimetry.

Suspecting that the problem was associated with the phosphorus trioxide oxidations, we oxidized and analyzed a sample of phosphorous acid using IC. As expected, a large phosphate ion peak appeared, but a 6.50 peak also appeared, as shown in figure 6 (g) vs. 6 (h). A stronger oxidation process was used, which yielded a complicated IC spectrum, figure 6 (i). Of significance in this spectrum is the appearance of the 6.50 peak.

Whether or not the 4.38 and 6.38 peaks contain phosphorus has not yet been determined. It is believed that they do. However, they may simply be different ionic species of the phosphite ion and as such have different retention times in the IC.

There was a possibility that the IC peaks at 4.38 and 6.38 represent the compound that yielded the positive blue green color with the molybdate ion without a reducing agent. Because of the kind of phosphorus compounds that have this ability have -P-P- bonding, an effort was made to detect this bonding using P-31 NMR.

A plasma PSG film was analyzed. The results, which reveal only two peaks, are shown in Figure 7. However, according to Crutchfield et al. (6), when analyzing oxoacids of phosphorus by NMR, the pH is critical (see Figure 8). Since the original NMR solution had a pH of 3, the sample was analyzed in a basic solution. Again only two peaks were obtained whose values were identical to those found for known solutions of phosphate and phosphite.

This preliminary NMR study needs considerable refining in terms of both controlled pH and detection limits. Without knowing what phosphorus compound may be present in plasma PSG films, this is difficult. The plasma films are our only source of the suspected compound, and because its concentration is very low, questions concerning NMR data could not be resolved.

Using colorimetric reactions developed to identify various species of phosphorus compounds in plasma PSG films, we obtained quantitative results on a variety of samples from three sources (identified as A, B, C). The results are listed in Table III. No attempt has been made to date to quantitate the suspected fourth compound.

Table III. Weight Percent Phosphorus in Plasma Doped Oxides

No.	Total	P_2O_5	P_2O_3	PH_3
A-1	9.5	6.2 (65.3)*	2.9 (30.5)	0.4 (4.2)
A-2	5.6	4.0 (71.4)	1.1 (19.6)	0.5 (9.0)
A-3	5.2	3.9 (75.0)	0.9 (17.3)	0.4 (7.7)
A-4	5.0	4.1 (82.0)	0.4 (8.0)	0.5 (10.0)
A-5	5.4	3.9 (72.2)	1.0 (18.5)	0.5 (9.3)
B-1	4.1	3.3 (80.5)	0.6 (14.6)	0.2 (4.9)
B-2	4.1	3.4 (82.9)	0.5 (12.2)	0.2 (4.9)
C-1	5.0	3.8 (76.0)	1.1 (22.0)	<0.1 (<2.0)
C-2	2.3	1.9 (82.6)	0.4 (17.4)	<0.05(<1.0)
C-3	7.5	6.2 (82.7)	1.3 (17.3)	<0.05(<1.0)

*Numbers in parentheses are relative percents of weight percent phosphorus.

Many LTO/LPO samples have also been studied. The results indicate that they have characteristics similar to plasma PSG. However, the quantity of the partially or unoxidized species is smaller. Like plasma PSG, the LTO/LPO also contain phosphorus pentoxide, trioxide, and phosphine. The 4.38 IC peak was not found in most of these films; however, when it was found, it appeared as a very slight trace.

One interesting result on one set of "atmospheric CVD" PSG films is worth mentioning. This set gave higher colorimetric results on oxidation indicating the presence of suboxides of phosphorus. Not having seen suboxides of phosphorus in atmospheric CVD films, we

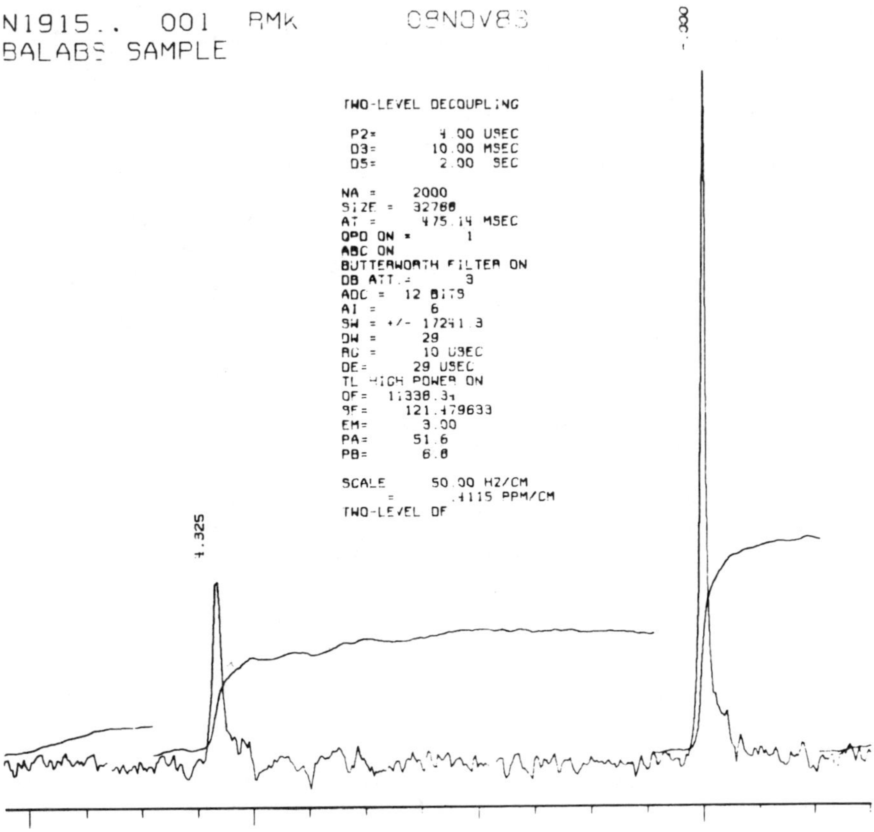

Figure 7. P-31 NMR of plasma PSG film.

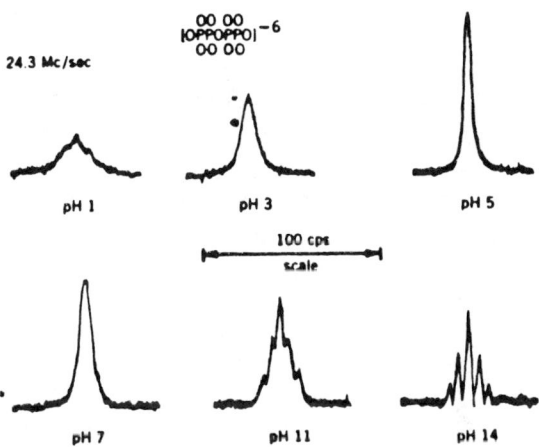

Figure 8. High resolution P-31 NMR.

questioned the process used to make these films. The oxides were grown in an atmospheric system using a silane/oxygen ratio of 1/17 and various amounts of 5% phosphine in argon. The system, however, had been connected to a vacuum system and the temperature used was 410°C. In other words, the system had been converted to an LTO/LPO system and gave results accordingly.

In conclusion, it has been shown that plasma, plasma enhanced, and LTO/LPO PSG and BPSG films contain phosphorus pentoxide, phosphorus trioxide, and phosphine. Although not discussed, we have found that BPSG films act the same as PSG films in this regard. There is strong evidence by both colorimetry and ion chromatography that a fourth compound exists. The fourth phosphorus compound that is indicated by colorimetry may or may not be the same as that causing the 4.38 peak in ion chromatography. The data to date on the presence and identification of this compound are ambiguous and need further study.

The known compounds in PSG films can be quantitatively measured colorimetrically. The ion chromatographic data based on the measurement of phosphate and phosphite ion do not give accurate results for total weight percent of phosphorus. The difference between the IC results and colorimetric results is believed to be due to unidentified or missing peaks in the IC that contain phosphorus. Until these compounds are identified, the most reliable method for measuring the known phosphorus compounds and total weight percent phosphorus appears to be colorimetry.

Literature Cited

1. See for example Electronic News, Sept. 6, 1983; Microelectronics Measurements and Testing, Oct. 1983; Semiconductor International, Nov. 1983 p 44; Solid State Technology, Nov. 1983 p 32.

2. Yoza, N.; Ohashi, S. Bull. Chem. Soc. Japan 37 (1) 33-40 (1964).

3. Ohashi, S.; Yoza, N. Bull. Chem. Soc. Japan 36 (6) 707-712, (1963).

4. Small, H.; Stevens, T.S.; Bauman, W.C. Anal. Chemistry 47, 1801, (1975).

5. Marmion, D.M. ASTM Standardization News 22, (1983).

6. Crutchfield, M.M.; Dungan, C.H.; Letcher, J.H.; Mark, V.; Van Wazer, J.R. "Topics in Phosphorus Chemistry, P-31 Nuclear Magnetic Resonance"; Interscience, Division of John Wiley and Sons, New York, 1967; Vol 5, p 15.

RECEIVED November 12, 1985

21

Process Control of Vacuum-Deposited Nickel-Chromium for the Fabrication of Reproducible Thin-Film Resistors

Vineet S. Dharmadhikari

Harris Semiconductor, Melbourne, FL 32901

As a part of ongoing efforts to establish a manufacturing technology for the production of thin film resistors, a statistical control of the materials and the manufacturing process was developed. Crucial to this goal was the thorough understanding of the nichrome thin films and the process development. A detailed study of the material properties which influence the resistor performance, such as deposition rate, film thickness, temperature coefficient of resistance and composition was performed. The process development encompassed the optimization of the process equipment parameters of power, ramp, deposition geometry and the base pressure. The reproducibility and stability of the process was confirmed by monitoring the sheet resistance and composition of the deposited nichrome layer from batch to batch. Statistical techniques were used to detect the sources of variation in the process and determine the degree of adjustment required.

Electrical properties of thin films have been studied for several decades and it's not surprising that a number of industrial processes and end products have been developed to exploit the characteristics of thin films. One such development led to the use of nickel-chromium (Ni-Cr) thin films in the manufacture of thin film precision resistors on the surface of a monolithic integrated circuit [1-2]. This transition metal alloy exhibits wide range of resistivity, low temperature coefficient of resistance and high stability [1-12], however it is difficult to mass produce. This is due, primarily, to the difficulty of obtaining well-defined reproducible structures by normal methods of production, such as thermal evaporation. Additionally, the added complications of surface scattering, film morphology, contamination of residual gases, alloy composition and purity do not represent a well-defined state and may also encourage a time dependent phenomena.

0097-6156/86/0295-0333$06.00/0
© 1986 American Chemical Society

Although the electrical response to thin nichrome films represent many interesting problems, this paper is mainly concerned with the statistical control and optimization of the manufacturing process which has received very little technological attention, even though the statistical concept and methodology involved in manufacturing are undeniably high. There exists a large amount of experimental data on the different aspects of resistor fabrication and characterization (3-12), but there is no coherent presentation of the totality of the information nor even of significant parts of it.

We present here an effective way to manufacture a reproducible thin film resistor, using accepted statistical techniques. The process requirements, experimental techniques, process development and characterization, examples of control charts, their interpretation for rapid appraisal and the impact of statistical techniques on the production yields and future developments will be described.

Process Background and Requirements

In monolithic integrated circuit technology, all circuit elements are fabricated on, or within, a single crystal piece of semiconductor material by means of a sequential series of diffusion, photolithography and thin film processes. As a production method, generally where thin film resistors are required, the following process sequence is defined:
1. photoresist application and patterning on single-crystal semiconductor substrates with the active devices already formed;
2. thin film deposition;
3. photoresist etching or lifting;
4. contact metal deposition;
5. contact pattern delineation, photoresist metal etch;
6. high temperature stability bake, 300-400°C, 30-60 min.

Thin Film Deposition Procedure. The nickel-chromium films were thermally evaporated in a cryopumped vacuum system. The composition of the initial charge was 80 Ni, 20 Cr. The literature indicated a difficulty in controlling the composition of the films from batch to batch, mainly because of the differences in the vapor pressure of nickel and chromium. Another problem indicated was the severe reactivity of nickel at high temperatures with tungsten boats. Initial efforts indicated the need for an in-situ-resistance monitoring capability to ensure repeatable results. This capability was provided as well as a shutter to control the composition. Pre-cleaned glass substrates with gold contacts were used as monitors. Once initiated, deposition continued until the desired value was indicated by the resistance monitor, at which time the shutter was closed and power turned off. The chamber was then brought up to atmosphere pressure by back filling with nitrogen. The product wafers or substrates were then removed and the monitor resistance (pre-bake) measurements carried out. Since the thin film resistors undergo a high temperature stabilization bake (300 to 400°C) during processing, to simulate the time and temperature of the subsequent process steps, the monitors were also baked for 30 minutes at 300°C, and the sheet resistance (post-bake) measured. The composition of the films was measured by x-ray fluorescence technique and the thickness of the films was determined by a Talystep profilometer.

The required specifications for our thin film resistor parameters are listed in Table I. Any product that deviates from the limits has to be reworked or scrapped.

Process History. The state of the process before any statistical process control existed is shown in Table II. As can be noted, over a period of five months only 44% of the time on an average was actually spent on production and of that about 10% of the product wafers were reworked. Rework required additional material, labor and time and at times, affected the production schedules adversely. Because of this unpredictable nature of the process, an accepted statistical technique was badly needed to establish the process and aid in producing quality product from batch to batch.

Statistical Analysis. One of the greatest contributions of the statistical method was in identifying the sources of problems. It was realized that 80% of the problem could be traced directly to the process or the evaporation system. Only 20% of the problem was related to special causes which were traceable to the individual operator. Figure 1 is the pareto chart analysis of the different variables that actually caused the percent (%) reworks reported in Table II. Post-bake sheet resistance and chromium percentage (% Cr) contributed the most, followed by the deposition time and pre-bake sheet resistance. The respective histograms for these variables are shown in Figures 2 and 3. A number of features are apparent. First, except for the pre-bake histogram, which is skewed towards the lower end, none of the histograms look normal. Secondly, the histograms are offset from the expected average value and the variability (i.e. the base of the histogram) is very large. Clearly much emphasis was needed in a process development and characterization.

Process Development and Characterization

The effect of the major process variables on the resistor performance was investigated. We would like to mention that since this was a production oriented investigation, every effort was made to optimize the process in terms of production efficiency. The most important parameter seemed to be the control of the film composition. In Figure 4, the variation in pre-bake and post-bake sheet resistance with % Cr is shown. Several features can be seen. The pre-bake shows a minimum with % Cr in the vicinity of 50%. For % Cr less than ~50%, the pre-bake increases with a negative slope, whereas for % Cr greater than ~50%, it increases with a positive slope. In contrast, the post-bake shows an entirely different trend. For % Cr greater than ~51%, the post-bake is nearly constant within experimental errors, whereas for % Cr less than ~51%, it decreases linearly with decrease in % Cr from ~51 to ~38%. This data was fed into a stepwise regression program in order to build an equation relating % Cr to post-bake. The specific equation arrived at for % Cr and post-bake was

$$\text{Post-bake} = 172\,(1-e^{-0.153\,(\%\ Cr - 31.64)})$$

Where Post-bake is in ohms/□ & % Cr in Wt. %.

Table I. Thin Film Resistor Requirements

Parameter	Value	Units
Sheet Resistance	200 ± 20	Ohms/sq.
TCR	$\leq \pm 50$	ppm/°C
Ratio Tracking TCR *	± 1	ppm/°C
Initial Match at 25°C *	± 0.25	%
Long Term Stability 1000 hr/125°C	± 0.1	%
Long Term Ratio Tracking *	± 24	ppm

* over 3σ limits

Table II. Summary of the Nickel-Chromium Process before Process Control

Months	Total No. of Evaporations	Production Wafers Shot (%)	Wafers Reworked (%)
11	192	46.88	10.61
12	335	48.06	12.54
01	178	51.68	9.74
02	226	32.32	7.21
03	315	38.73	9.92
		$\bar{x} = 43.53 \pm 7.86$	$\bar{x} = 10.01 \pm 1.92$

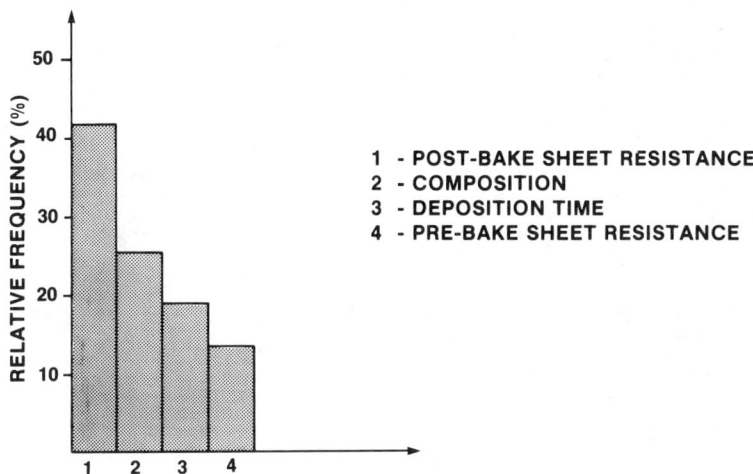

Figure 1. Pareto chart analysis of variables that caused rework reported in Table II.

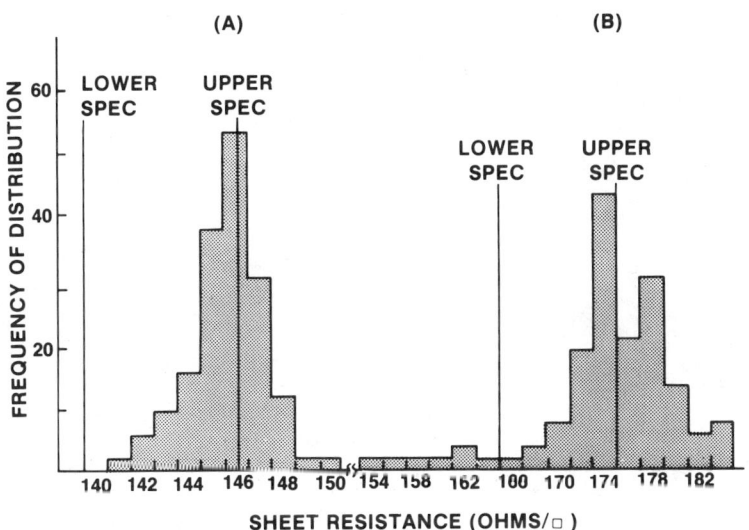

Figure 2. Distribution of sheet resistance of thin films before process control for a period of one month. (A) pre-bake and (B) post-bake.

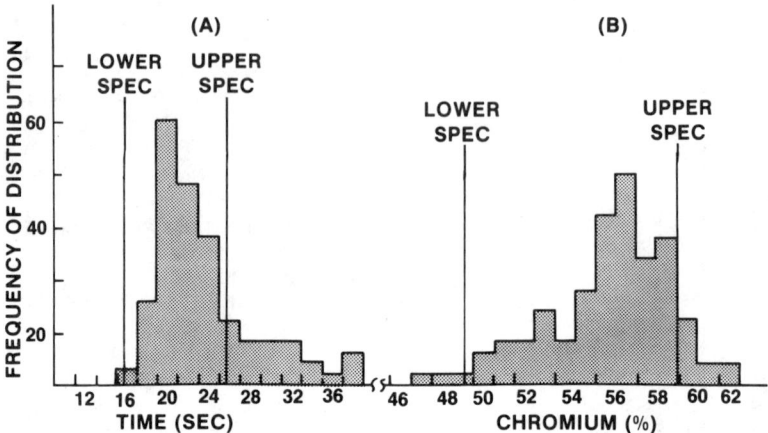

Figure 3. Distribution of (A) deposition time and (B) percent chromium for a period of one month, before process control.

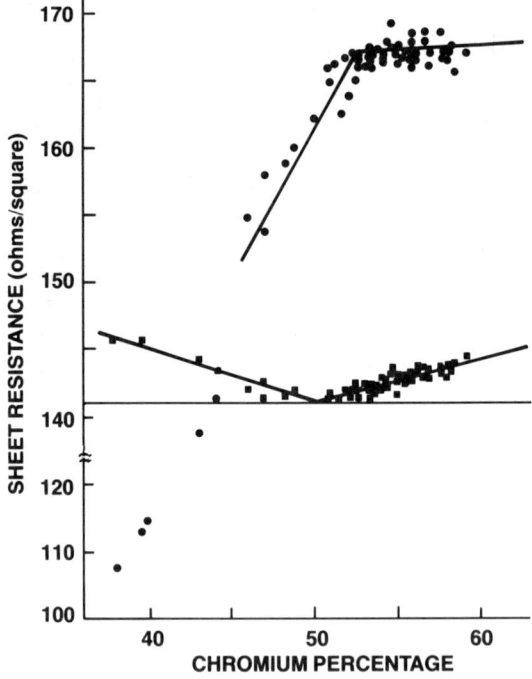

Figure 4. Variation in pre-bake (■) and post-bake (●) sheet resistance with percent chromium for a fixed value of the resistance monitor. The films were baked at 300°C for 30 min. in air.

The behavior of temperature coefficient of resistance with % Cr is shown in Figure 5. The curve can be divided into three regions: i) with % Cr greater than 52%, ii) % Cr between 52 to 48%, and iii) % Cr less than 48%. In region i) the TCR is practically independent of % Cr and varies between -5 to -30 ppm/°C. In region ii) again the TCR is low and as the % Cr decreases, the TCR changes sign. This change takes place in the vicinity of 51 to 52% Cr. In region iii), however, the TCR abruptly increases with little decrease in % Cr. These results again demonstrate that to be in specification, it is better to have % Cr between 50 to 60%.

The effect of oxidation on the sheet resistance and TCR is shown in Figure 6. For a Ni-Cr film with ~54.5% Cr and baked at 300°C, the sheet resistance increases by ~22%, while the TCR decreases from +52 to -10 ppm/°C for a 30 min. anneal in air. In contrast to the decrease in TCR caused by oxidation, for a Ni-Cr film with ~45% Cr (not shown here), the oxidation increased the TCR. For example, the sheet resistance decreased by 5%, whereas the TCR increased from +78 to +146 ppm/°C.

The second factor of critical importance observed is maintaining the chemistry of the film which is being deposited onto the substrates, i.e. to control the deposition rate and the partial pressure of the gases in the deposition chamber. The simplest means of controlling the partial pressure of residual gases is to use the same deposition pressure for each evaporation.

In addition to this, a check on the total residual gas composition before, during and after evaporation was occasionally performed using quadrapole residual gas analyzer (RGA) traces. The main components of the RGA spectra at a base pressure of $\sim 2 \times 10^{-6}$ torr were water, atomic and molecular hydrogen, carbon monoxide, carbondioxide, nitrogen and oxygen. Initial test runs showed a correlation of the film properties with residual partial pressures. Thus, over a limited pressure range, the sheet resistance varied as a power function of the base pressure. However, the effect on the post-bake as compared to pre-bake sheet resistance was more pronounced.

The changes in the above mentioned variables have produced enough information to obtain an indication of trends. This information was used for developing a suitable process control for the Ni-Cr thin films:
o Deposition pressure had a definite effect on the film properties. A good starting point was to use the same base pressure for each run.
o The results of % Cr variation indicated workable limits for meeting the specification on sheet resistance and TCR.
o The pre and post-bake sheet resistance behavior indicated a different trend. The results showed that post-bake - % Cr interaction was a better choice for process monitoring.
o The deposition time which was affected by the power, which also affected the % Cr had to be controlled. Further refinement was obtained by ramping the power to the desired value.

Based on these results, a set of process limits termed as "Engineering Tolerance" limits were chosen to apply the statistical techniques and begin control charts on these parameters, which had reasonable characteristics of a production process.

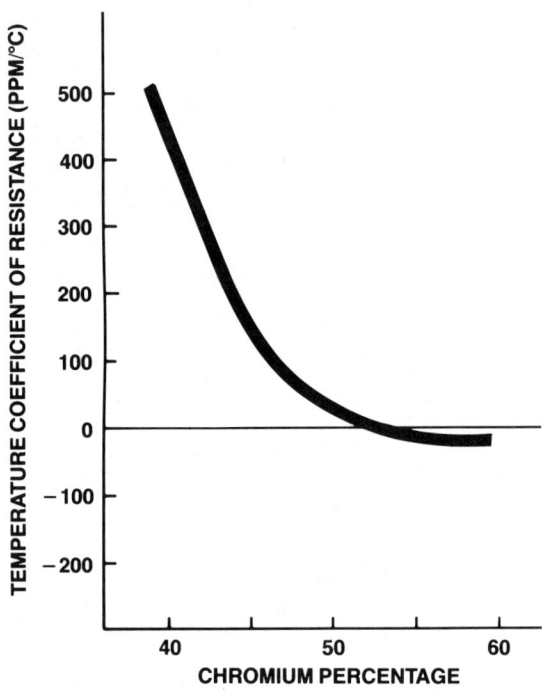

Figure 5. Change in TCR with percent chromium.

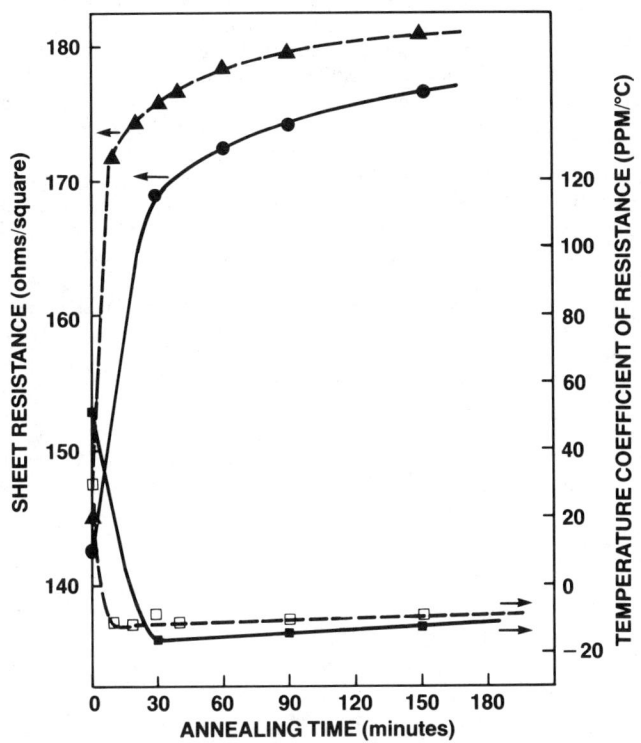

Figure 6. Effect of annealing time on sheet resistance and TCR for films with percent chromium of 54.5% (●,■) and 55.5% (▲,□) respectively.

Control Charts

One important tool in statistical quality control is the shewhart control charts (13-14). The power of the shewhart techniques lies not only in separating out assignable causes of quality variations, but they tell us when to leave a process alone and thus prevent unnecessarily frequent adjustments that tend to increase the variability of the process rather than to decrease it. However, the often troublesome part in any organization during the introduction of the control chart techniques is the choice of variables and types of charts. In selecting variables for applications of control chart techniques, it is important to choose those variables which can prevent scraps and will be recognized by anyone in a manufacturing or management capacity as having economic significance.

As discussed in section 2, the post-bake sheet resistance and % Cr caused the most reworks and scraps. Therefore, control charts were initiated to control these variables. Once the variables were selected, the next important step was to select a subgroup size. Generally speaking, subgroups should be selected in a way that makes each subgroup as homogenous as possible and that gives the maximum opportunity for variations from one subgoup to another. There is a trade off between subgroup size and sensitivity (15). The smaller the subgroups size the sooner the information obtained will provide a basis for action. However, the assurance of whether the basis is sound or not will be less. Moreover, where measurements are available slowly and only one measurement is obtained at a time (as is the case in the present study), it's a natural desire to use a minimum subgroup size. Hence, in present case, a subgroup size of two was used to calculate the control limits for the average ($\bar{x}$) and range (R) charts. The method for calculating the limits is shown below:

$$\text{(UCL) Upper Control Limit}_{\bar{x}} = \bar{\bar{x}} + A_2 \bar{R} \qquad (1)$$

$$\text{(LCL) Lower Control Limit}_{\bar{x}} = \bar{\bar{x}} - A_2 \bar{R} \qquad (2)$$

where $\bar{\bar{x}}$ is the average of $\bar{x}$ values, $\bar{R}$ is the average of the range values between the measurements, and A_2 is a constant dependent on sample size (13-15).

Similarly the control limits for the process variability, i.e. R chart, are

$$UCL_R = D_4 \bar{R} \qquad (3)$$

$$LCL_R = D_3 \bar{R} \qquad (4)$$

where D_3 and D_4 are constant dependent on sample size (13-15).

Using these equations, control charts for $\bar{x}$ and R were generated for all the dependent variables mentioned in section 2. However, for lack of space only the control chart for % Cr is presented here. Figure 7 is the $\bar{x}$ and R chart for % Cr. These data are from a batch taken a few weeks after the process control was started. There is no evidence of the process being out of control on the R chart. This shows that the process variability or dispersion is in control. However, on $\bar{x}$ chart, even though there are no points outside the 3σ-control limits, the distribution of the points about the central

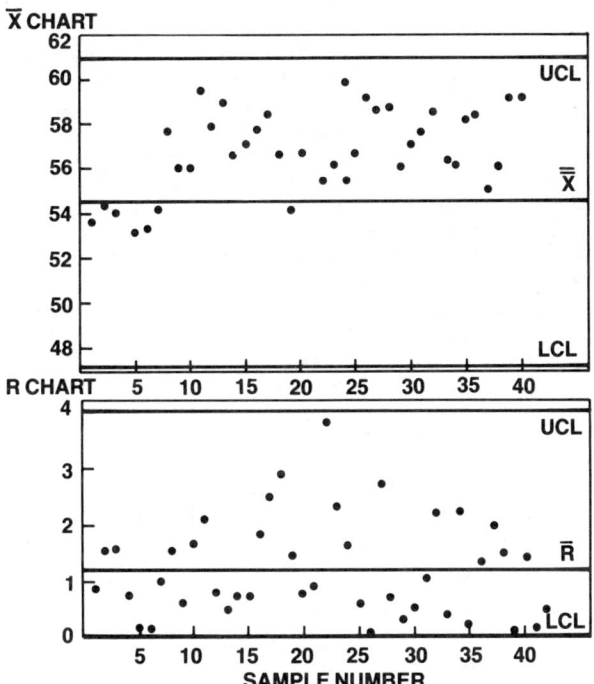

Figure 7. A segment of the control chart for x and R showing the changes in percent chromium from shot to shot. This data taken after process control was initiated.

line is not even, i.e. this chart fails the runs test (13-16), showing that the process is out of control on the average. The histogram in Figure 8(a) also shows the same features. The process average has shifted and is above the expected mean.

These results demonstrate that there are production processes that tend to have relatively uniform dispersion even though the process mean changes from time to time. In such processes, lack of statistical control will be detected from the $\bar{x}$ chart. Further, Figure 7 also shows that processes have an inherent natural variation. Attempts to adjust the process based on the results of each subgroup will drastically increase the process variability. This is shown in Figure 8(b). The process standard deviation increased from 1.08 (Fig. 8a) to 2.02 (Fig. 8b).

However, once a state of control was established, the next problem was maintaining such a state of control over time. At this stage, it was also appropriate to do the process capability analysis (17) to see if the process was capable of meeting the specifications. This is usually done by comparing the engineering tolerance (ET) limits with the natural tolerance (NT) limits. Taking the data presented in Figure 8(b) as an example, the NT obtained ~12.14 is greater than the ET (~10), showing that the process is not capable of meeting the specification. On an average, about 5% of the parts were out of specification. As mentioned earlier, a number of factors are responsible. Possibilities are that the process average has shifted, or the process variability needs to be reduced or a need exists to change the specifications. We opted to reduce the process variability. The influence of this is shown in Figure 9.

Process Influence and Improvements

After the process was brought into statistical control, the next task of the control chart was to help continue this state of control over time. This involved simply leaving the process alone as long as it stayed in control and when lack of control was indicated, hunting for and removing assignable causes of variation.

The improvement brought about by using the statistical techniques to establish a process is shown in Figure 10, where % defects out of total number of wafers produced before and after process control was instituted is shown. It is clear that the % defects, after the process control was introduced have reduced dramatically.
Table III summarizes a few of the other influences on the process control. Before process control (see Table II) on an average only 44% of the time was spent on actual production, whereas the rest of the time was spent in qualifying and calibrating the system for production. Even then 10% (on average) of the product wafers were reworked. After the statistical process control was initiated not only the reworks went down (~2.5%), but ~63% of the time was now spent on the actual production. Thus, resulting in a straight increase of ~55% in the production. Achieving this objective not only helped in meeting the production schedules, but also helped in minimizing the cost of maintaining the process. Less machine adjustments were required. The operators could control the process on their own. It also gave an identity to the process and made the engineering time available for other research and development.

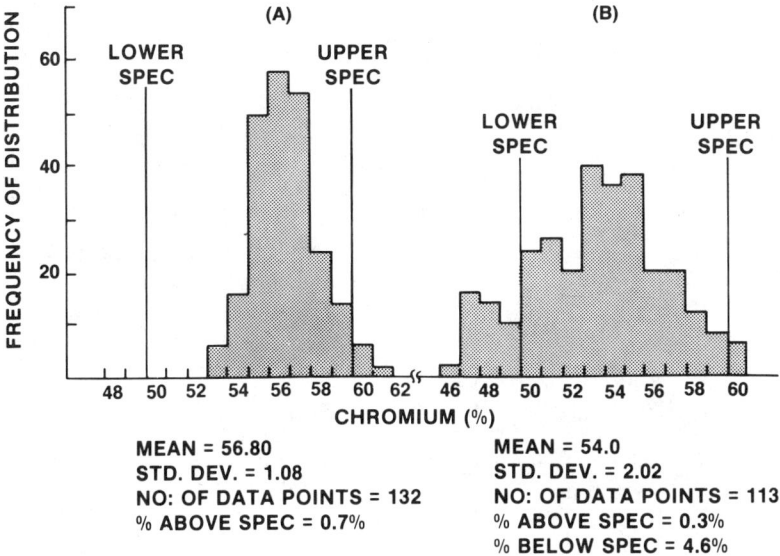

Figure 8. Distribution of percent chromium. (A) data of Figure 7 and (B) showing the influence on the distribution when too frequent adjustments are made in the process settings.

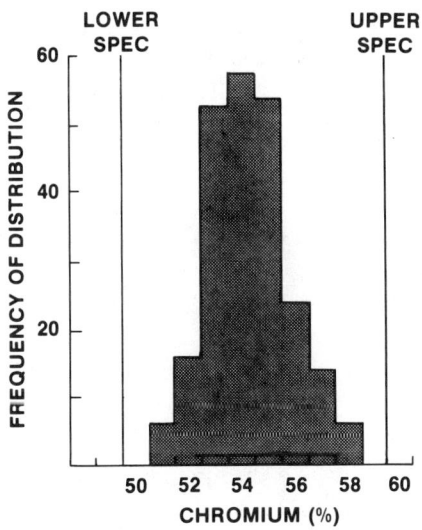

Figure 9. Distribution of percent chromium when a good process control was established.

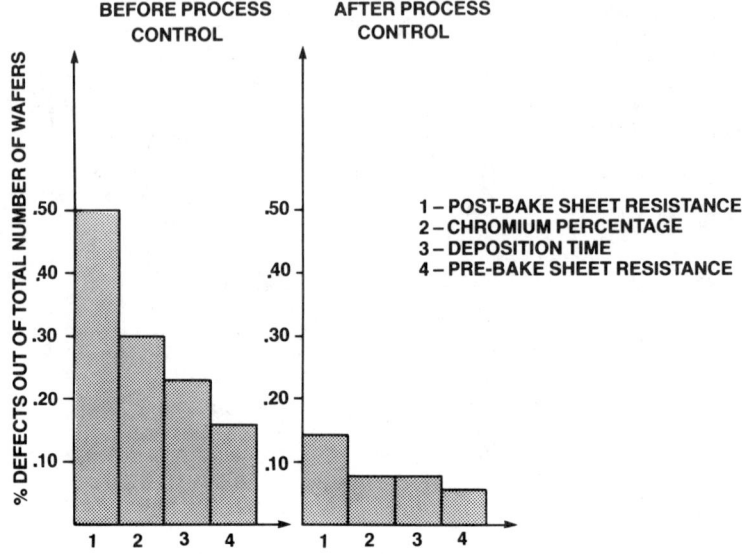

Figure 10. Pareto chart analysis of the percent defects out of total number of wafers produced before and after process control.

Table III. Summary of the Nickel-Chromium Process after Process Control

Months	Total No. of Evaporations	Production Wafers Shot (%)	Wafers Reworked (%)
2	170	60.9	5.00
3	216	69.91	3.40
4	181	62.43	3.18
6	131	64.12	0.0
8	311	57.23	0.18
		$\bar{x} = 62.74 \pm 4.78$	$\bar{x} = 2.44 \pm 2.23$

In addition to the above mentioned merits, it also helped in identifying new variables that were not detected before, such as controlling the power output, initial charge and tungsten boat dimensions used for each evaporation, system geometry and preventive maintenance schedules.

It should be noted that if further refinement or improvement in the process variability is needed, the precise contribution of the above mentioned starting material to the process variability must be understood. Unfortunately, during the course of this study the process engineering group had very little control on this, since most of the vendors were unaware of the statistical concept. However, it is hoped that such work in the future will promote a dialog between vendors and processing which will result in a mutually satisfactory resolution of this problem.

Concluding Remarks

There is no doubt that the electrical response in thin Ni-Cr films represent a wide range of characteristics. There is still the possibility of establishing a manufacturing process for Ni-Cr using thermal evaporation which is amply confirmed by the present analyses. A good understanding of the fundamentals of Ni-Cr process together with the statistically controlled process has given us an entirely new understanding of the Ni-Cr response, which previously appeared as a complex and seemingly arbitrary process.

Throughout the present treatment, we have been stressing the use of statistical techniques and control charts. This does not imply that control charts can change processes or solve the problems. They are tools that provide a road map to assist in solving problems at a low cost to products and services. It is people working in conjunction with these tools that change and improve the processes.

Finally, it has been the intent of this article to present some simple statistical techniques used in manufacturing of Ni-Cr process. Actual data was used to provide examples for their application. Although this paper explicitly dealt with Ni-Cr thin films, it is felt that the basic experimental strategy used here can be applied to any IC manufacturing process of thin films.

Acknowledgments

I would like to thank Harris Semiconductor for giving me the permission to publish this research report. An acknowledgment also is due to W.M. Connor for his constant encouragement throughout the work, and to B.V. Phillips for his help in the instrumentation and to A. Grabow for typing the manuscript.

Literature Cited

1. Moore, G.E. "Microelectronics" Keonjin, E., Ed.; McGraw-Hill: New York, 1963; p. 316.
2. Polata, B. Proc. Third Microelectronics Symposium IEEE, 1964, 111-C-1 (1964).
3. Maissel, L.I.; Glang, R. "Handbook of Thin Film Technology"; McGraw-Hill: New York, 1970; chap. 18.

4. Lassak, L.; Hieber, K. Thin Solid Films, 1973, 17, 105-111.
5. Campbell, D.S.; Hendry, B. B.J. Appl. Phys., 1965, 16, 1719-25.
6. Tavzes, R.; Kansky, E.; Banovec, A.; Zalar, A.; Gregon, M.; Barna, A.; Pozsgai, I. Thin Solid Films, 1976, 36, 33-41.
7. Dhere, N.G.; Vauide, D.G. Thin Solid Films, 1979, 59, 379-382.
8. Nocerino, G.; Singer, K.E. J. Vac. Science and Technology, 1979, 16, 147-150.
9. Hegner, F. Thin Solid Films, 1979, 57, 359-62.
10. Nahar, R.K.; Devashrayee, N.M. Electrocomponent Science and Technology, 1983, 11, 43-51.
11. Rolke, J. Proc. International Microelectronics Conference, Tokyo, 1980, p. 12-19.
12. Thiel, R.A.; Maures, E.M. IEEE Trans. on Components, Hybrids & Manuf. Technology, 1979, CHMT-2, 467-75.
13. Grant, E.L.; Leavenworth, R.S. "Statistical Quality Control"; McGraw-Hill: New York, 1980.
14. Shewhart, W.A. "The Economic Control of Quality of Manufacturing Product"; Van Nostrand, 1931; American Society for Quality Control, 1980.
15. Duncan, A.J. "Quality Control & Industrial Statistics"; R.D. Irwin, Inc.; Homewood, Ill.; 1974.
16. Roberts, S.W. Bell System Technical Journal, 1958, 37, 83-114.
17. Clifford, P.C. J. Quality Technology, 1971, 3.

RECEIVED June 28, 1985

Characterization of Spin-On Glass Films as a Planarizing Dielectric

Satish K. Gupta and Roland L. Chin

Electronic Chemical Products, Allied Chemicals, Buffalo, NY 14210

Substrate planarization is a critical requirement in the fabrication technology of state-of-the-art integrated circuits (ICs). After a brief review of the existing planarization techniques the use of spin-on glass (SOG) films, which have the inherent quality of planarizing underlying topography, as an interlevel dielectric is described. The physical, chemical, and dielectric properties of two SOG materials, ACCUGLASS 203 and 305, were investigated. It is shown that SOG films of 2-3000Å thickness yield adequate smoothing of one micron high features and that their material and processing characteristics are compatible with IC processes. The chemical and structural characteristics of SOG 203 were further studied with a variety of spectroscopic techniques. The SOG/silicon interface was found to be structurally similar to that between thermal SiO_2 and silicon.

The fabrication of high density LSI and VLSI circuits entails the use of multilevel interconnect systems with concomitant isolation of metallization levels by dielectric layers. Conventionally, chemical vapor deposition (CVD) of undoped or phosphorus-doped SiO_2 glass films has been employed for interlevel dielectric insulation ([1-2]). CVD dielectric films provide only a conformal, at best, coverage of substrate features and often form retrograde step profiles. Therefore, they are not conducive to continuous and uniform step coverage by the overlying metallization layer. The poor step coverage results in discontinuities and thin spots in the conductor lines causing degradation of metallization yields as well as device reliability problems ([3]). The problem becomes worse as the lateral dimensions shrink further and sidewalls become steeper to accommodate the higher packing densities of the advanced ICs. The higher aspect ratio of the interconnect lines, additionally, places severe demands on the subsequent microlithographic processing for fine-line definition.

0097-6156/86/0295-0349$06.00/0
© 1986 American Chemical Society

Thus, substrate planarization has become an essential requirement in the fabrication technology of the state-of-the-art IC chips.

Over the years, a number of techniques for planarizing interlevel dielectric layers have been developed. The so-called reflow glass process (4) consists in depositing a 6-8 w/o P phosphosilicate glass (PSG) and annealing at a temperature of 1000-1100°C. Smoothing of sharp step profiles occurs as a result of glass flow caused by the lowering of the softening point of SiO_2 with the incorporation of phosphorus. In order to minimize dopant redistribution in the shallow source/drain junction employed in the fabrication of advanced IC devices, glass flow temperatures well below 1000°C are required. Since the concentration of phosphorus in PSG can not be increased much above 8 w/o P without causing aluminum corrosion problems, the use of phosphogermanosilicate glasses (5) or borophosphosilicate glasses (6) having flow temperatures as low as 800°C has recently been proposed. The reflow technique can not be applied on aluminum metallization, however, where processing temperatures below 500°C only are permitted. Several other planarization techniques available utilize plasma or sputter etching processes. The "etch back" method (7) is based on the inherent property of spun-on films to planarize underlying substrate. It consists in coating the substrate with a sacrificial layer of an organic polymer, such as polyimide or positive photoresist, and curing appropriately to obtain a planar surface followed by plasma, reactive ion, sputter, or ion-beam etching. The etch conditions are selected so that the dielectric and the organic layers are removed at the same rate. The sputter-etching planarization technique (8) exploits the angular dependence of sputter yields and the fact that sputter yield is maximum at step edges. The principle has also been applied in using bias sputtered quartz (9) as a planarization layer. The latter group of planarization techniques suffer from the drawbacks of low throughput and dependence on complex and expensive equipment.

Spin-On Dielectrics

Spin-on glass (SOG) films are desirable as an interlevel dielectric because of their inherent ability to planarize underlying topography. Depending upon the film thickness obtainable and the material characteristics, SOG films can be utilized either as a stand-alone interlevel insulation layer or as a smoothing layer in conjunction with conventional CVD dielectric layers. The various schemes possible with the application of SOG films in interlevel insulation are illustrated in Figure 1.

In the stand-alone application (Figure 1a) the SOG film must have good dielectric characteristics and a thickness range of 1-2 microns. No such SOG materials are currently commercially available. Polyimides have adequate electrical characteristics and can be spin-applied into films up to several microns thick. However, they have not been widely adopted because of difficult processing required in their use and certain inherent limitation in material characteristics. The smoothing layers schemes (b), (c), and (d) of Figure 1 require a thickness of only 1000-3000Å for the SOG film. The electrical properties of the SOG film in these structures are relatively less critical since the bulk of the dielectric function is

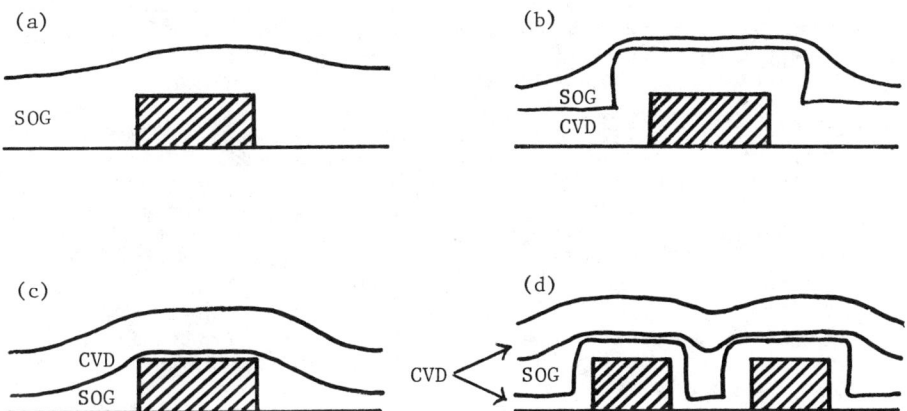

Figure 1. Spin-on glass films as planarization layers.

provided by the thicker CVD layer. Although the schemes (b) and (c) apparently yield equivalent results, the processing characteristics of the two layers, such as etch rate and adhesion, determine which scheme should be employed in a given situation.

We have developed several SOG materials for application as a smoothing dielectric. In this paper the physical, chemical, and processing characteristics of two of these materials, ACCUGLASS 203 and ACCUGLASS 305, are described. Chemically, SOG 203 films consist of essentially pure SiO_2 while those of SOG 305 are classified as siloxane polymer.

Results and Discussion

<u>General SOG Characteristics</u>. Both the SOG products are a solution of hydrolyzed alkoxysilanes in common organic solvents such as alcohols and ketones. The solutions are filtered through 0.2 micron membrane filters and do not form particles or crystals on standing for long periods of time or during the spin-on process. Trace metal contamination levels of the SOGs were determined by atomic absorption spectrophotometry to be well below 0.5 ppm for K, Fe, Cr, Cu, Ni, and Mn and less than 0.1 ppm for Na.

<u>Spin-on and Cure Procedure</u>. The spin-on applications of SOG were carried out with a manual photoresist spinner (Headway, Carrolton, Texas) under a laminar flow hood. In order to minimize the 'splash back' problem, spray of the substrate by the mist or droplets formed by the SOG solution thrown off against the spinner bowl wall, it was necessary to apply a downward exhaust through the bottom of the bowl. The films were spun on 3 in. or 4 in. diameter single-crystal silicon wafers using an acceleration of 20,000 rpm/sec^2 and a spin-time of 20 sec. at the desired speed. For most of the characterizations, the SOG film was cured at 100°C for 15 min. followed by 400°C for 60 min. in ambient air.

<u>Physical Characteristics</u>. The cured film thicknesses as a function of the spin speed are listed in Table I. Thickness was measured with a Rudolph Research AutoEL-III ellipsometer and a Nanometrics Nanometer thickness analyzer.

Table I. Film Thickness as a Function of Spin Speed

Spin Speed, rpm	Film Thickness (400°C cure)*, Å	
	ACCUGLASS 203	ACCUGLASS 305
2000	2930	3830
3000	2380	2840
4000	2110	2430
5000	1900	2180
6000	1760	2030

*Each value listed is the average of 5 measurements at different points on the wafer. The average standard deviation is 2%.

The cured SOG films are crack-free and highly amorphous with smooth grain-free surfaces as determined by scanning electron microscopy (SEM). Selected physical characteristics of films cured at 400°C are given in Table II.

Table II. Physical Characteristics of SOG Films

Property, units	Value
Thickness variation, %	<2
Radial striations, % of thickness	± 5 (max.)
Film density, g/cc	2.1 ± 0.1 (SOG 203 only)
Refractive index	1.43 ± 0.01
Pinhole density, /cm^2	<1
Particulate density, /cm^2	<1

The pinhole density was determined by the electrolysis method while the particulate density was determined by examination under an optical microscope. The low pinhole density of the SOG film should help improve the insulation integrity of the CVD dielectric layers. The film density and refractive index of 203 are comparable to those of low temperature (325-475°C) CVD SiO_2 films (2). Upon densification at 800-900°C in air/oxygen ambients, the value of refractive index of 203 film increased to that for thermal SiO_2, 1.46, similar to the behavior shown by CVD SiO_2 films. As will be discussed later, these density characteristics are also reflected in the SOG etch rates in various media. Attempt to densify 305 film under the same conditions resulted in a film with a refractive index of 1.41 ± 0.01. This is attributed to the presence of Si-R groups in SOG 305, the thermal decomposition of which interferes with the densification mechanism (viscous sintering).

Electrical Characteristics. The dielectric constant, volume resistivity, and electric breakdown strength of the SOG films cured at 400°C and those cured at 800°C were determined by fabricating MOS capacitor structures. The measurements were carried out using a wafer probe station housed in a shielded enclosure, HP Model 4275A LCR meter, Keithley Model 616 electrometer, voltage ramp generator, and a Keithley microvoltmeter. The results are given in Table III. Corresponding values for thermal SiO_2 have also been listed for comparison purposes.

Table III. Dielectric Properties of SOG Films

Property	305 (400°C)	203 (400°C)	203 (800°C)	Thermal SiO_2
Dielectric constant (1 MHz)	4.7±0.5	8±1	4.2	3.9
Volume resistivity, ohm cm	1×10^{12}	1×10^{13}	1×10^{16}	1×10^{16}
Breakdown field, V/cm	10^5–10^6	10^5–10^6	1×10^6	2–8×10^6

The dielectric properties of the SOG films cured at 400°C are, in general, inferior to those of thermal SiO_2 or low temperature CVD SiO_2 films. The high dielectric constant of 203 indicates the presence of a significant amount of polarizable material in the SOG film. This polarizable species is H_2O that is adsorbed into the micropores of the film, the adsorption being facilitated by presence of silanol (≡Si-OH) groups in the material. The variations in the dielectric constant values arise due to the reversible adsorption/ desorption of H_2O in the film due to heat produced by the applied electric fields. The low values of volume resistivity are attributed to protonic conductivity. The dielectric properties of 203 films densified at 800°C become quite comparable to those of thermal SiO_2.

Using the CV flatband voltage shift under bias-temperature stress measurements, the mobile ion (Na^+) density in the SOG film was determined to be about 1×10^{11} cm^{-2}.

Film Etch Rates. The etch rates of the SOG films were determined in various wet and dry media. The results are summarized in Table IV.

Table IV. Wet and Dry Etch Rate of SOG Films

Etch Medium	Etch Rate, Å/min.[1]			
	305(400°C)	203(400°C)	203(900°C)	Thermal SiO_2
100:1 BOE^2, 19.5°C	800	250	50	50
50:1 BOE^2, 20.4°C	-	650	120	120
$CF_4 + O_2$ plasma	680	570	500	490
$CF_4 + O_2$ RIE	750	680	-	450

[1] Estimated error in etch rates is ± 10%.
[2] BOE stands for buffered oxide etch made by mixing 100 or 50 parts by volume of 40% NH_4F solution in water with 1 part of 49% HF.

Wet etch rates, or chemical solution rates, are a function of density for films of same chemical composition. Therefore, it is not surprising that the BOE etch rates of densified 203 films are identical to those of thermal SiO_2. Corresponding wet etch data for CVD SiO_2 films are not available. However, the etch rates of 325-475°C CVD SiO_2 films in 6:1 BOE (at 25°C) are 5500Å/min. and 12-1500Å/min. respectively for as-deposited films and for films densified at 800°C (2). Thus the ratios of the etch rate for the undensified film to that of the densified film are roughly the same for the SOG and CVD films. Assuming that the densified films etch at equal rates, the BOE etch rates of a 203 film cured at 400°C are similar to that of an as-deposited CVD SiO_2 film. Thus, the application of smoothing layers of 203 SOG either before or after the deposition of the CVD dielectric would be compatible with wet etch processing for via (contact) holes. On the other hand, because of their relatively larger etch rates in BOE the SOG 305 films would optimally be used on top of the CVD dielectric to avoid the formation of undercut via holes.

The plasma etch rates of the SOG films were determined in an MRC Model 51 RIE parallel plate etcher which had 6 in. diameter electrodes. The electrode spacing was 2 in. and the substrates were placed on the powered electrode which was water cooled. The chamber pressure was about 200 mTorr for plasma etching and 10-15 mTorr for reaction ion etching (RIE). A net total power of 50W was used in each experiment. In contrast to their wet etch behavior, the SOG films etch only slightly faster than thermal SiO_2 in $CF_4 + O_2$ plasmas. This is so since plasma etch rates are determined to a greater extent by the chemical nature than by the density of the film. When using plasma etching of via holes, therefore, the SOG smoothing layers can be utilized in either configuration (Figure 1 (b) and (c)).

Planarization and Adhesion Characteristics. The adhesion of the SOG films on unpatterned surfaces of silicon, thermal SiO_2 and aluminum was qualitatively evaluated by placing cured SOG films on the various substrates in boiling water for 30 min. No detachment or lifting-off of either film from any substrate was observed. However, it was found that SOG 203 spun on aluminized substrates had a tendency to form cracks (craze) upon cure at 400°C indicating poor adhesions, or, more likely, mismatch of thermal expansion coefficients of the two materials.

The planarization efficiency of the SOG films was evaluated on patterned layers of thermal SiO_2 and aluminum. The thermal SiO_2 substrates were patterned by RIE while wet etching was employed for aluminum. The SOG films were spun on at 3000 rpm and cured at 400°C. The SEM micrographs of one micron thick thermal SiO_2 lines before and after smoothing by a 203 film are shown in Figure 2. The 90° angle of sidewalls is clearly seen to be reduced to about 45°. Similar results were obtained by the application of 305 film on thermal SiO_2 substrates. Figure 3 shows the smoothing of a 0.75 micron thick aluminum line by a film of SOG 305. Again, similar smoothing was obtained with a 203 film. Results were less satisfactory when smoothing of 1.5 micron thick anisotropically etched aluminum features was attempted with 203 films. In addition to a significantly smaller reduction of the sidewall angles, some separation of the SOG film near the top edges of the features was observed. The reduced planarization in this case, of course, is a result of the greater step height since the degree of planarization obtained with a given SOG material is known to be a function of the line and space-widths as well as the step height of substrate features (10).

Film Composition and Structure. It was of interest to compare the characteristics of the SiO_2-like films of the SOG 203 to other SiO_2 films. Therefore, the chemical composition, stoichiometry, bulk structure of the SOG 203 films as well as the SOG-single crystal silicon interface characteristics were investigated using a number of spectroscopic techniques. Compositional and structural changes were studied as a function of cure temperature. Rutherford Back-Scattering (RBS) analyses on SOG films cured at 100°C, 400°C, and at 900°C (performed at Charles Evans & Associates, San Mateo, Cal.) indicated that these films were essentially pure SiO_2 - the oxygen to silicon ratio in each case was determined to be 2.0 ± 0.1. The results obtained from AES (Auger Electron Spectroscopy), ESCA (Electron Spectroscopy for Chemical Analysis), IR (Infrared Spectroscopy), and SIMS (Secondary Ion Mass Spectrometry) are discussed in the following paragraphs.

IR Spectra. The IR spectra of the SOG film were obtained with a Perkin Elmer Model 683 Infrared Spectrophotometer in transmission mode by passing radiation through the film coated on a single crystal silicon wafer. In order to enhance the spectral signal, three layers

 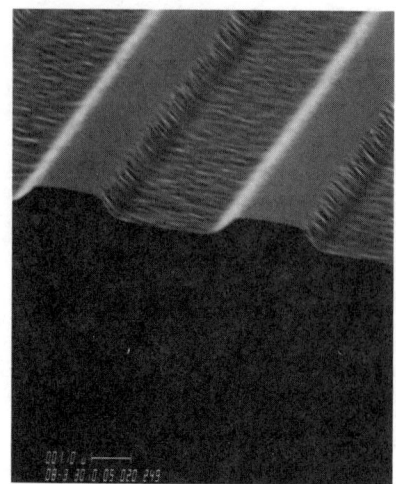

Figure 2. Smoothing of thermal SiO$_2$ lines by SOG 203.

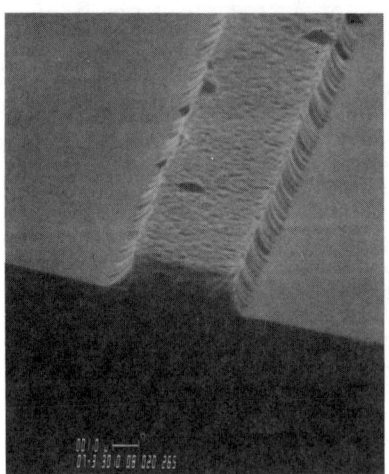 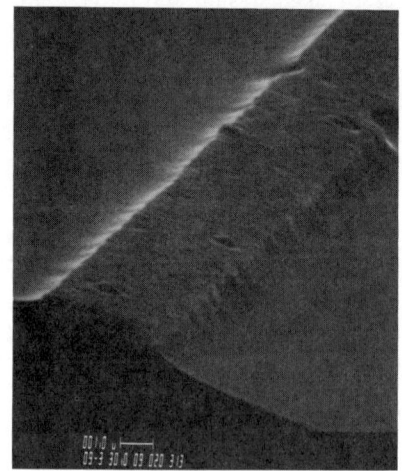

Figure 3. Smoothing of aluminum line by SOG 305.

of the SOG were applied with only a 100°C/15 min. treatment between successive applications. The spectra observed for the composite films cured at 100°C, 400°C, and at 900°C are shown in Figure 4.

All three films show the well-known characteristic absorption bands of SiO_2 films at 1080 cm^{-1} (Si-O stretch), 800 cm^{-1} (O-Si-O bending), and 450 cm^{-1} (Si-O-Si rocking vibration) bands. The 100°C and 400°C films show broad absorption bands in the 3000-3700 cm^{-1} region due to silanol ($\equiv$Si-OH), hydrogen bonded silanol and absorbed water contents characteristic of low temperature porous SiO_2 films (11). The H_2O deformation band near 1625 cm^{-1} is also clearly visible. The 100°C film also shows a small but sharp absorption peak at just below 3000 cm^{-1}. This we attribute to aliphatic C-H stretch arising from residual alcohol solvent or from Si-OR groups in the film. The C-H absorption is reduced to nearly zero as a result of the 400°C cure. However, the silanol and water absorption bands remain. The 900°C spectrum shows no absorption band due to $\equiv$Si-OH, $\equiv$Si-OR, or H_2O indicating that the film is highly densified and has little propensity for absorbing H_2O.

The Si-O stretch band in the 1050-1100 cm^{-1} region is seen to shift to higher frequencies and become narrower with increasing temperature of film cure. Such energy shifts and band broadening in SiO_2 films have been attributed (11) to changes in stoichiometry, porosity, and chemical bonding or bond strain. Since the stoichiometry of these films was determined (by RBS analysis) to be essentially that of SiO_2, we believe that the observed spectra reflect changes in the bond strain or stress in these SOG films due to different cure temperatures.

Auger, ESCA and SIMS Studies. The ESCA and Auger spectra were obtained with a Physical Electronics Model 560 spectrometer, equipped with a double pass cylindrical mirror analyzer employing single channel detection. Typical operating pressure was 1×10^{-9} torr except during AES depth profiling. In the ESCA mode, Mg Kα radiation (1253.6eV) was utilized as the excitation source at a pass energy of 25eV. Under these conditions, the full width at half maximum (FWHM) of the $Au4f_{7/2}$ line was 1.1eV. For AES, a 70nA electron beam at 5.0KeV was used with the analyzer set at an energy resolution of 0.4%.

In the ESCA studies of the SOG films cured at 100°C, 400°C and 900°C, no alkoxy ($-OCH_2CH_3$) groups could be detected on the surface. The interior of the films was found to be virtually free of carbon in AES depth profiling experiments. This is in agreement with the results of RBS analysis discussed earlier. However, because of the relatively poor sensitivity of the Auger and ESCA techniques in detecting minor components (about 0.1 at.%), the results obtained do not rule out the presence of alkoxy or other carbon-containing species in these films. Therefore, the SOG films were probed by SIMS. Trace quantities of $Si-OCH_2CH_3$ (O-Et) were detected in the 100°C and 400°C cured films. Although quantitative information could not be obtained, the 400°C film showed a smaller amount of Si-OEt contamination.

SOG/Silicon Interface Studies. Interface widths were measured by AES depth profiling using the 10%/90% method (12-13). The films

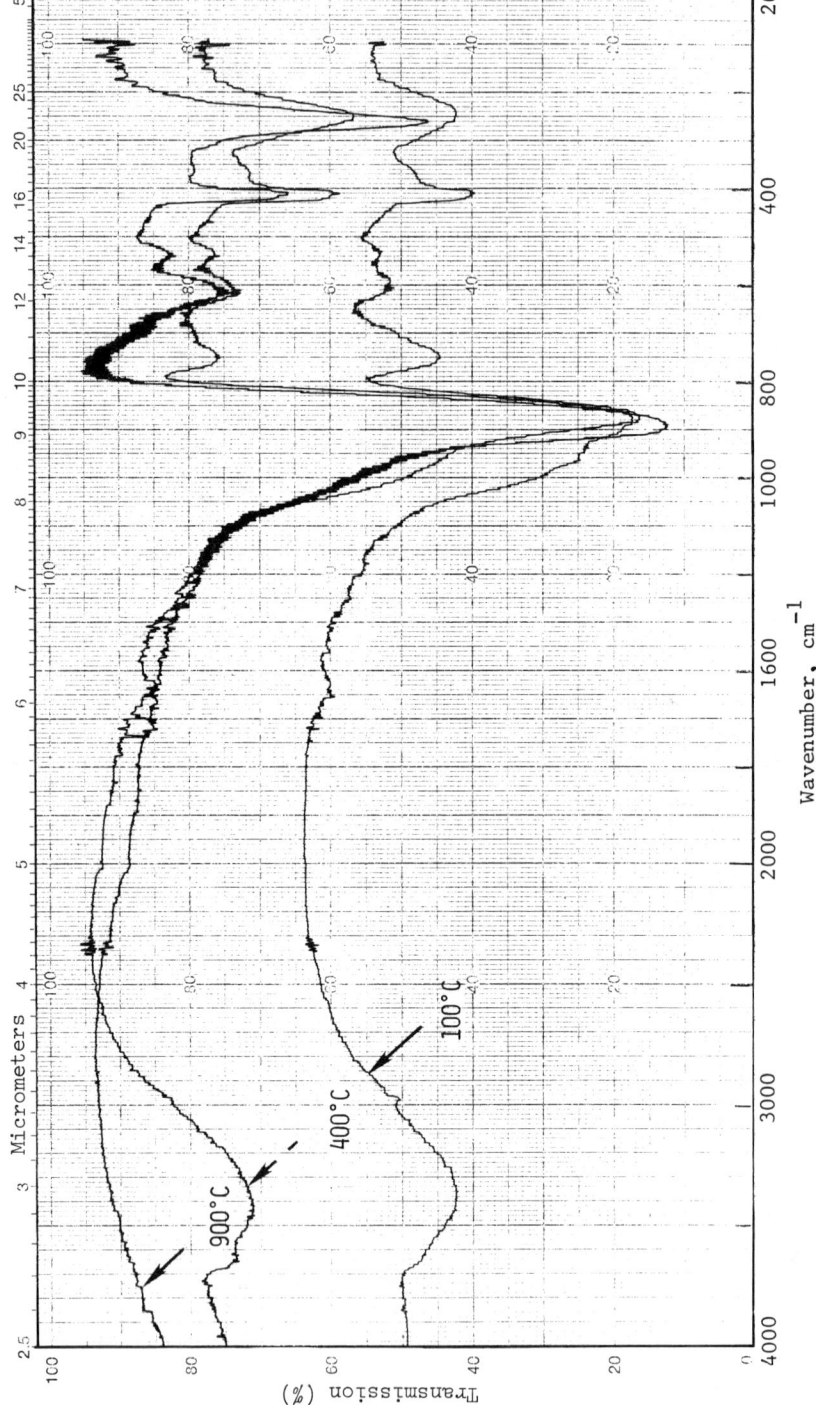

Figure 4. IR spectra of SOG 203 films as a function of cure.

were sputtered with a 1.5KeV Ar^+ beam rastered over a $4mm^2$ area. Under these conditions, a sputter rate of about 5Å min.$^{-1}$ was obtained. Both the low energy LVV (designated Si1) and high energy KLL (Si2) Si Auger transitions were monitored as well as the KLL lines of C and O. The depth profile obtained for 100Å thermal oxide film is shown in Figure 5. The calculated value of 25Å agrees well with those reported by others using the same technique (14-16).

Preparation of the thin SOG film required special precautions to eliminate any effects due to the presence of native oxide which may be initially on the Si substrate. The Si wafer was first etched with HF vapors to remove any oxide followed by immediate (15-28 sec. delay) application of the SOG film. After a bake at 100°C for 15 min. the sample was cured for 1 hour at 400°C. Despite these precautions, regrowth of native oxide on the Si substrate may have occurred during the elapsed time between the HF etch and application of the SOG film or after application due to the presence of H_2O and O_2 in the uncured film. This possibility could not be verified. The depth profile for the SOG film is shown in Figure 6.

The accuracy with which the interface width can be determined is highly dependent on not only the precise measurement of the original film thickness but also uncertainties associated with the interaction of the ion beam with the sample surface. The sputtering process may induce broadening of the interface by a mechanism referred to as ion knock-on mixing in which atoms at the interface are "knocked" across the interface. Additionally, broadening may result from nonuniform ion etching which causes roughening of the surface. These effects can be minimized by the use of thin films (<100Å) and low primary ion beam energies (<2KeV).

For this study a cured SOG film of 65±10Å was used. The relatively large uncertainty associated with this measurement is chiefly due to instrumental limitations. Based on this value of 65Å, the calculated width of the SOG/Si interface is 23±3.5Å. The error limits were based on the method described by Taubenblatt and Helms which takes into account the broadening mechanisms described above (13).

Although the experimental uncertainty is high, it is significant that the SOG/Si interface is not abrupt. If a thin layer of native oxide existed on the Si substrate before the application of the SOG, its thickness would be much less than that of the measured interface width. The results would indicate that the SOG forms an extension of the native oxide structure with interface characteristics much like that of thermally grown oxide. Whether a native oxide was initially present or not, the measured width of the SOG/Si interface is, within experimental error, similar to thermal oxide.

Film Stress. AES spectra in the low kinetic energy valence band region were employed to study the bonding characteristics of silicon in 100Å thick thermal SiO_2 and thin (<100Å) SOG films cured at 100°C and 400°C. These spectra are presented in Figure 7. The principal Auger feature from thermal oxide appears at 76eV and originates from a $L_{2,3}VV$ transition. Note a slight shoulder is present on the high kinetic energy side of this peak at around 79eV. This shoulder has previously been attributed to a change in stoichiometry which leads to a decrease in the surface charge and is in good agreement with

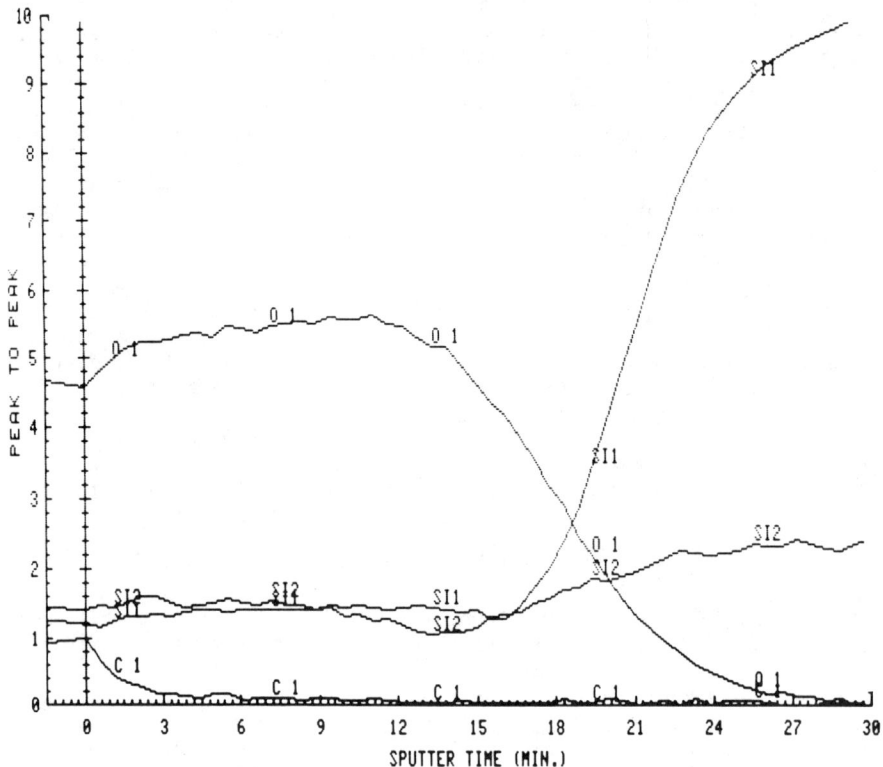

Figure 5. Auger depth profile of 100Å thick thermal oxide film.

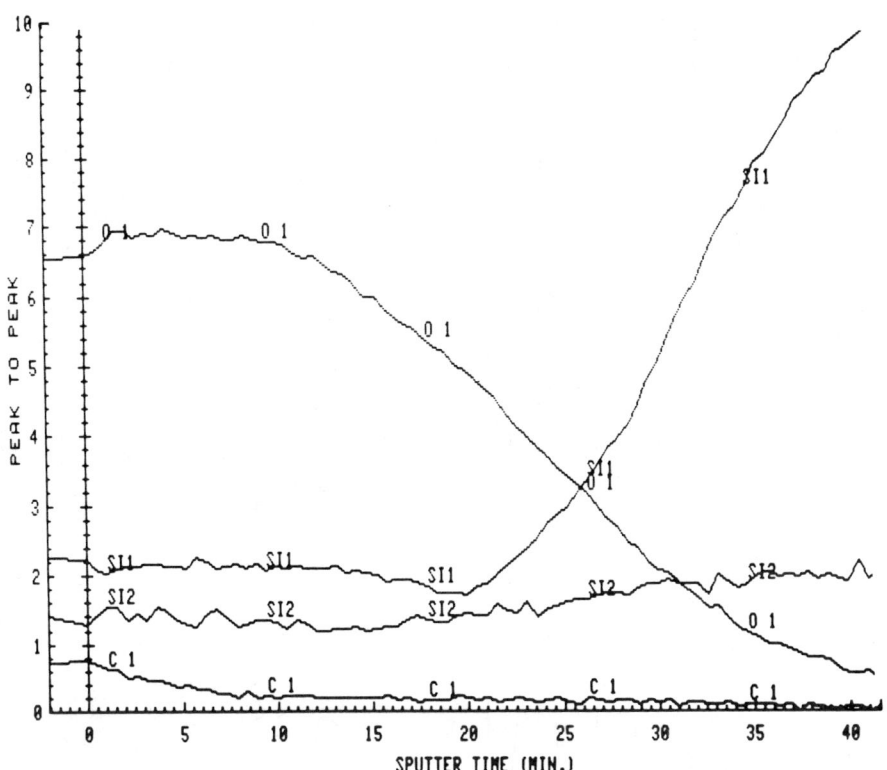

Figure 6. Auger depth profile of 65Å thick spin-on glass film.

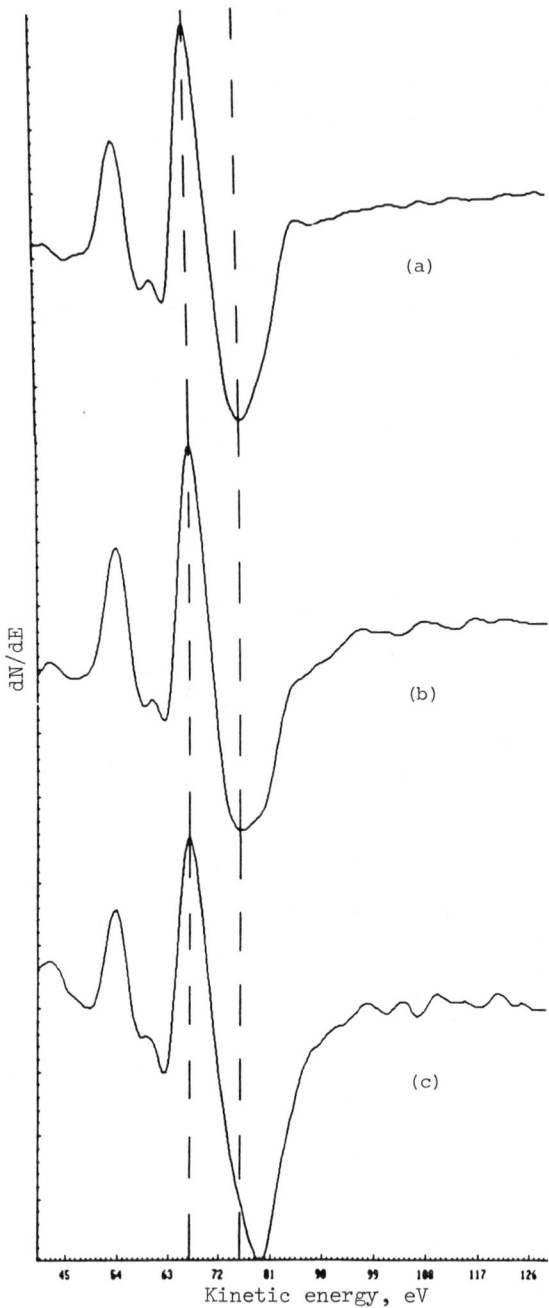

Figure 7. Si LVV Auger transition. a) thermal oxide, b) SOG cured at 400°C, c) SOG cured at 100°C.

published data (7-8). However, any change which leads to a change in the electron charge distribution around the Si atoms comprising the sample would be expected to show a similar effect.

The spectrum of the SOG film cured at 400°C is similar to that of thermal oxide; however, a very slight increase in energy of the maximum negative excursion of the $L_{2,3}VV$ transition is observed as well as a pronounced increase in the intensity of the high kinetic energy shoulder. Under the low primary beam currents (70nA) employed in this study, the spectrum remained unchanged after several hours of exposure; thus, eliminating electron beam induced modification of the sample surface as a factor for this change.

For the 100°C SOG film, the $L_{2,3}VV$ peak is markedly different from that obtained for the 400°C film. The peak has shifted in energy to around 79eV and a shoulder appears on the low kinetic energy side. It would appear that the dominant feature for SOG (100°) corresponds to the shoulder found in the spectrum of SOG (400°).

The observed differences in the spectra of the two SOG films may be interpreted in terms of differences in the film stress which leads to a displacement of the charge density around the Si atoms due to bond bending. The film cured at 400°C is expected to be under greater stress than the film cured at 100°C because of the thermally induced crosslinking. Thus, the 79eV peak of the 100°C film may be associated with Si atoms with lower bond strain. In order to test this hypothesis, an AES spectrum in the low kinetic energy valence band region was also obtained for a 5% P-doped SOG film cured at 400°C. This spectrum along with that for the undoped SOG film is presented in Figure 8. The magnitude of the shift and resultant peak shape of the P-doped SOG spectrum are similar to those observed for the 100°C SOG film. It is well known that incorporation of phosphorus in (CVD) SiO_2 films reduces film stress. Thus, AES spectra in the low kinetic energy valence band region may provide a means for characterization of film stress.

Summary and Conclusions

Thin spin-on glass films in the thickness range of 2-3000Å can be employed to planarize a CVD dielectric in the fabrication of integrated circuits. The physical, dielectric, and processing characteristics of the two SOG investigated, ACCUGLASS 203 and 305, are suitable for their use as a smoothing layer in interlevel dielectric insulation. Films of SOG 203, cured at 400°C, show several material characteristics common with low temperature CVD SiO_2 films. The SOG/Si interface is structurally similar to that between thermal SiO_2 and silicon.

Acknowledgments

The authors wish to thank Ms. S. A. Ferguson for assistance in SEM and spectroscopic studies. The permission by Allied Chemical to publish this work is also appreciated.

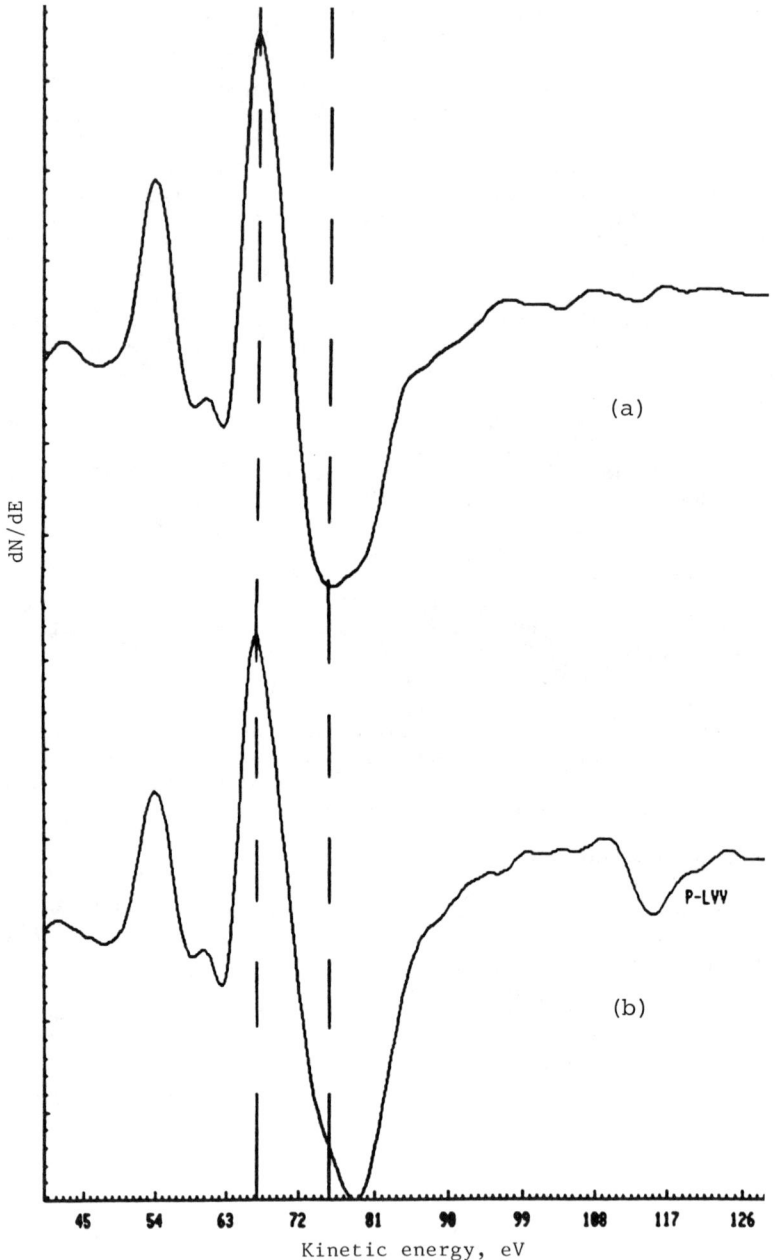

Figure 8. Si LVV Auger transition. a) SOG cured at 400°C, b) 5% P-doped SOG cured at 400°C.

Literature Cited

1. Kern, W. *Semiconductor International* 1982, 3, 89-103.
2. Mattron, B. *Solid State Technology* 1980, 23 (1), 60-64.
3. Kern, W.; Vossen, J. L., Schnable, G. L. *11th Ann. Proc. Reliability Physics, IEEE* 1973, 214-223.
4. Armstrong, W. E. and Tolliver, D. C., *J. Electrochem. Soc.* 1974, 121, 307.
5. Iqbal, A.; Lehrer, W. I.; and Pierce, J. M., *ECS Extend Abstracts*, Abst. 228, Washington, D.C. Meeting, Fall 1983.
6. Kern, W. and Schnable, L. *RCA Review*, 1982, 43, 423-457.
7. Hom-ma, Y.; Harada, S.; and Kaji, T., *J. Electrochem. Soc.* 1979, 125, 1531.
8. Kotani, H.; Yakushiji, H.; Harada, H.; Tsukamoto, K.; and Nishioka, T., *J. Electrochem. Soc.* 1983, 130, 645.
9. Desbiens, D., *J. Electrochem. Soc. Ext. Abst.* 1984, 84-2, 605.
10. White, L. K., *J. Electrochem. Soc.* 1983, 130, 1543.
11. Pliskin, W. A., *J. Vac. Sci. Technol.* 1977, 14(5), 1064-1081.
12. Cook, C. F.; Helms, C. R.; Fox, D. C. *T. Vac. Sci. Technol.* 1980, 17, 44.
13. Taubenblatt, M. A.; Helms, C. R. *J. Appl. Phys.* 1983, 54, 2667.
14. Helms, C. R.; Johnson, N. M.; Spicer, W. E., *Solid State Commun.* 1978, 25, 673.
15. Wagner, J. F.; Wilsem, C. W., *J. Appl. Phys.* 1979, 50, 874.
16. Helms, C. R.; Johnson, N. M.; Schwarz, S. A.; Spicer, W. E., J. Appl. Phys. 1979, 50, 7007.
17. Smith, J. N.; Thomas, S.; Ritchie, K. *J. Electrochem. Soc.* 1974, 121, 827.
18. Carriere, B.; Lang, B., *Surf. Sci.* 1977, 64, 209.

RECEIVED June 24, 1985

23

Effects of Various Chemistries on Silicon-Wafer Cleaning

D. Scott Becker, William R. Schmidt, Charlie A. Peterson, and Don C. Burkman

FSI Corporation, Chaska, MN 55318

This paper reviews the current knowledge on aqueous cleaning of silicon wafers. Some new information regarding the chemical action of the most common cleaning solutions is revealed. Information on individual solutions as well as the sequential use of more than one solution is presented and discussed. Through the use of Secondary Ion Mass Spectrometry and Laser Defect Scanning, the best overall cleaning was accomplished through the sequential use of the four cleaning solutions (mixed in parts per volume) $4H_2SO_4(96\%):1H_2O_2(31\%)$, $100H_2O:1HF(99\%)$, $5H_2O:1H_2O_2(31\%):1NH_4OH(29\%)$, and $5H_2O:1H_2O_2(31\%):1HCl(38\%)$ in that order. It was also found that reversing the order of use of the hydrofluoric acid and ammonium hydroxide hydrogen peroxide solutions yielded a wafer surface lower in metal contamination yet higher in particle contamination.

The performance of semiconductor devices may be altered by the presence of contamination before, during and after device fabrication. Because of this, it is necessary to achieve the highest degree of cleanliness possible during semiconductor wafer processing. There are many cleaning procedures and the most desirable process will depend on which step in device fabrication it follows and which step it precedes. This paper reviews the current understanding of the most common silicon wafer cleaning procedures and reveals new information on their chemical behavior. Most processes involve sequential use of several cleaning solutions. The sequencing is of great significance and will be discussed in detail.

Contaminants

Contaminants on a surface may be classified as molecular, ionic or atomic. The molecular contamination, primarily organic, could be oils, waxes (from mounting, slicing, etc.), photoresist, particles from wafer carriers and boxes and airborne hydrocarbons. Complications created by these contaminants will depend on the process that would follow. These contaminants could result in poor adhesion of deposited materials, changes in oxidation and EPI deposition rates as well as lead to the formation of crystalline defects. In addition, the hydrophobic nature of some types of organic material inhibits the interaction of aqueous cleaning solutions with the wafer surface, therefore, limiting removal of ionic and atomic contaminants. For these reasons, the removal of organic contamination is usually the first step of a cleaning process. The presence of an oxide layer on a silicon wafer is inevitable if the wafer has been exposed to an oxygen or water atmosphere. Oxidizing agents such as H_2O_2 will also oxidize a silicon wafer. It might be necessary to remove some of this oxide in order to ease the removal of entrapped contaminants. Presence of contamination can be detrimental to a processing step. An example would be the interference that particles cause during EPI deposition (1,12,13).

Ionic and atomic contamination may come from the environment or result from the use of processing chemicals. The adsorption of halide ions (2,3) and metal cations (3) from common processing reagents have been investigated. It was found that ions may be adsorbed in concentrations as great as 10^{17} ions/cm^2. Oxidized wafers adsorbed more Na+ than did bare silicon wafers (2). Adsorption of some metals such as gold and silver were more dramatic in HF. Because a wafer without an oxidized surface is in good electrical contact with an aqueous solution it is capable of electrochemical reduction of some metal ions. This reverse plating results in atomic metal contamination. This idea was reinforced when it was found that the adsorption of gold on silicon was three to four orders of magnitude greater than on silica (3).

Ions and atomic metals contamination can be detrimental to device performance for a variety of reasons. Small ions may migrate under the influence of an electric field or at high temperature. Metals will also diffuse at high temperatures. Ionic contamination can result in drift currents and surface potential drifting (4,5). Heavy metals can affect minority carrier lifetimes and surface recombination velocities (6,7). Both types of contamination can affect the formation of accumulation and depletion layers.

Cleaning Chemistries

Feder and Koontz (8) and Peters and Deckert (9) have evaluated various photoresist stripping methods. Their evaluations were based on ability to remove photoresist as well as the ability to leave a critically clean surface. Some of the methods evaluated were 1) oxidizing agents such as H_2O_2, HCl, H_2SO_4, HNO_3, $HCl-H_2O_2$, $H_2SO_4-H_2O_2$; 2) caustics such as NH_4OH, $H_2O_2-NH_4OH$; 3) solvent systems such as CCl_4, J-100, A-20, CH_3COCH_3, $CH_3COC_2H_5$ and 4) ashing (air) and plasma stripping. Both reports found that some systems were good for photoresist stripping and others were good for critical cleaning.

The organic solvent systems were good for stripping positive resist but did not leave a clean surface. Ashing and plasma stripping were very effective on negative resist but they too did not yield a clean surface after positive resist stripping. The aqueous cleaning methods yielded various results of which only two were found to be effective. The $CrO_3-H_2SO_4$ and $H_2SO_4-H_2O_2$ were very effective in photoresist stripping but did not leave a clean surface. A problem from using chromic acid solutions is that it can leave chromium ions on a silicon wafer surface. This should not be surprising since it is well known to chemists that after chromic acid has been used to clean glassware it is extremely difficult to extract all of the chromium ions out of the glass. As mentioned earlier, the presence of these ions can be detrimental to device performance.

The only solution which was very effective in cleaning a stripped surface was the $NH_4OH-H_2O_2$ solution, even though this solution was not effective for stripping photoresist. Because stripping methods do not leave a good clean surface and require further cleaning in an aqueous solution it might be easier to strip the resist and clean the surface in aqueous solutions. Based on the discussions above, the best organic cleaning procedure would be to use $H_2SO_4-H_2O_2$ to remove any heavy organic contamination and follow with a $NH_4OH-H_2O_2$ cleaning step.

An oxide layer on a silicon wafer is most commonly removed with HF or NH_4F-HF (BOE) solutions. The reaction results in formation of silicon hexafluoride which is water soluble (10). The buffered HF is used when a constant oxide etch rate is required. Complication due to NH_4F precipitation can occur. Because of NH_4F precipitation and economic considerations, HF is the preferred reagent for oxide removal as part of a cleaning procedure.

Two problems can arise from the HF treatment of silicon wafers. The reverse plating of metals such as gold has already been mentioned. The other problem is that HF-treated silicon is very susceptible to hydrocarbon contamination. Henderson (11) found that when HF-treated wafers were placed into an oven at high temperatures, carbon containing adsorbants on the silicon surface decomposed to form β-SiC particles. These particles are known to create problems such as nucleate polycrystalline regions during the growth of epitaxial silicon (12,13). Because of these two problems it is recommended that after HF the wafers should be cleaned to remove the heavy metals and carbon containing adsorbants.

Cationic contamination may be lowered by rinsing in water, acids, bases, acidic peroxide, basic peroxide or solution with chelating agents (2,3,14,15). Water rinsing of sodium contaminated silicon wafers was effective. A 6N HCl solution was found to be more effective than the water rinsing. If the sodium was chemically adsorbed the action of HCl can be viewed as a simple cation exchange reaction. In general, HCl solutions were very effective in removing metals due to this as well as the complexing ability of chloride ions. It was also reported that sodium contaminated wafers that were stored in room air had lower desorption rates in both water and HCl than did the fresh samples.

Use of an NH_4OH solution can be effective if the contaminant has a tendency to form amino complexes, such as $Cu(NH_3)_4^{+2}$. However, some cations such as Mg^{+2}, Al^{+3} and Fe^{+3} will form insoluble hydroxide complexes in basic solutions. Because of this, metals need to be removed by acidic solutions. Chelating agents are capable of forming water soluble complexes with many metal ions. Yet, chelating agents were ineffective in preventing metal contamination in etchant solution (2). Metals that reverse plated onto the surface were ineffectively removed by most single component cleaning solutions. It was found that it was necessary to oxidize these metals before they could be removed. From the investigations of Kern (14) it was revealed that the most effective cleaning was accomplished with H_2O_2 based solutions containing HCl or NH_4OH. Basic peroxide solutions, typically 5:1:1 ppv in $H_2O:H_2O_2(31\%):NH_4OH(29\%)$, are effective in removing organic contamination as well as group IB and group IIB metals. Preliminary reports indicate that this basic peroxide mixture may be effective in removing gold as well. The acidic hydrogen peroxide solutions, typically 5:1:1 ppv in $H_2O:H_2O_2(31\%):HCl(38\%)$, were very effective in removing all of the metals that were investigated.

One drawback of hydrogen peroxide solutions is that they readily decompose. Decomposition is greater at higher pH's and the presence of metals will accelerate the decomposition. Another potential problem is when the peroxide concentration gets too low, in a basic solution, the solution will etch a silicon surface (14). Solution decomposition and concentration variation may be eliminated by use of centrifugal spray processing systems (15). In these systems the cleaning solutions are mixed on-line, in controlled ratios immediately before being sprayed onto the wafer surface.

Chemical Sequencing

Most cleaning procedures require the sequential use of two or more cleaning solutions. This may be necessary when most cleaning procedures require the sequential use of two or more cleaning solutions. Wafers are contaminated with more than one type of contamination. Also, a cleaning solution can remove some contaminants while leaving a different contaminant on the wafer surface (e.g. HF). It has also been mentioned that one type of contaminant could interfere with the removal of another type. Chemical sequencing has been discussed in great detail. Amick (17) and Burkman, et al (16) suggested that a cleaning sequence should be 1) removal of organic material, 2) removal of oxide layers and 3) removal of metallic and ionic contaminants. From the data available on silicon wafer cleaning, step 1) could be accomplished with SPM and/or APM, step 2) could be accomplished with an HF solution, and step 3) with HPM. SPM refers to a 4:1 ppv mixture of $H_2SO_4(96\%):H_2O_2(31\%)$. APM refers to a 5:1:1 ppv mixture of $H_2O:H_2O_2(31\%):NH_4OH(29\%)$. HPM is typically a 5:1:1 ppv mixture of $H_2O:H_2O_2(31\%):HCl(38\%)$.

Experimental Procedures

The metallic and ionic cleaning procedures studied by Kern were evaluated by radiotracer methods. This technique requires the use of radionuclide enriched reagents. Recently, a procedure utilizing secondary ion mass spectrometry (SIMS) has been developed for studying very low levels of contamination on silicon surfaces (18). This technique does not require radionuclide enriched reagents so normal reagents may be used. The SIMS technique also has the advantage that it is sensitive enough to detect contamination on a wafer after cleaning with the best cleaning procedures. The high sensitivity of SIMS can be seen in Figure 1. The Auger Electron Spectrum (AES) of the silicon wafer only shows Si and O. A SIMS spectrum of the same sample shows the presence of several contaminants not detected in the AES spectrum. There are, however, some limitations in the SIMS tech-

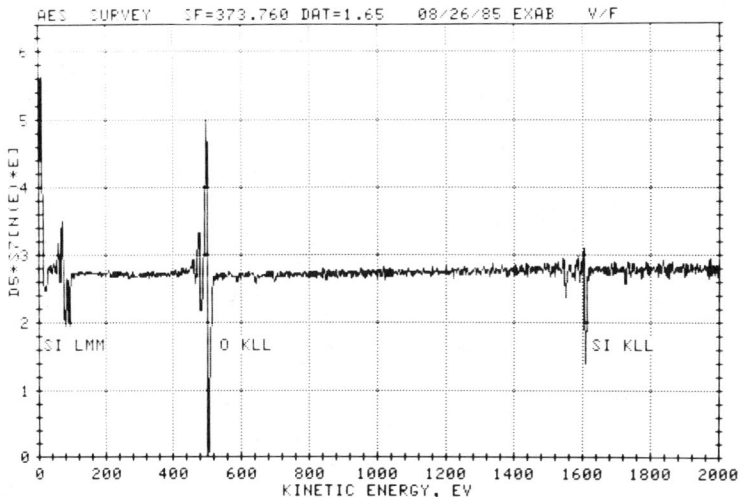

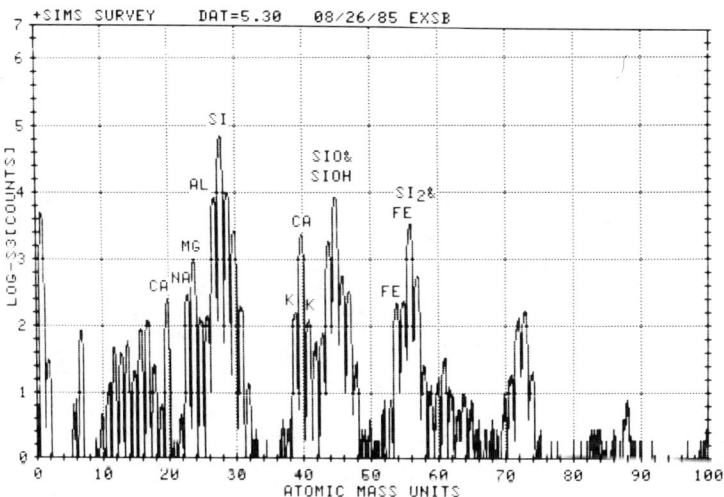

Figure 1. AES and SIMS spectra obtained from the same spot on a contaminated wafer.

nique. The technique is not suitable for detecting iron or heavy metals on silicon. The most intense iron peaks Fe^{+2} and Fe^+ occur where the silicon peaks due to Si^+ and Si^+_2 occur. Heavy metals such as gold have low ion yields which gives them a high detection limit.

All of the cleans to be discussed were evaluated for metal ion contamination by using the SIMS procedure previously described by Phillips, et al (18). Particle contamination was determined using a Tencor Surfscan set at maximum sensitivity (1µm) with an edge exclusion setting of 7. All of the samples were prepared and cleaned as previously described (19) unless stated otherwise.

Results and Discussion

An experiment that evaluated the necessity of oxide removal as a step in cleaning was performed (19). Contaminated wafers were cleaned using the two sequence of SPM, HF, APM, HPM or SPM, APM, HPM. Table I gives the SIMS results that were obtained.

Table I. The Influence of Oxide Removal on Cleaning

Cleaning Process	Contaminant/Silicon Peak x 10^{-6}		
	Na(23)	K(39)	Cu(63)
SPM, HF, APM, HPM	26	156	23
SPM, APM, HPM	90	246	66

These results indicated that removal of the native oxide with HF affords a cleaner surface. The same conclusion was ascertained by Beyer (10).

Further evidence that the HF step is good for removing some metal ion contamination can be seen from the experiment that compared wafers cleaned by APM and APM, HF. Table II shows the results of the SIMS analysis.

Table II. Removal of Metal Ions With HF

Cleaning Process	Contaminant Peak/Silicon (28) Peak x 10^{-6}				
	Na(23)	Mg(24)	Al(27)	K(39)	Ca(40)
APM	38	46	18807	49	79
APM, HF	13	7	97	30	48

It can be seen that omission of the HF step results in a surface with greater metal ion contamination. The most significant observation was that the aluminum content was over two orders of magnitude greater on the sample without an HF step.

The HF step was effective in lowering the levels of those metals shown above in Table II. A different experiment was performed in order to evaluate the particle contamination due to HF treatment. In Table III it can be seen that treatment of a wafer with SPM then HF changes the particle level on four-inch wafers from three to 80. When the HF step was followed by APM then HPM the particle levels after the clean were approximately the same as before the clean.

Table III. Total Point Defects on Four-inch Wafers Before and After Cleaning

	SPM, HF		SPM, HF, APM, HPM	
	Before	After	Before	After
	3	80	4	5

From this experiment it was concluded that the APM, HPM cycles removed the particulate contamination that resulted from the HF step.

It was pointed out earlier that the best clean for hydrocarbon contamination was SPM followed by APM. It was also shown that after an HF step another step is required to remove residual particulate contamination, suggesting that the APM step should follow the HF step. An experiment was performed in which highly contaminated wafers were cleaned by the two sequences of SPM, APM, HF, HPM and SPM, HF, APM, HPM (19). In Table IV the results of that experiment are shown.

Table IV. The Effect of APM and HF Sequencing

Cleaning Process	Contaminant Peak/Silicon Peak X 10^{-6}							Total Point Defects
	Na (23)	K (39)	Ca (40)	Mg (24)	Cr (52)	Cu (63)	Al (27)	
SPM,APM, HF,HPM	27	32	70	ND	ND	ND	ND	236
SPM,HF, APM,HPM	30	25	134	131	ND	ND	645	86

ND = none detected

The results clearly show that the sequence of SPM, APM, HF, HPM was the more effective in removing metals, but it was the less effective for particle reduction.

The particle levels can be explained on the basis that HF leaves a surface ridden with particles that can be removed more efficiently by APM, HPM than by HPM alone. The data in Table II showed that the metals of interest were present in only trace amounts after an HF step. So the difference in metal contamination between the two cleaning processes in Table IV must reflect one of the differences between the use of APM, HPM and only HPM after an HF step. The results suggest it would be advantageous to use APM last in a cleaning sequence to yield a particulate clean surface. Because of this we evaluated the metal contamination on wafers cleaned by HF, APM, HPM and HF, HPM, APM. Figure 2 shows the SIMS spectra of these cleans. It can be seen from the spectra that metal ion contamination resulted from using the APM step last as evidenced by the presence of a large alumi-

374 MICROELECTRONICS PROCESSING: INORGANIC MATERIALS CHARACTERIZATION

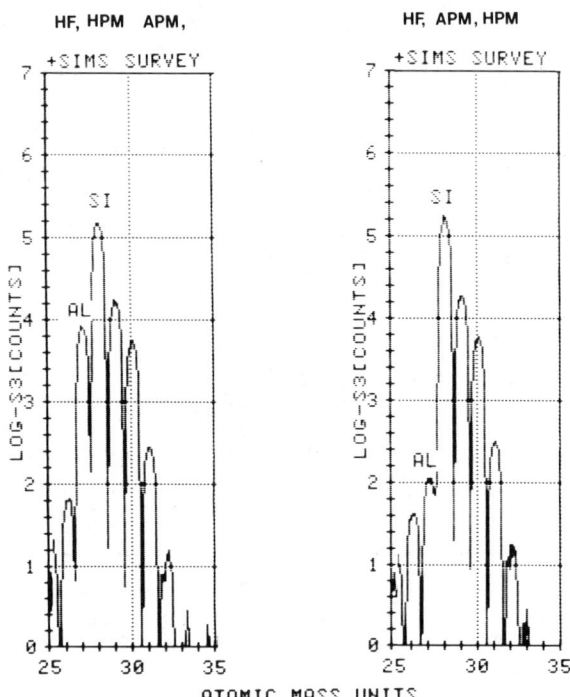

Figure 2. SIMS spectra taken from silicon wafers processed by the cleaning sequences of HF, APM, HPM, and HF, HPM, APM.

num peak. This indicated that metal ion contamination (Al, Ca, and Mg) actually increased from the use of APM. Since aluminum is insoluble in NH_4OH the most probable source in the APM solution would be the H_2O_2.

There are a variety of cleaning processes used in the microelectronics industry. The results of an experiment which compared the cleaning sequence of SPM,APM,HF,HPM to some of the more common cleaning sequences are listed in Table V (19).

Table V. Chemical Sequence Comparison by SIMS

Cleaning Process	Contaminant Peak/Silicon Peak X 10^{-6}						
	Na (23)	Mg (24)	Al (27)	K (39)	Ca (40)	Cr (52)	Cu (63)
SPM,APM,HF,HPM	15	57	ND	103	270	80	14
SPM,HF	30	30	ND	210	360	80	20
HCl/HNO_3,HF,HCl/HNO_3	38	160	171	ND	440	57	340
Fuming Nitric	179	93	380	1763	560	160	246

It can be seen that the lowest levels of metallic contamination resulted from cleaning with SPM, APM, HF, HPM, and SPM, HF. However, this SIMS data did not reveal the problems of heavy metal and particle contamination after using HF. Hence, of these four cleans, the best overall cleaning was accomplished using the SPM, APM, HF, HPM sequence.

Conclusion

It was revealed that some metal and particle contamination may result from the use of certain cleaning solutions. In turn, these contaminants may be removed by other cleaning solutions. Because of this, as the purity of processing chemicals change so will the most effective cleaning process. The most effective cleaning process will also depend on what contaminants are present and what contaminants must be removed. Based on the results that were obtained the best cleaning sequence for metallic contamination was SPM, APM, HF, HPM. Reversing the order of the APM and HF steps was more effective for particle reduction and slightly less effective for metal ion contamination.

Literature Cited

1. Batsford, K. O.; Thomas, D. D. Elect. Comm. 1963, 38(3), 354.
2. Kern, W. RCA Review 1970, 31, 207.
3. Kern, W. RCA Review 1970, 31, 234.
4. Hofstein, S. R. IEEE Trans. on Elect. Dev. 1967, ED-14, 749.
5. Snow, E. H.; Grove, A. S.; Deal, B. E.; Sah, C. T. J. Appl.Phys. 1965, 36, 1664.

6. Goetzberger, A.; Shockley, W. *J. Appl. Phys.* 1960, 31, 1821.
7. Bullis, W. M. *Solid-State Elect.* 1966, 9, 143.
8. Feder, D. O.; Koontz, D. E. *Symposium on Cleaning of Electronic Device Components* 1959, ASTM STP No. 246, 40.
9. Peters, D. A.; Deckert, C. A. *J. Electrochem. Soc.* 1979, 126(5), 883.
10. Beyer, K. D.; Kastl, R. H. *J. Electrochem. Soc.* 1982, 129(5), 1027.
11. Henderson, R. C. *J. Electrochem. Soc.* 1972, 119(6), 772.
12. Joyce, B. A.; Neave, J. H.; Watts, B. E. *Surf. Sci.* 1969, 15, 1.
13. Thomas, D. J. D. *Phys. Status Solidi* 1966, 13, 359.
14. Kern, W.; Poutinen, D. A. *RCA Review* 1970, 31, 187.
15. Burkman, D. C.; Peterson, C. A.; Schmidt, W. R. in "Treatise on Clean Surface Technology"; Mittal, , Ed.; Plenum Press: New York, to be published.
16. Burkman, D. C. *Semiconductor International* 1981, 4(7), 103.
17. Amick, J.A. *Solid State Technol.* 1976, 19(11), 47.
18. Phillips, B. F.; Burkman, D. C.; Schmidt, W. R.; Peterson, C. A. *J. Vac. Sci. Technol.* 1983, A1(2), 646.
19. Burkman, D. C.; Schmidt, W. R.; Peterson, C. A.; Phillips, B. F., *Proc. Semiconductor 83 International*, Birmhingham, England Sept. 1983. Also available from the authors as FSI TR217.

RECEIVED September 13, 1985

24

Monitoring of Particles in Gases with a Laser Counter

C. E. Nowakowski and J. V. Martinez de Pinillos

Air Products and Chemicals, Inc., Allentown, PA 18105

> The effect of various sampling techniques on the particle count and distribution sampled from gas streams, has been determined, as a function of gas stream Reynolds number and particle size (from 0.176 to 1.091 μm diameter). Sampling through a tee in a piping system has been shown to give higher particle counts than isokinetic sampling, indicating a "worst case" and thus, a conservative method of determining the purity of a gas system. The importance of characterizing pressure reducing devices has also been demonstrated.

There has been increasing interest in the semiconductor industry in monitoring the particle content of process gases. Advances in VLSI manufacturing processes have resulted in increased purity demands, not only with respect to gaseous contaminants but also particulate matter (1-4). Two types of particle counters have been preferentially used to monitor submicron particles: those based on coherent light sources and those based on incoherent light sources (5-6). The incoherent light scattering instruments (tungsten lamps) can detect particles with diameters of 0.3 micron or larger. The currently available coherent light devices (LASERS) can detect particles as small as 0.09 micron. A study was made to determine sampling conditions using a LASER counter.

Experimental, Results and Discussion

Laser Aerosol Spectrometer

All measurements were made using a Particle Measuring Systems Inc, Laser Aerosol Spectrometer Model LAS-X. The LAS-X is a light scattering instrument which uses a He-Ne laser. The gas sampling rate is adjustable from 0.4-3.0 cc/sec. The sample gas stream is surrounded by a filtered, 20 cc/sec, sheath flow which confines the particles to the center of the stream, at the focal point of the laser beam as can be seen in Figure 1. Particles present in the sample stream will deflect laser light. This scattered light is collected by a parabolic mirror, as presented in Figure 2, and detected by a photodiode which converts the collected light into a photocurrent. The voltage of this photocurrent, which is proportional to the amount of light collected, can then be interpretted as a particle size.

The LAS-X can detect particles as small as 0.09 micron diameter. Several size ranges, within the range 0.09 to 3.05 micron, are available with both differential and cumulative modes. For any size range a sixteen channel size distribution is obtained.

Number of Samples

To determine the number of samples required to achieve an experimental standard deviation (S) equal to or less than the Poisson distribution standard deviation (σ), a 1 cc/sec gas flow rate and four minute samples were used. The particles were generated with a PMS, Inc. particle generator model PG-100, providing a particle count of 2000 particles/ft^3 or greater.

The σ of a Poisson distribution can be shown to be equal to $(N)^{1/2}$, where N is the number of particles counted. An experimental standard deviation (S) for counts in all channels less than or equal to σ, was chosen as the criterion for number of samples to indicate that the error was due to counting statistics rather than any other experimental parameter. The results are presented in Table I and a channel size identification in Table II. It can be seen that the above criterion was met when the number of samples was at least 6. It should be understood that when the total number of particles is small, increased counting times and number of samples are needed to attain statistical significance.

Pressure Reduction Diffuser

The device used to sample high pressure lines, and to reduce the pressure from the line value to atmospheric, was a PMS diffuser. A photograph of this device is presented in Figure 3. The gas line is attached at the narrow end and gas is allowed to flow and expand into the larger diameter end (the isokinetic sampling cone). From the sampling cone, gas escapes either through the side holes or through the coupling between the diffuser and the instrument. A sampling needle projects from the instrument into the diffuser and introduces the sample from the isokinetic sampling cone into the interrogation zone of the instrument.

The minimum pressure needed to assure that particles from the outside environment were not being introduced into the isokinetic sampling cone, was studied. (Intake of unwanted ambient air could

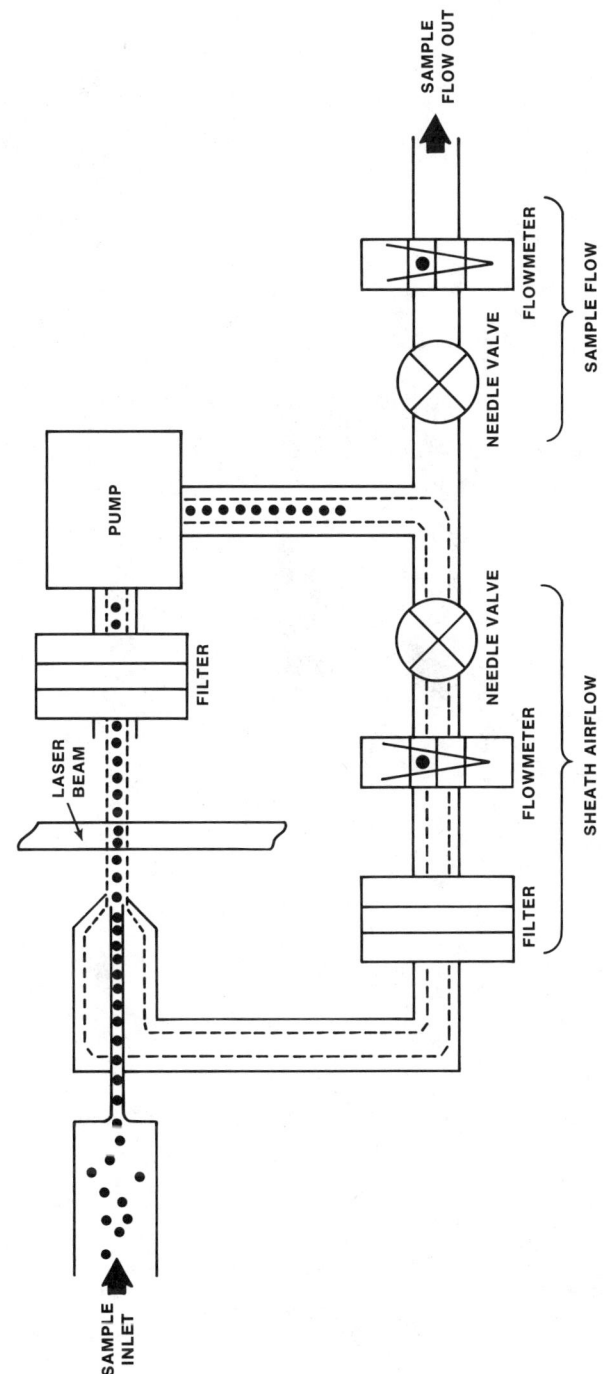

Figure 1. PMS LAS-X Flow Diagram. (Reproduced with permission. Copyright Particle Measuring Systems Inc.)

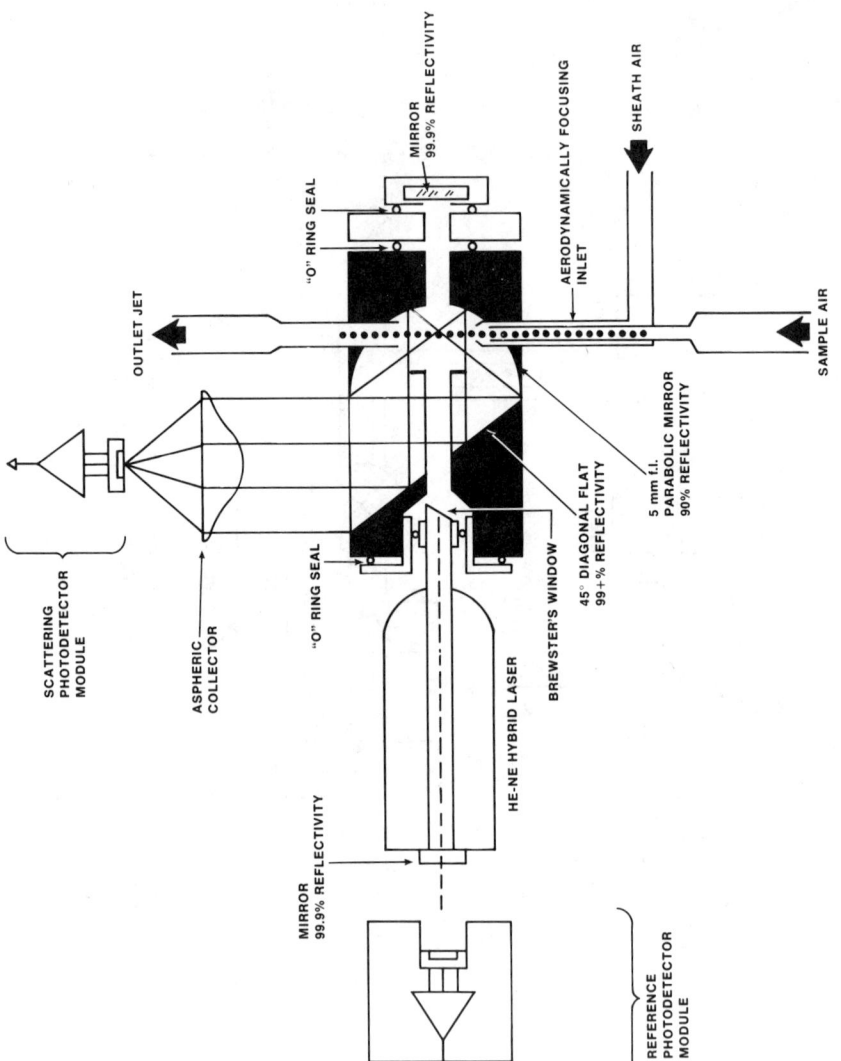

Figure 2. PMS LAS-X Optics Diagram. (Reproduced with permission. Copyright Particle Measuring Systems Inc.)

Table I

Number of Samples Needed to Achieve Statistical Significance

Counts X 10^{-3}

Channel	1	2	3	4	5	6	7	8
N								
1	2.14	1.88	2.92	6.93	19.7	44.2	78.1	142.
2	2.22	1.84	2.96	6.80	19.4	44.1	78.0	143.
3	2.10	1.88	2.85	7.00	19.7	44.1	78.3	142.
$X_3 (X10^{-3})$	2.16	1.86	2.91	6.91	19.6	44.1	78.1	143.
S_3	62.9	24.0	53.2	103.	163.	84.4	166.	481.
$\sigma_{Poisson}$ (3)	46.4	43.2	54.0	83.1	140.	40.1	280.	378.
4	2.12	1.87	2.95	6.82	19.7	44.1	77.8	143.
$X_4 (X10^{-3})$	2.15	1.87	2.92	6.89	19.6	44.1	78.0	143.
S_4	54.7	20.1	47.8	95.3	145.	69.3	211.	393.
$\sigma_{Poisson}$ (4)	46.3	43.2	54.0	83.0	140.	210.	279.	378.
5	2.17	1.84	2.91	6.87	19.5	44.1	78.1	143.
$X_5 (X10^{-3})$	2.15	1.86	2.92	6.88	19.6	44.1	78.1	143.
S_5	48.2	21.7	41.6	83.1	139.	61.1	186.	397.
$\sigma_{Poisson}$ (5)	46.4	43.1	54.0	83.0	140.	210.	279.	378.
6	2.14	1.85	2.90	6.85	19.6	44.2	77.8	143.
$X_6 (X10^{-3})$	2.15	1.86	2.92	6.88	19.6	44.1	78.0	143.
S_6	43.4	20.4	38.2	75.3	124.	64.8	201.	358.
$\sigma_{Poisson}$ (6)	46.3	43.1	54.0	82.9	140.	210.	279.	378.

Channel	9	10	11	12	13	14	15	16
N								
1	67.4	24.9	22.2	18.5	13.8	10.7	8.42	22.9
2	67.4	24.8	22.0	18.5	14.0	10.8	8.42	23.1
3	66.8	25.1	22.2	18.2	13.7	10.8	8.29	22.9
$X_3 (X10^{-3})$	67.2	24.9	22.1	18.4	13.8	10.8	8.38	23.0
S_3	348.	155.	68.1	159.	153.	97.0	76.8	126.
$\sigma_{Poisson}$ (3)	259.	158.	149.	136.	118.	104.0	91.5	152.
4	67.2	24.8	22.1	18.5	13.8	10.6	8.38	23.0
$X_4 (X10^{-3})$	67.2	24.9	22.1	18.4	13.8	10.8	8.38	23.0
S_4	284.	151.	56.0	140.	125.	105.	62.7	104.
$\sigma_{Poisson}$ (4)	259.	158.	149.	136.	118.	104.	91.5	152.
5	67.3	24.8	22.1	18.4	13.7	10.7	8.35	23.2
$X_5 (X10^{-3})$	67.2	24.9	22.1	18.4	13.8	10.7	8.37	23.0
S_5	253.	137.	48.6	122.	132.	91.4	55.5	124.
$\sigma_{Poisson}$ (5)	259.	158.	149.	136.	118.	104.	91.5	152.
6	67.4	25.0	22.2	18.3	13.7	10.6	8.42	23.2
$X_6 (X10^{-3})$	67.2	24.9	22.1	18.4	13.8	10.7	8.38	23.0
S_6	237.	133.	46.0	119.	123.	92.0	52.4	128.
$\sigma_{Poisson}$ (6)	259.	158.	149.	136.	117.	104.	91.5	152.

Table II

Channel Number and Particle Diameter Range for PMS LAS-X Counter

Number	Range (μm)
1	0.09 - 0.11
2	0.11 - 0.15
3	0.15 - 0.20
4	0.20 - 0.25
5	0.25 - 0.30
6	0.30 - 0.40
7	0.40 - 0.50
8	0.50 - 0.65
9	0.65 - 0.80
10	0.80 - 1.00
11	1.00 - 1.25
12	1.25 - 1.50
13	1.50 - 2.00
14	2.00 - 2.50
15	2.50 - 3.00
16	> - 3.00

occur due to the flow outwardly of the gas though the holes in the diffuser creating air currents that will force contaminated ambient air into the diffuser.) A polypropylene vessel was constructed that enclosed the diffuser as presented in Figure 4. A pressure gauge at the diffuser inlet monitored the gas pressure into the diffuser while the gas was injected from a cylinder. The vessel was maintained at atmospheric pressure or slightly higher. Particles of known diameter were injected into the vessel using the PMS particle generator. Particle counts were obtained with and without particles introduced in the vessel surrounding the diffuser. The gas pressure at the diffuser inlet was increased until no statistically significant differences were observed in the proper instrument channels in the two modes of operation.

Figure 5 presents the data obtained with a diffuser-called diffuser number 1 – for particles generated at 1.09, 0.460 and 0.176 micron. It can be seen that to exclude the smaller particles, pressures greater than 85 psig were required while 60 psig were needed to exclude the 1.09 micron particles. A new diffuser – number 2 – was tested and it was found to require pressures of approximately 20 psig to exclude all ambient particles as shown in Figure 6.

Isokineticity Requirements

Isokinetic sampling requires that the velocity of a gas through a sampling tube and into a measuring device, be equal to the velocity in the main gas stream. Under turbulent conditions, true isokinetic sampling is not possible. However, throughout this paper, sampling under turbulent conditions, when the velocity at the center of the gas stream is equivalent to the average velocity in the sample probe, will be referred to as isokinetic or pseudo-isokinetic. The velocity at the center of the gas stream for turbulent conditions was calculated from known relationships between Reynolds' number and the ratio of average velocity to maximum velocity in a smooth circular pipe (7). Under laminar conditions, the maximum velocity was calculated as two times the average velocity with average velocities being based on volumetric flow measurements.

The error obtained when anisokinetic sampling conditions are chosen can be caused by two factors: misalignment of the sampling probe and different velocity in the sample tube than in the bulk stream. The magnitudes of both types of error are dependent on the particle diameter through Stokes number. The Stokes number for the sampling tube inlet is defined by:

$$Stk = \frac{\tau V_M}{D}$$

where τ is the relaxation time or the time a particle takes to relax to a new velocity due to a new condition of forces; V_M, the gas velocity in the main gas stream; and D, the diameter of

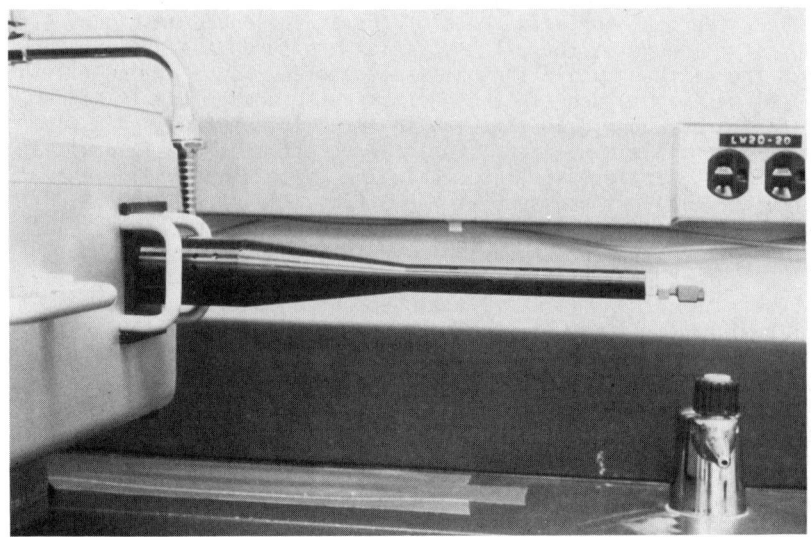

Figure 3. PMS High Pressure Diffuser.

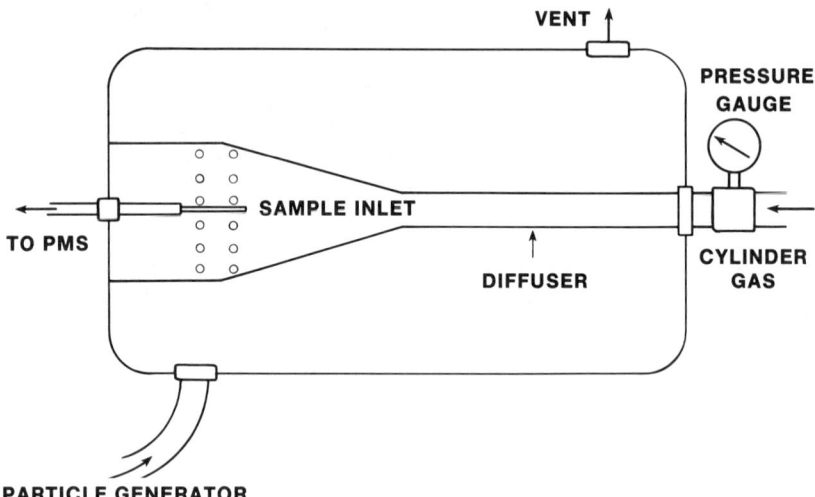

Figure 4. Apparatus for Determination of Minimum Pressure Requirements.

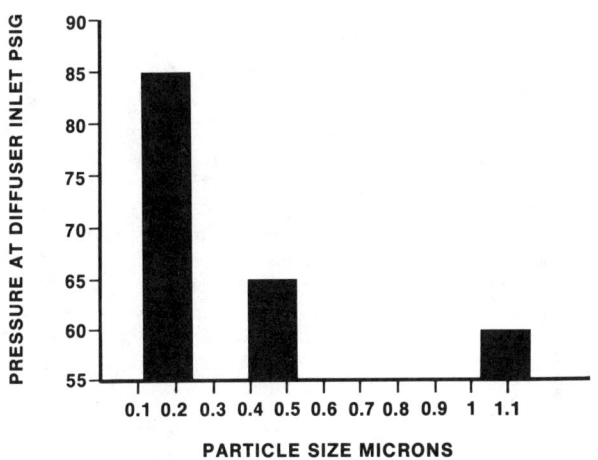

Figure 5.   Pressure Requirements for Diffuser #1.

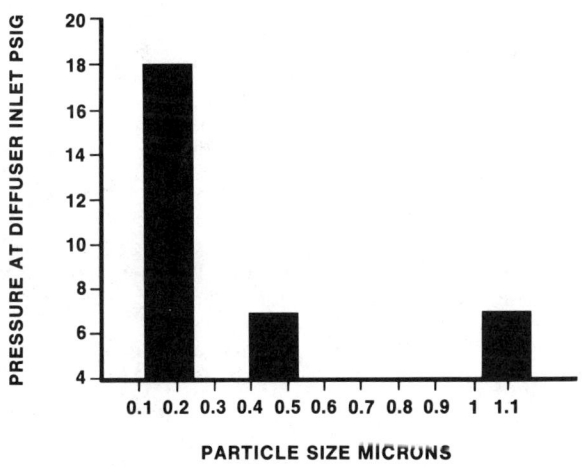

Figure 6.   Pressure Requirements for Diffuser #2.

the sampling tube. It can be shown (8) that in terms of particle diameter, for a spherical particle, τ is given by:

$$\tau = \frac{\rho_p \; d^2 \; C_c}{18 \; \eta}$$

where ρ_p is the particle density; d, its diameter, η, the gas viscosity; and Cc, the slip correction factor. Cc corrects for the assumption made in the derivation of Stokes law that the relative velocity of the gas at the surface of the sphere is zero. This assumption is not met for small particles whose size approaches the mean free path of the gas. The particle diameter, d, is proportioned to $(Stk)^{1/2}$. If the sample probe diameter is 1 cm and the gas velocity is 1000 cm/sec, the Stokes number for 1.0 μm particles is less than 0.1 and decreases rapidly for decreasing diameters.

The fractional error C/Co, where C is the measured particle concentration when the true gas concentration is Co can be shown (9) to be equal to:

$$\frac{C}{C_o} = 1 + (\cos \theta - 1) \left(1 - \frac{1}{1 + 0.55 \; (Stk') \; \exp(0.25 \; Stk')}\right)$$

where θ is the angle between the streamlines in the manifold and the axis of the sampling tube. Stk' is defined by:

$$Stk' = Stk \; \exp(0.022 \; \theta)$$

Figure 7 shows the effect of misalignment of the sampling probe on the concentration if the velocity, V_S, in the sampling tube is equal to V_M. The abcissa is $(Stk)^{1/2}$ which is proportional to particle diameter. For misalignment of less than 15° the error in measured concentration is negligible for particles smaller than about 5 microns. However this error increases to about 10% for misalignment of 45°, and more than 30% for misalignment of 90°.

When the probe is correctly aligned and $V_S = V_M$, the fractional error can be shown (9) to be equal to:

$$\frac{C}{C_o} = 1 + \left(\frac{V_M}{V_S} - 1\right) \left(1 - \frac{1}{1 + (2 + 0.62 \; V_S/V_M) \; Stk}\right)$$

Figure 8 shows the result of the calculation for different V_M/V_S. It can be seen that for particles with Stokes numbers less than 0.1, the error due to variations in sampling velocity relative to the manifold velocity, is negligible.

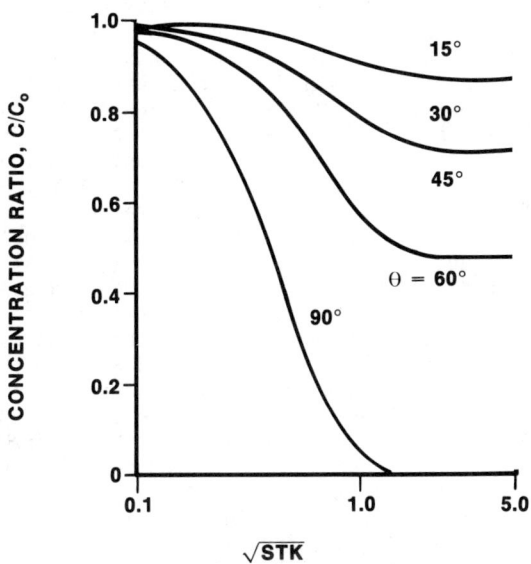

Figure 7. Effect of Probe Misalignment on Concentration Ratio, $V_S = V_M$. (Reproduced with permission from Ref. 8. Copyright 1982, John Wiley.)

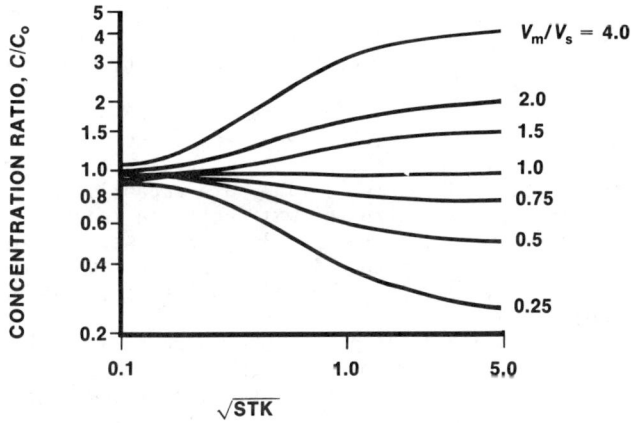

Figure 8. Concentration Ratio Versus the Square Root of the Stokes Number for Several Values of Velocity Ratio, $\theta = 0°$. (Reproduced with permission from Ref. 8. Copyright 1982, John Wiley.)

Particle counters can be calibrated for particle size by passing polystyrene microspheres of known diameter through the interrogation zone, interpreting the intensity of scattered light as a diameter, and placing an added count in a selected channel. There is no method available to calibrate these instruments with respect to absolute particle count. The number produced by a particle counter must be interpreted as a figure of merit, or a qualitative rather than absolute quantitative number. For the purpose of this study, ratios of particle counts between two different probes, or of the same probe under two different conditions will be used to show comparative rather than absolute quantitative results. It will be assumed that the condition, or the probe, that counts the highest number density is at least, a worst possible condition, if not the "true" number density. Since we are interested in qualifying clean rooms or gas streams, and are unable to determine absolute particle counts, the "worst case" conditions will be considered as the most reliable.

The laboratory manifold used to test different sampling techniques, is shown in Figure 9. The tubing used was 316L Process Type C chempolished 3/4 inch tube purchased from Tube Service, Inc. in Milpitas, CA. Two ports, numbered 1 and 3, were fitted with 0.47 mm I.D. chamfered isokinetic probes, as shown in Figure 10, that extended 4.5 inches into the tube. Another port, 2, was installed by welding a Cajon fitting on the tube wall forming a tee, with no isokinetic sampling capabilities. This type of port is typically found in field installations. The gas, obtained from a nitrogen lab source, was passed through a Tylan, Corp. mass flow controller and a Pall Corp. 0.1 µm electronic grade filter. An inlet port was provided to inject polystyrene particles of known diameter. A high pressure particle generator was built based on Figure 11. Gas from a cylinder was filtered through a Balston 0.1 µm filter and passed through a rotameter prior to the particle generator. The particles used were a mixture of Dow's polystyrene microspheres of 0.176, 0.305, 0.460 and 1.091 µm diameter.

Three manifold flow conditions were chosen with Reynold's numbers of 1800, 6000, and 10500. The first is definitely laminar, the latter is turbulent and the mid range corresponds to the mid-range of the PMS counter's flow rate capabilities. The pumping speed in the PMS instrument was chosen so that the linear velocity in the tubing and in the sampling probe were equal, resulting in isokinetic or pseudo-isokinetic sampling conditions. The raw data are presented in Table III and the calculated count ratios in Table IV.

Measurements were made at port number 3 under isokinetic and anisokinetic conditions. Two modes of anisokinetic sampling were used - where the velocity in the sampling tube, V_S, was greater than the velocity in the manifold, V_M, and viceversa. The results appear in Figure 12 where the ratio of the isokinetic to the anisokinetic counts is plotted against the particle diameter, at different Reynold's numbers. Under laminar conditions, one measurement was made ($V_S > V_M$) since the PMS instrument cannot pump consistently at a lower rate than 0.6 cc/sec. Under laminar conditions, a higher count is observed under anisokinetic

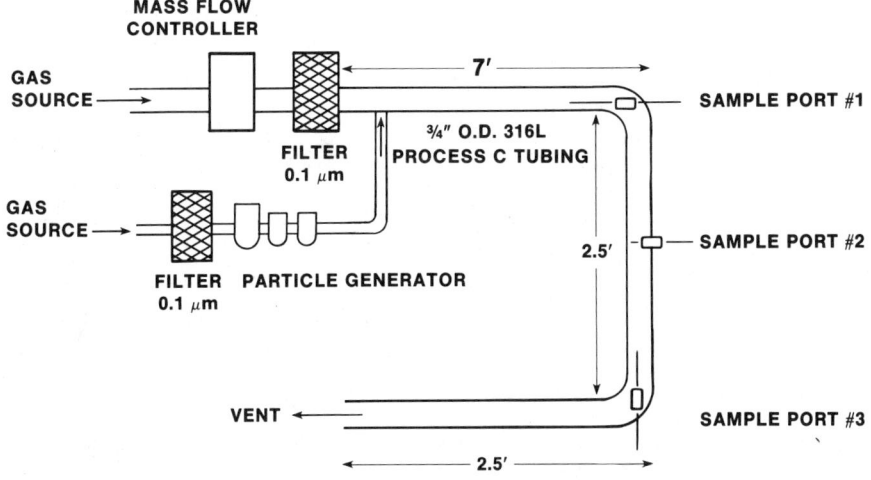

Figure 9. Laboratory Manifold.

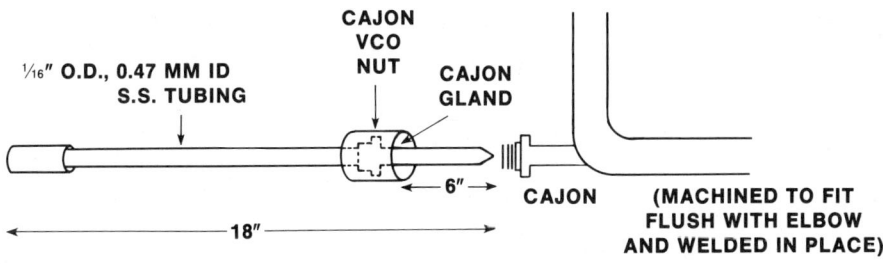

Figure 10. Isokinetic Sample Probe.

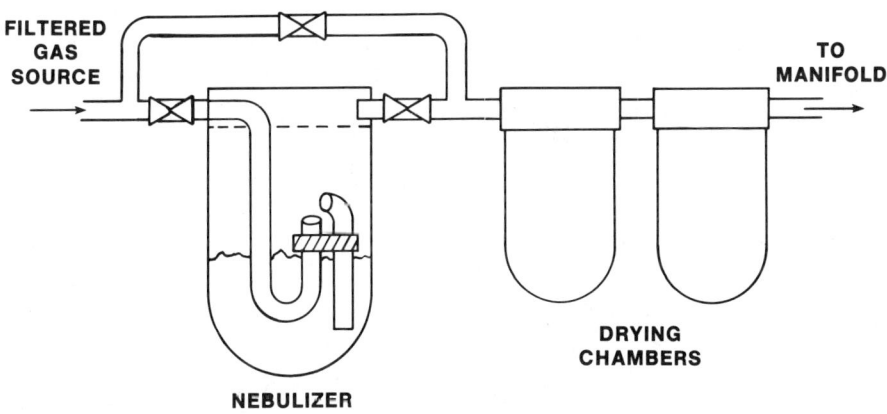

Figure 11. High Pressure Particle Generator.

Table III

Particle Counts

Particle Size (μm)	Counts X 10^{-3}			
	0.176	0.305	0.460	1.091
Port 3, Vs = Vm				
Turbulent	101. ± 1.71	34.4 ± 0.380	7.75 ± 0.098	0.160 ± 0.007
Mid-Range	19.5 ± 0.595	10.2 ± 0.313	3.83 ± 0.154	0.341 ± 0.011
Laminar	9.11 ± 0.633	5.55 ± 0.375	2.52 ± 0.162	0.291 ± 0.031
Port 3, Vs > Vm				
Turbulent	106. ± 0.897	35.3 ± 0.218	7.60 ± 0.062	0.151 ± 0.014
Mid-Range	19.9 ± 0.567	10.3 ± 0.392	4.10 ± 0.192	0.411 ± 0.024
Laminar	10.8 ± 0.261	6.96 ± 0.153	3.69 ± 0.100	0.526 ± 0.021
Port 3, Vs < Vm				
Turbulent	107. ± 0.809	38.1 ± 0.298	8.11 ± 0.071	0.181 ± 0.011
Mid-Range	25.1 ± 0.348	13.2 ± 0.285	5.03 ± 0.075	0.463 ± 0.033
Port 2, Partial Flow Thru Tee				
Turbulent	131. ± 5.82	47.1 ± 2.23	9.69 ± 0.468	0.231 ± 0.010
Mid-Range	31.0 ± 0.421	18.0 ± 0.329	6.91 ± 0.188	0.652 ± 0.017
Laminar	8.49 ± 0.150	5.96 ± 0.121	2.96 ± 0.057	0.396 ± 0.025
Port 1, Vs = Vm				
Turbulent	145. ± 2.64	47.7 ± 0.304	9.24 ± 0.071	0.131 ± 0.004
Mid-Range	21.4 ± 0.285	11.1 ± 0.227	4.20 ± 0.053	0.347 ± 0.010
Laminar	10.5 ± 0.297	7.16 ± 0.110	3.28 ± 0.027	0.480 ± 0.017

Mean ± $ts/(N)^{1/2}$, t = 2.015, N = 6

Table IV

Calculated Count Ratios

Particle Size (μm)	0.176	0.305	0.460	1.091
Port 3, Isokinetic **Port 3, Anisokinetic, Vs > Vm**				
Turbulent	0.952 ± 0.008	0.975 ± 0.005	1.020 ± 0.002	1.060 ± 0.046
Mid-Range	0.979 ± 0.002	0.987 ± 0.007	0.934 ± 0.006	0.830 ± 0.020
Laminar	0.841 ± 0.036	0.797 ± 0.035	0.684 ± 0.024	0.553 ± 0.034
Port 3, Isokinetic **Port 3, Anisokinetic, Vs < Vm**				
Turbulent	0.942 ± 0.009	0.902 ± 0.003	0.955 ± 0.002	0.884 ± 0.014
Mid-Range	0.777 ± 0.013	0.770 ± 0.007	0.762 ± 0.019	0.737 ± 0.026
Port 3, Isokinetic **Port 2, Partial Flow Thru Tee**				
Turbulent	0.768 ± 0.019	0.730 ± 0.024	0.800 ± 0.028	0.693 ± 0.001
Mid-Range	0.628 ± 0.010	0.565 ± 0.007	0.555 ± 0.007	0.523 ± 0.002
Laminar	1.072 ± 0.054	0.930 ± 0.042	0.854 ± 0.037	0.735 ± 0.029
Port 3, Isokinetic **Port 1, Isokinetic**				
Turbulent	0.693 ± 0.001	0.721 ± 0.003	0.839 ± 0.002	1.221 ± 0.011
Mid-Range	0.910 ± 0.015	0.919 ± 0.009	0.913 ± 0.024	0.983 ± 0.003
Laminar	0.868 ± 0.034	0.776 ± 0.039	0.768 ± 0.042	0.606 ± 0.040

$$\frac{X}{Y} \pm \frac{Y\Delta X - X\Delta Y}{(Y+\Delta Y)^2}, \text{ where } \Delta X = ts_x/(N)^{1/2}, \text{ and } \Delta Y = ts_y/(N)^{1/2}$$

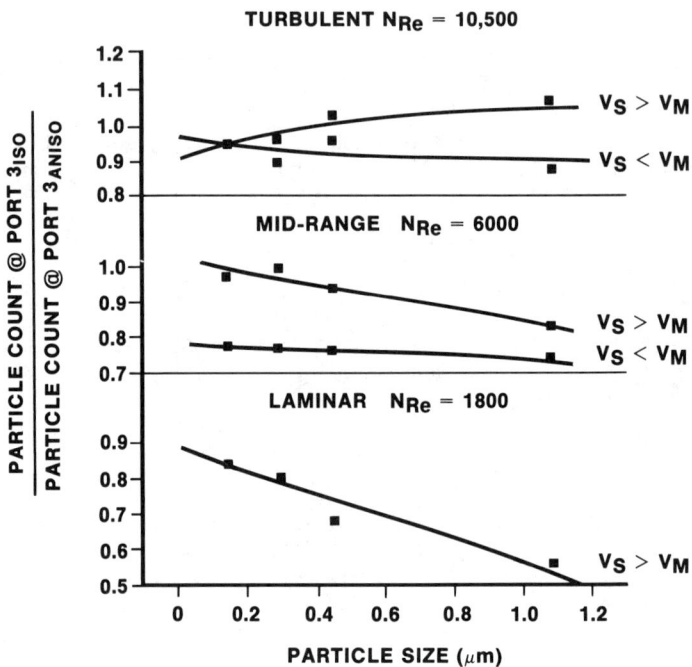

Figure 12. Effect of Anisokinetic Sampling on Particle Concentration.

conditions. These results can be understood if it is assumed that, under laminar conditions, segregation of the particles by particle size occurs, with the larger particles moving towards the wall. Isokinetic sampling will sample the center of the manifold, and, therefore, will sample a greater number of smaller size particles. When $V_S > V_M$, some local turbulence is created, resulting in better mixing and a more even distribution of particles in all size ranges. Under a strictly turbulent regime with $V_S < V_M$, linear behavior is observed, particle size is not a significant factor, and higher counts are seen under anisokinetic conditions in all sizes. For $V_S > V_M$, particle size is not a significant factor either. However, higher counts are seen under isokinetic conditions for particles larger than 0.4µm, with smaller counts seen for smaller particles under similar conditions. In the mid-range velocity, when $V_S > V_M$, the trend parallels laminar flow with higher counts observed under anisokinetic conditions. This is probably due to some segregation of the particles by particle size as discussed earlier for laminar flow, indicating that even though the Reynold's number is greater than 4000, the flow characteristics may be closer to laminar than strictly turbulent. When $V_S < V_M$, the trend is similar to that when $V_S > V_M$, however particle size is less significant, and anisokinetic conditions produce even larger counts than when $V_S > V_M$. This may be due to a better mixing when $V_S < V_M$ since the vortex lines create more turbulence under these conditions than when $V_M > V_S$.

Figure 13 shows the comparison of counts obtained at port 3 under isokinetic conditions, with those at port 2. Diverting part of the gas stream through a tee to port 2 produces an even greater disturbance than that caused by a 90° bend in the system. Impaction at the disturbance causes particles to be deposited on the tube wall, with deposition rates being lower for larger particles. This reduced deposition rate may be due to the larger particles having sufficient momentum to bounce back and become re-entrained in the gas stream. The substantial disturbance caused by the tee prevents redevelopment of the velocity profile, minimizing the stagnant boundary layer at the tube wall. This effect will reduce the number of particles that can remain deposited on the wall, thus causing the counts at port 2 to be greater than at port 3. The flow rate which should be laminar will also behave more like the turbulent flow rates, because the laminar velocity profile will not be redeveloped.

As seen in Figure 14, if the counts obtained under isokinetic and laminar conditions at ports 1 and 3 are compared, it is found that the counts in port 1 are larger than those in port 3 and increase as the particle diameter increases. This phenomenon could be caused by flow disturbances produced by the 90° bend. Many of the larger diameter particles will lose momentum in collisions with the wall at the bend and become deposited on the tube wall. The smaller particles will remain suspended in the gas with minimal disturbance due to the bend. On the other hand, this effect may also be caused by a segregation of particle diameters in the bulk stream, with particles having a larger diameter migrating away from the center of the stream. Particles with

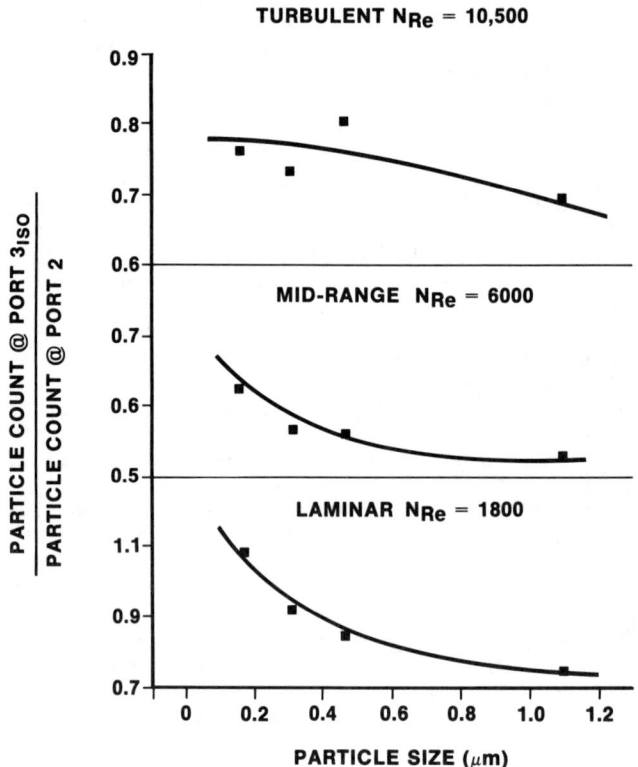

Figure 13. Effect on Particle Concentration of Sampling Through a Tee.

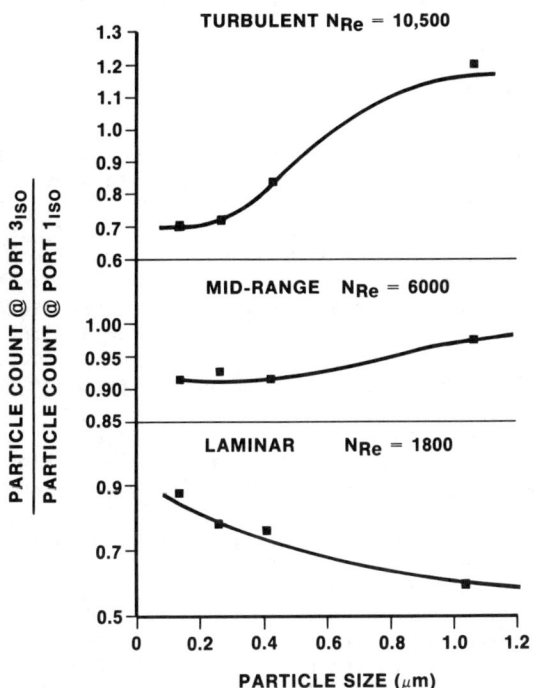

Figure 14. Effect of Piping Bends on Particle Concentration.

diameters less than approximately 0.5 microns move under a Brownian motion regime and those with large diameters are affected more by gravitational forces as described by Allen (10). Under turbulent conditions, particles are again being deposited on the tube wall after the bend. However, the trend is reversed, with deposition being greater for smaller particles. The larger particles may have enough momentum to bounce back and become re-entrained into the gas stream, while smaller particles may remain deposited on the wall. The difference in count between the two probes is a minimum at larger particle diameters. This indicates that the losses of the larger particles are minimal under turbulent conditions. At mid-range velocities, there is little dependence of counts on particle size at either probe. It appears that these conditions are best suited for particle transfer with minimal losses.

Conclusions

The number of samples needed to obtain an experimental standard deviation less than or equal to the Poisson standard deviation was found to be six when particle counts were 2000 particles/ft^3 or greater. If the particle concentration is lower, increased counting times and number of samples will be required to attain the same level of statistical significance.

The importance of characterizing any pressure reducing or sampling device with respect to the conditions required to eliminate intake of ambient particulates has been demonstrated. Significant differences in condition requirements are possible between seemingly identical devices.

The effect of sampling technique on total particle count and particle distribution has also been demonstrated. Since there is no standard method of calibrating particle counters and no definite way to determine the "true" particle count or distribution in a gas stream, for the purposes of qualifying clean rooms or gas streams, "worst case" or highest particle counts should be considered preferable. Under both laminar and turbulent conditions, counts increase over those under isokinetic sampling conditions, when flow is partially diverted through a tee to sample. If flow is disturbed by piping elbows particle counts decrease due to deposition on the pipe walls. This effect is greater for larger particles under laminar flow conditions, and greater for smaller particles under turbulent conditions. Particle counts are also increased by anisokinetic sampling except under turbulent conditions, when the sampling velocity is greater than the main gas velocity for particles larger than about 0.4 μm diameter. Therefore, for particles with diameters less than or equal to 1.0 μm, turbulent flow conditions and anisokinetic sampling are preferred to yield the highest counts.

Literature Cited

1. Burgraaf, P. S.; Semiconductor Int., July, 1982, 35.

2. Francis, T.; ASTM Standardization News, May, 1983, 16.

3. Francis, T.; Ind. R&D, August, 1982, 121.

4. Brummett, R., and Sulavick, R. V.; J. of Env. Sc., 1982, 25, 16-20.

5. Davies, R. "Laboratory Instrumentation Microscopy Pollution Analysis", Series II, Vol. III, Microscopy - Pollution Analysis, International Scientific Comm., 1977; p. 100-120B.

6. Liu, B. Y. H., Berglund, R. N., and Agarwal, J. K.; Atmospheric Env., 1974, 8, 717.

7. Perry, R. H., Chilton, C. H. "Chemical Engineers Handbook", McGraw-Hill, NY, 1973; p. 5-10.

8. Hinds, William C. "Aerosol Technology: Properties, Behavior, and Measurement of Airborne Particles", John Wiley & Sons, NY, 1982; p. 104.

9. Ibid., p. 187-193.

10. Allen, T. "Particle Size Measurement", Chapman and Hall, NY, 1981; p. 223.

RECEIVED July 8, 1985

25

Microelectronics Processing Problem Solving: The Synergism of Complementary Techniques

J. N. Ramsey

IBM Corporation, Hopewell Junction, NY 12533

It has been almost 25 years since Prof. David Wittry presented the first application of the electron probe microanalyzer to the solution of semiconductor/microelectronics problems(1). The need for small area chemical analysis has been strong in this field and is growing because of continuing miniaturization(2,3). The electron probe microanalyzer (and later the scanning electron microscope with energy dispersive analysis capability) has developed markedly over these 25 years. This small area analysis, almost exclusively elemental, has been essential for process development and problem solving. Elemental information, however, is often not sufficient. For example, finding hydrogen, carbon, oxygen and nitrogen in a particle does not add measurably to the identification of a material and its elimination from a process. Thus, the search has been for small area molecular analysis techniques to complement the elemental analysis.

X-ray or electron diffraction allows identification of crystalline species by the long-used Hannawalt-Dow-ASTM-JCPDS system(4). Small particles can be removed for analysis in a small rotating specimen X-ray powder camera, or by extraction replication and selected area diffraction in a Transmission Electron Microscope (5). For those specimens where a residue of reactant or corrosion product is too adherent, the material may be removed for analysis by micro-bulldozing (with a microhardness indentor), micro-jack hammering (with a needle attached to a small piezoelectric crystal on a pencil-like rod), and micro-boring (with a precision controlled dental drill)(5). While these techniques are very useful, they fail with amorphous materials, organics and polymers.

Polarized light microscopy has long been used in chemical and mineralogical studies, and was raised to a high level of applicability by Walter McCrone and associates, culminating in the extensive Particle Atlas(6), which included innumerable glasses, organics and polymers.

Small area molecular spectroscopies are obviously desirable, but there have been problems of coupling a microscope and a spectrometer and retaining high efficiency energy transfer. This was solved for Raman spectroscopy by Dalhaye and Dhamelincourt(7), which resulted in 1 micron lateral-dimension

Raman imaging capability in an instrument (the MOLE) produced by Jobin-Yvon(8). This technique has been widely applied (several symposia have been held)(9,10) including the field of microelectronics(11,12). Small area infrared analysis has been made possible with the introduction of all-reflecting infrared microscope optics for use with either dispersive or FT infrared instruments(13). We have applied this technique extensively to process control problems in microelectronics manufacturing(14-16).

The vibrational transition energies of the molecules, while in the infrared portion of the spectrum, are "seen" as the difference (Delta) between the impinging visible laser light and the scattered Raman light, also in the visible. Both infrared and Raman give not only a fingerprint of each molecule, but also information on its functional units and structure. However, because the two methods respond to different quantum mechanical selection rules, the intensities of functional groups and structures can vary, giving sometimes markedly different fingerprints of the same material. The use of visible light in Raman to stimulate infrared transitions offers several advantages (11,17):
- Smaller analytical area (because the diffraction limit (focussing) of visible light less than 1µm vs. about 15-20µm for infrared)
- Heavy metal compounds with very low vibrational frequencies can be seen (because with Raman, the range extends downward to below $100cm^{-1}$, whereas infrared stops at about $500-600cm^{-1}$)
- The spectrum can usually be examined in-situ (because the Raman can be in the reflective mode, whereas with infrared, in the transmission mode*, the specimen must be transferred to an infrared transparent substrate - fortunately for silicon device work, silicon is transparent and usually particles can be analyzed in situ).
[*NOTE: ADDED IN PROOF: Since this conference, considerable progress has been made in reflection microscope attachments on FTIR instruments(30).]

Raman, however, also has several disadvantages theat need to be worked around. Raman scattering is weak, requiring high laser optical power densities that may "burn" or volatilize the sample (although most specimens can be immersed in a drop of water, under a cover glass(17)). In addition, the specimen may fluorosece, which will overwhelm the Raman signal, and all materials do not have Raman-active vibrational bands. Thus, Raman and infrared are very complementary techniques.

Laser desorption/ionization and mass analysis has been available in the transmission mode for several years (18,19). This technique has recently been extended to the much more useful reflection mode as the LAMMA 1000 by Leybold Heraeus(18,20) and the LIMA by Cambridge Mass Spectroscopy Co.(21,22) In these instruments, the region of interest is located under a microscope. The laser beam is focussed and pulsed through a microscope, producing a plasma that fragments and ionizes the molecules, which are then analyzed in a time of flight mass spectrometer. One of the major parameters to be controlled and measured is the laser power, to get reproducible fragmentation:

the fragment pattern should be recognizable to an organic mass spectroscopist and close to those from electron impact so that the extensive literature and spectra libraries can be utilized. If the laser power is too high and fragmentation extends to H, C, O and N, then, of course, little molecular information is obtained.

The most productive interaction of an analytical/ characterization group and the process development and manufacturing people who have the problem, is to operate in a problem-solving mode. When yields, reliability, in-processing testing, or some other critical product parameter start to drift, whether it is in development or manufacturing, the analyst should not accept a large number of samples for exhaustive elemental analysis. It should be determined what the nature of the problem is, or what attributes of the process have shifted, or what changes were made recently. (Every solution to a problem carries within it the seeds of the next problem.)(3) The attitude of the analyst should be "We are here to help you by getting to know your process, and its interactions with previous and subsequent process steps". The analyst decides the specimens most likely to show differences and the proper form for the techniques to be employed (for example, Auger or ESCA should not be performed on a sample that has been carried around in a someone's hands). Constant and iterative interaction is generally required, leading to design of experiments, clarifications, more (and different) analyses, until the process is understood and the problem is solved. Sometimes the procedure is short; more often, with the complex process interactions so common in the microelectronics field, it is not.

This problem solving approach is always productive because everyone "wins": the process engineer has more understanding for better process control, and the analyst is not just an analyst but, by using broad science and engineering concepts, the analyst is an integral part of the processing team.

Examples of Problem Solving by Molecular Analysis

Analysis of Intermetallics - Need for Molecular Analysis.

The IBM chip joining system uses solder bonding in a new large module, called Thermal Cooled Module, which requires an ability to remove and replace chips for engineering changes. Stress tests made during the development period of the program required in-situ analysis and characterization of the phases formed in the roughly 50 micron diameter vias on the chips. Figure 1 shows electron microprobe video scans of the intermetallic phases remaining on the Cr/Cu/Au bond pad via metallurgy after several reflows and subsequent etching away of the solder. There appear to be two types of intermetallics, with different morphologies and different levels of Cu and Sn (it is not possible to do a quantitative analysis on regions as small and irregular as these particles since the electron beam cannot be contained in a particle to match conditions with elemental standards).

Miniaturization is progressing with X-ray diffraction: a locatable 10-micron diameter area can now be analyzed with a microdiffractometer(23,24). Figure 2 shows that both Cu_3Sn and

Figure 1. Intermetallics after solder reflow-- elemental analysis by electron probe.

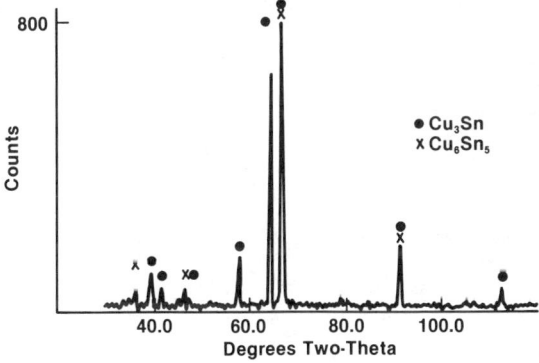

Figure 2. Different Cu-Sn intermetallics formed on solder reflow-- X-ray diffraction.

Cu_6Sn_5 are present. The Au in the Cr/Cu/Au film attachment scheme on the Si chip also forms Au-Sn intermetallics, which mainly float into the solder and have been etched away in preparing these samples. In addition to phase analysis, the departure from equilibrium lattice positions can be interpreted as a measure of the stress in a crystalline substance. Figure 3 shows the decrease in stress in the Cu_3Sn layer as a function of the number of reflows: the decrease in stress was attributed to stress-relief by cracking and spalling into the solder (which was verified by SEM).

Extraction Replication (Crystals in Fired Moly). The multilayer ceramic package that is used by IBM has a sintering operation in which the alumina/glass body undergoes sintering-with-a-liquid phase simultaneously with the firing of the molybdenum lands. This is a complex process, requiring close control over incoming materials and processes(25). The molybdenum surface is electroless nickel plated and electroless "flash" gold plated to provide a solderable or brazeable surface. In the early days of the program, there was an occurrence of blisters of the Ni/Au plating after a subsequent thermal operation. Figure 4a shows an SEM micrograph of such a blister, which has been opened and laid back with a needle to expose material for analysis. A smaller blister is seen towards 12 o'clock from the big blister. Figure 4b, an enlargement, shows a cavity network under the blister and Figure 4c shows that the cavity is lined with needle or rod crystallites and some platelets. Energy dispersive X-ray analysis showed primarily aluminum, magnesium and silicon. The crystallites were extracted on a replica(5), and electron diffraction was done: the needles (rods) are mullite while the plates are anorthite. These unexpected phases were necessary clues to institute process development changes and controls to eliminate the voids, which had trapped plating solution. These, in turn, expanded and loosened or burst the plating during the next thermal excursion. Thus, appropriate small area analysis gave a proper diagnosis (what appeared to be a plating problem was really a sintering problem) and led to greater process understanding and controls. Part of this was a tightening of the composition limits on the glass, which is controlled by a Materials Engineering Specification. Such specifications are seen to be "living" documents, being tightened or loosened as trade-offs occur in yield/cost of product/cost of materials/cost of controls.

Au-Sn Braze Material - Why Doesn't It Work? In the early development stages of a brazing operation, Au-Sn brazing alloy preforms were used to obtain a hermetic seal. A problem arose when leakers occurred periodically. The process, good and bad product and the materials (there was no a priori identification of good and bad) were examined using metallography, electron probe, Auger, ESCA, X-ray diffraction, thermal analysis (for melting point) and Plasma Chromatography/Mass Spectroscopy (to check for possible organic films, such as rolling and cutting

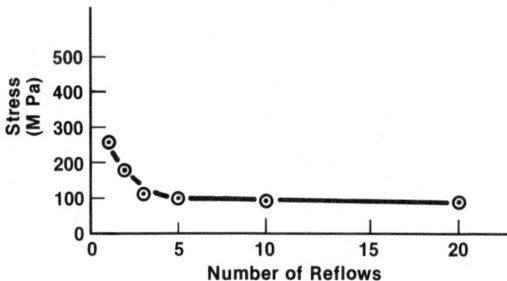

Figure 3. Cu_3Sn intermetallic stress vs. number of solder reflows-- X-ray diffraction.

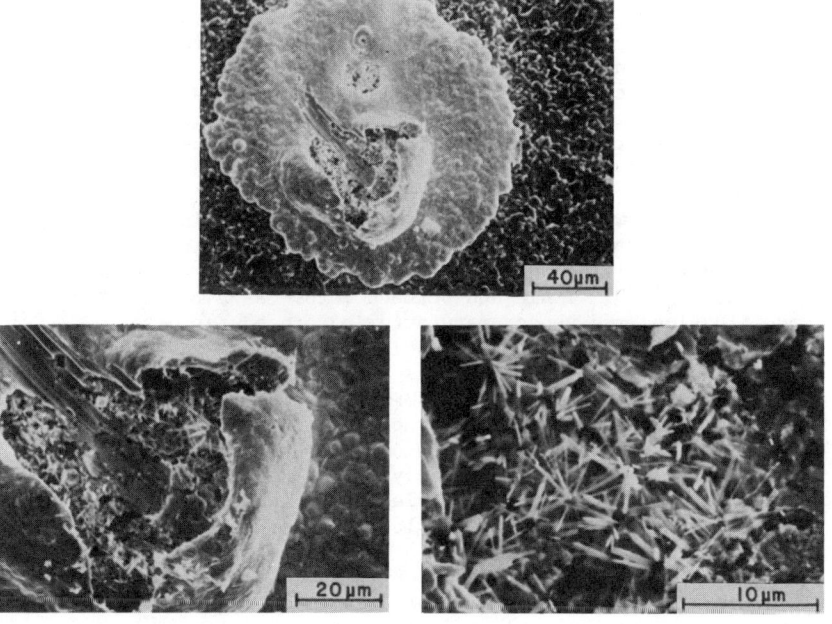

Figure 4. SEM of blistered nickel over fired molybdenum via-- opened with needle, showing mullite and anorthite.

oils that survived the cleaning operations). Nothing was found that indicated any type of problem, so it was recommended to clean up and optimize the process. The process was eventually put into manufacturing, where problems again occasionally developed but they were more grouped or clustered in time. Something new had been added: manufacturing line discipline. Process and materials specifications had been established, including vendor lot control on the preforms. Manufacturing process engineers could now observe that there was a lot-to-lot dependence on the preforms. Again the processes and materials were examined, and again nothing was found to confirm the observation or to assist the process engineers. However, a newly designed Theta-Theta X-ray diffraction unit(23) with hot stage, in which the specimen remained horizontal, allowed melting and re-melting studies to be preformed by cycling through the melting point(26) under controlled environments. This equipment allowed us to test the hypothesis that since there were not organics (by PC/MS) there must be tin oxide in the specimen. The preform was held a few degrees over the melting point in forming gas for 10 hours to allow any SnO_2 in the preform to float to the top (thermodynamically, SnO_2 should not decompose at 283°C), and then cooled to room temperature. Figure 5 shows X-ray diffraction scans of a "bad" preform before and after thermal cycling, in which the build-up of SnO_2 is quite evident, whereas there was no such build-up in the "good" preform. This can be seen in Figure 6 in which reflection electron diffraction shows significantly more SnO_2 in the "bad" (optical microscopy also showed a reddish color). Thus, the melted surface after pre-tin would indeed show lot-to-lot variations which could "tip a process over the edge" if it was not properly centered or optimized. This SnO_2 information showed that the vendor had been re-using scrap that had oxidized during rolling and stamping in the starting material for his next rolling and sampling operation, thus allowing variation of SnO_2 within and between lots. As a result, the specifications and vendor manufacturing procedures were tightened. Now that the SnO_2 "noise" in the brazing process was identified and eliminated, it was possible to unlayer to the next major process detractor (a contamination) which allowed a re-optimization of the entire process. Incidentally, a re-analysis showed that the previous problem parts in development had the same SnO_2. Thus, an early signal was not picked up due to the lack of manufacturing-style line discipline in the development environment and the lack of insight and capabilities fostered by the new in-situ hot stage XRD melting and wetting experiments.

There was a later problem of some lots of AuSn braze pastes not wetting satisfactorily. Investigation of line yield data revealed that the problems were actually AuSn powder-lot dependent, even though they were within specification. The previous success of X-ray diffraction dictated its application to this problem. The procedure was repeated, with the results in Figure 7, where a "bad" lot of powder showed no SnO_2 as-received, but showed SnO_2 after melting under forming gas. This did not

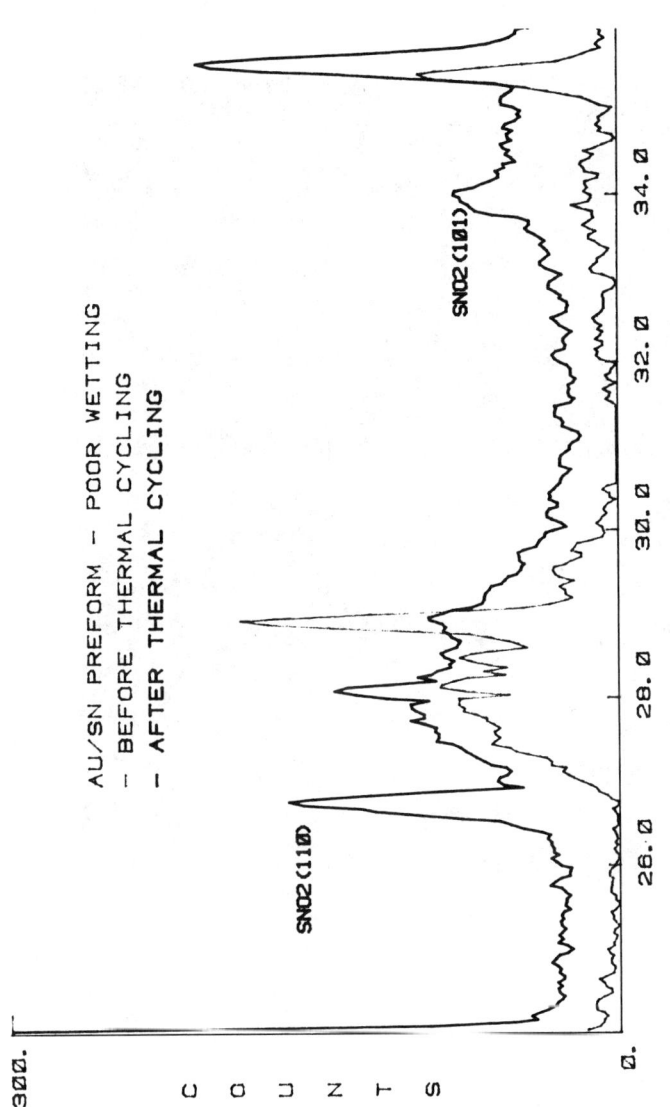

Figure 5. Poorly wetting Au-Sn preform: contains SnO_2--by X-ray diffraction.

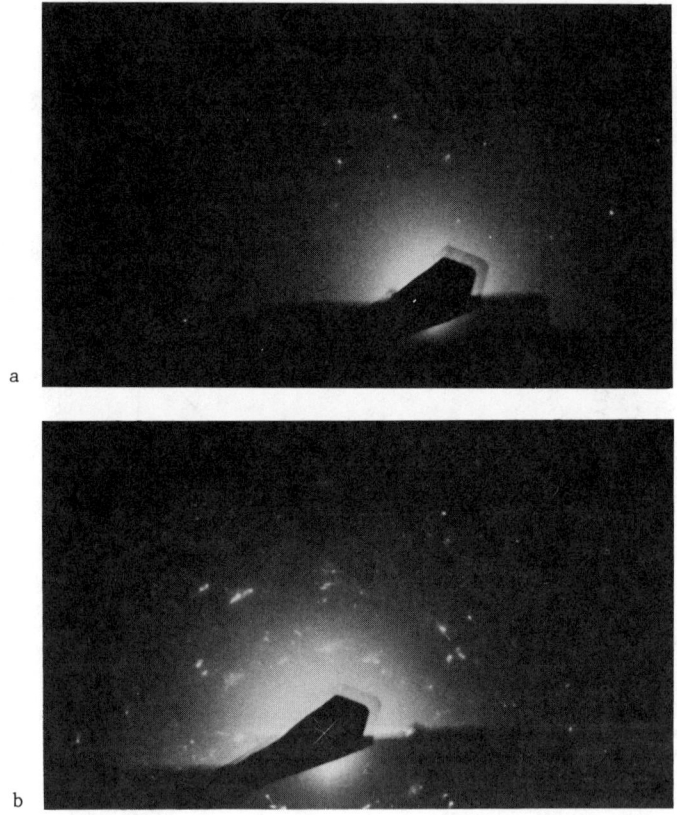

Figure 6. Reflection diffraction of Au-Sn preforms after melting. (a) "Good" preform shows Au-Sn plus minor SnO_2 reflections. (b) "Bad" preform shows Au-Sn plus major SnO_2 reflections.

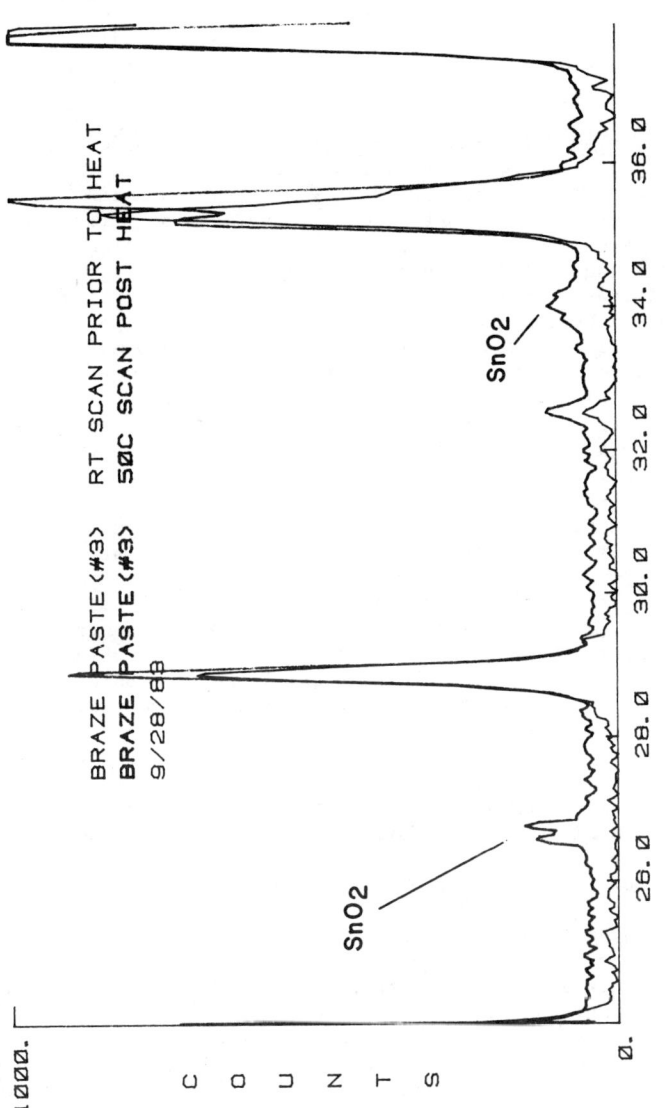

Figure 7. "Bad" Au-Sn braze paste shows SnO_2 after melting by X-ray diffraction.

happen with the "good" powder. ESCA confirmed an increase in SnO_2 on the "bad" powder, but because there wase some SnO_2 even on the "good" powder (remember ESCA is looking at only the outer 3.5-5.0nm) it lacked the definitive go/no-go distinction of Figure 7. The markedly different appearance of the two samples after the melting under forming gas in the X-ray θ-θ diffractometer provided an additional test for incoming lots of powder: the SnO_2 then floated to the surface upon melting provided a distinctive reddish visual cast even though it was difficult to photograph. The "bad" powder could, of course, be salvaged by etching.

Inadvertent E-Beam Polymerized Film. This is an example of the solution to a problem being the basis of future problems.

In electron optics columns, there are electron beam forming/shaping apertures that develop contamination films with use due to the polymerization of hydrocarbons by the electron beam in the vacuum. This insulating film has to be cleaned in order to eliminate the loss of resolution due to deflections of the beam caused by charging of the film. For several years, a solution for this problem of frequent down-time has been to heat the apertures to a high enough temperature to give a zero sticking coefficient. Thus, small thin film resistors were designed into an E-beam photoresist exposure system. The system was used quite successfully for several years by maintaining a schedule for all parts, including the aperture heaters. Recently, beam resolution and deflection problems began to appear. It was found that the shaping aperture plate had developed a film around the aperture in the characteristic circular beam pattern. It was postulated that a new photoresist was outgassing and providing the contamination film, and it would be necessary to know what material was involved. Reflectance Fourier Transform Infrared (FTIR) was required for a molecular analysis of the film. To check for possible surface polarization distortions in grazing angle spectra, two angles were used - as seen in Figure 8. The spectra are the same at 82° and 88°; the 1734cm^{-1} absorption is the carbonyl of ester, while the 1271 is the C-O stretch of aromatic ester. This indicates that the unknown material was an aryl ester, possibly a phthalate, or, more likely, an alkyd resin. While it was known that none of the various resists, solvents, cleaning solutions and vacuum system lubricants would match, spectra were run anyway on the chance that something was contaminated. There were no matches. When, after cleaning, the aperture plate again developed an insulative film, the E-beam unit was stripped down for cleaning, preventative maintenance, and further detective work. An alert, experienced technician noted that the aperture heater was brown instead of being clear, as usual. An isopropyl alcohol (IPA) extraction was analyzed by FTIR. The spectrum is shown in Figure 9, along with a spectrum from Figure 8 for comparison. The match is obvious. The material was outgassing at the operating temperature of the resistor heater.

25. RAMSEY *Microelectronics Processing Problem Solving* 409

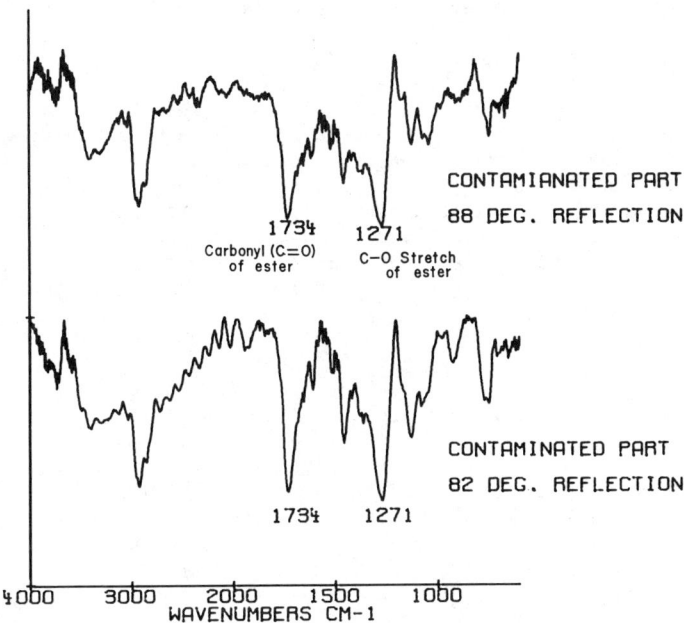

Figure 8. FTIR spectra of contamination layer on E-beam aperture at two reflective angles.

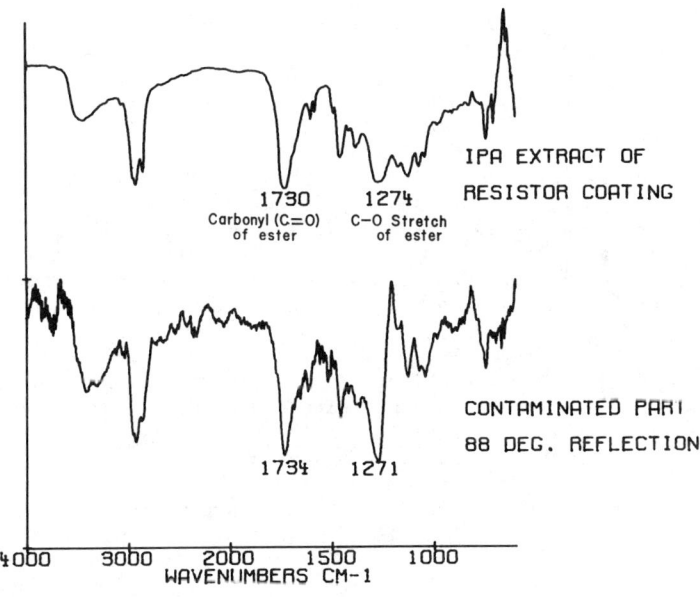

Figure 9. FTIR spectra of contamination layer on E-beam aperture, and IPA extract of new resistor coating.

What are the lessons to be learned from this example? The early question we ask is, "What have you changed in the process?" The answer was "nothing" -- and this was correct. The resistor heater vendor changed an SiO_2 coating to an organic coating as a cost reduction. We had qualified the heater with SiO_2 and had built purchase specifications around it. These were violated by the vendor who made process/material changes without our knowledge. Also, the change was not noticed by relatively new personnel doing the last two preventative maintenance operations. The high value of good observation by experienced people (noting the brownish heater) saved considerable analysis time of virtually everything in the E-beam column to track down the source of the vapor.

Is Your Solder Pad Joining "Hairy"? As part of a solder/flux/ cleaning procedure, the residue in Figure 10 (insert) was produced. The residue consists of 6x10µm leaf-like crystals on a ceramic substrate. The location of the residue prevented its analysis in-situ, so extraction replication(5) was used to remove some of the crystals for analysis by several small area techniques, as needed. Electron probe microanalysis showed lead, carbon and oxygen, which could indicate many possibilities. Small area infrared analysis produced the spectrum in Figure 10; the absorption at $1400cm^{-1}$ was interpreted as basic lead carbonate. This was also confirmed by electron diffraction of the same crystals. Raman microprobe gave conflicting data because the lead carbonate decomposed under the laser beam, giving a mixture of "daughter" products (mixtures constitute a "real world" analytical problem).

Who Exfoliated in the Cleaning Tank? A contamination problem developed in a device line cleaning operation: particulate contamination monitors (flamed Si wafers) showed many particulates, as in Figure 11. [(done in-situ because Si is 50% transmissive in the infrared.)] The in-situ small area infrared spectrum (indicates) zein, a corn-derived protein, which has over 100 commercial uses, many in paper processing. However, as bare paper is not allowed in this clean area, another source of zein had to be found. Zein is also widely used in cosmetics as a powder or "pancake" make-up base. Thus, in all likelihood, some operator had violated the rules regarding cosmetics in the clean room, and "exfoliated" over the cleaning station tanks.

Particles on Recording Surface. One of our development lines had a problem, with ~10µm particulate contamination on a recording surface, which caused, "dropped bits" on reading, and subsequent re-writing(27). In-situ small area Raman analysis (top two spectra of Figure 12) showed the material to be polypropylene (comparing it with the bottom spectrum). Polypropylene, we were told, was not a possible contaminant because there was no polypropylene in the materials of construction or packaging. A further look at the package revealed that a paper-like material

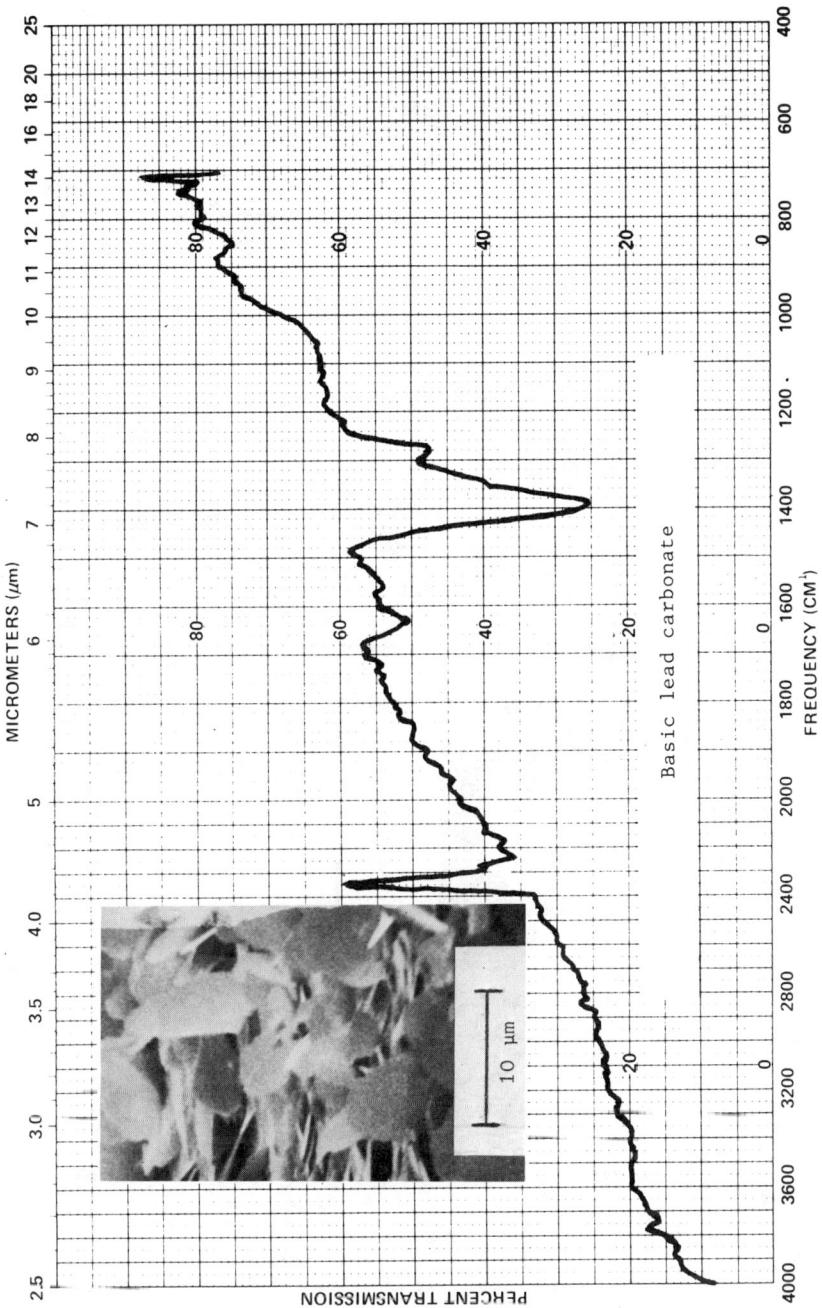

Figure 10. Residue from solder/flux/cleaning operation--by small area IR.

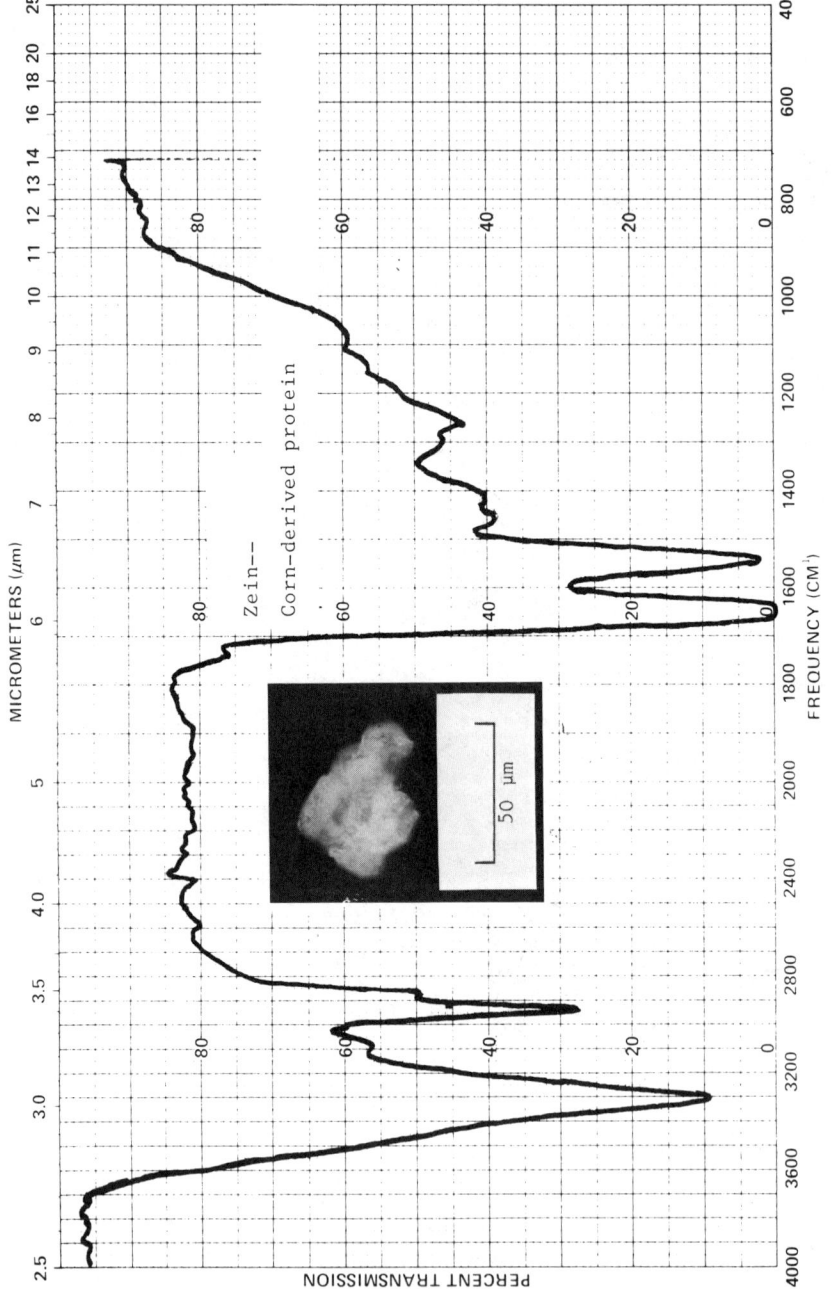

Figure 11. Residue on wafer after cleaning operation--by small area IR.

25. RAMSEY *Microelectronics Processing Problem Solving* 413

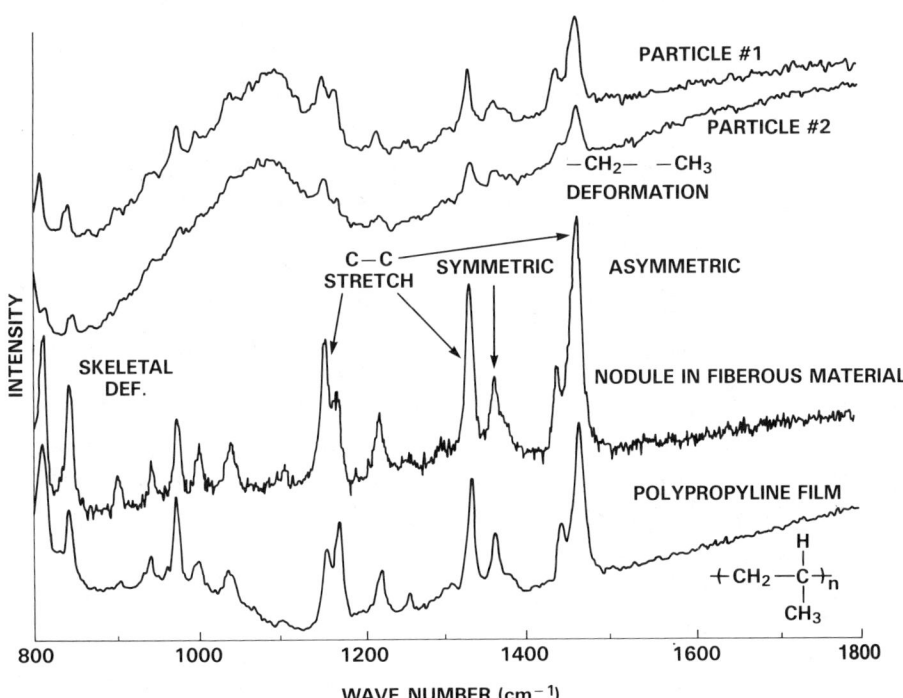

Figure 12. Particles on recording surface—by small area Raman. (Reproduced with permission from Ref. 27. Copyright 1983 International Conference on X-ray Optics and Microscopy.)

was used to "clean" the surface on command. This material was "teased" apart and examined microscopically (insert Figure 13). There are small nodules in the fibers to help hold the fibers together. The fibers are shown to be cellulosic by conventional infrared analysis and by polarized light techniques. The nodules are shown to be polypropylene by small area Raman (third spectrum, Figure 12) and by small area infrared (Figure 13). Was this indeed the source of the problem polypropylene particles? We could not be sure because, while this was the only known source, there were no distinguishing elements unique to these polypropylenes (electron microprobe and LAMMA showed Ti, but that is the usual catalyst in polypropylenes). The "proof" came when a different vendor's filter paper (without nodules of any type) was used without subsequent problems.

What difficulties did we encounter with these analyses? The particles were embedded in the recording medium and were difficult to remove, giving fragments which were difficult to collect together on an infrared transparent substrate for small area infrared. The fibers (approximately 10μm) were too small for small area infrared, and masses were too thick: thus, small area IR failed. LAMMA fragmented the fibers and nodules too much: it also failed (see later for LAMMA successes).

Particulate Analysis: Is it Nylon or Skin? Nylon and human skin are very common constituents of contamination in clean rooms (and not-so-clean rooms, too). The Nylon comes from garments and bushings/gears, while humans are constantly exfoliating skin flakes and fragments. We have had great success with small area infrared(14,15,16) and small area Raman(11,12). However, R.M. Scott pointed out(15) that, as Nylon and skin are both amides, their infrared spectra are quite similar, requiring full width at half maximum measurements on high resolution spectra to distinguish. Figure 14 is the IR spectrum of fluorescent hemispherical particles found on our modules after a brazing operation, showing that Nylon shreds or threads had melted. Figure 15 shows the IR spectrum of skin. Raman and polarized microscopy were explored as improvements, with the latter being especially successful. The very high bi-refringence of drawn Nylon (1.580 index of refraction axially, with 1.520 transverse) coupled with its distinctive hemispherical form with dimpled craters after melting ($\sim$265°C) allows unequivocal differentiation from skin, with its low bi-refringency ($\sim$1.530 index of refraction) with no change of shape (but darkening in color) on being heated to $\sim$265°C). In addition, the oils and salts in skin outgas at room temperature (and especially after $\sim$265°C).(27)

Analyses With The LAMMA 1000. As mentioned earlier, the degree of fragmentation of a molecule controls the amount of useful information to lead to a molecular analysis. I mentioned two examples of excessive fragmentation. Therefore, precise control and measurement of very low laser power applied to the spectrum is essential. Frank Anderson, analytical chemist of my laboratory, carried over a concept from his Transmission Electron

25. RAMSEY *Microelectronics Processing Problem Solving* 415

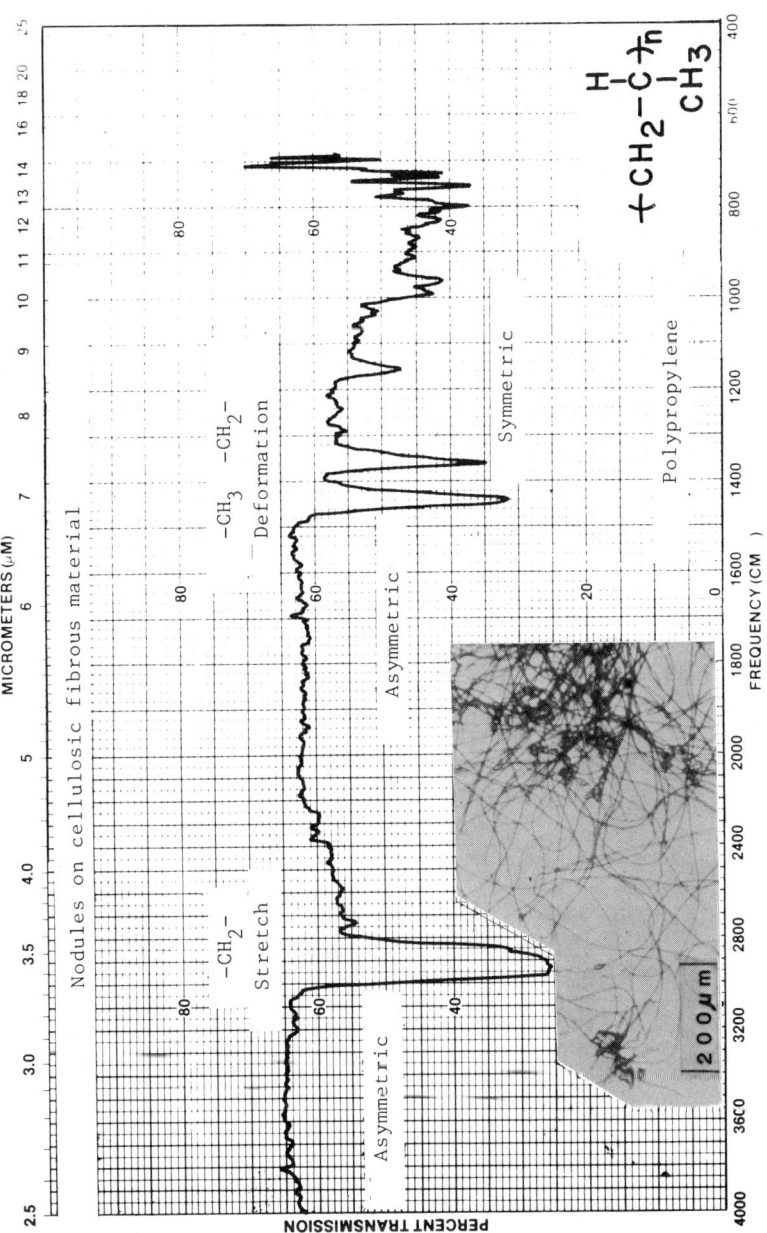

Figure 13. Nodules on cellulosic fibrous material--by small area IR. (Reproduced with permission from Ref. 27. Copyright 1983 International Conference on X-ray Optics and Microscopy.

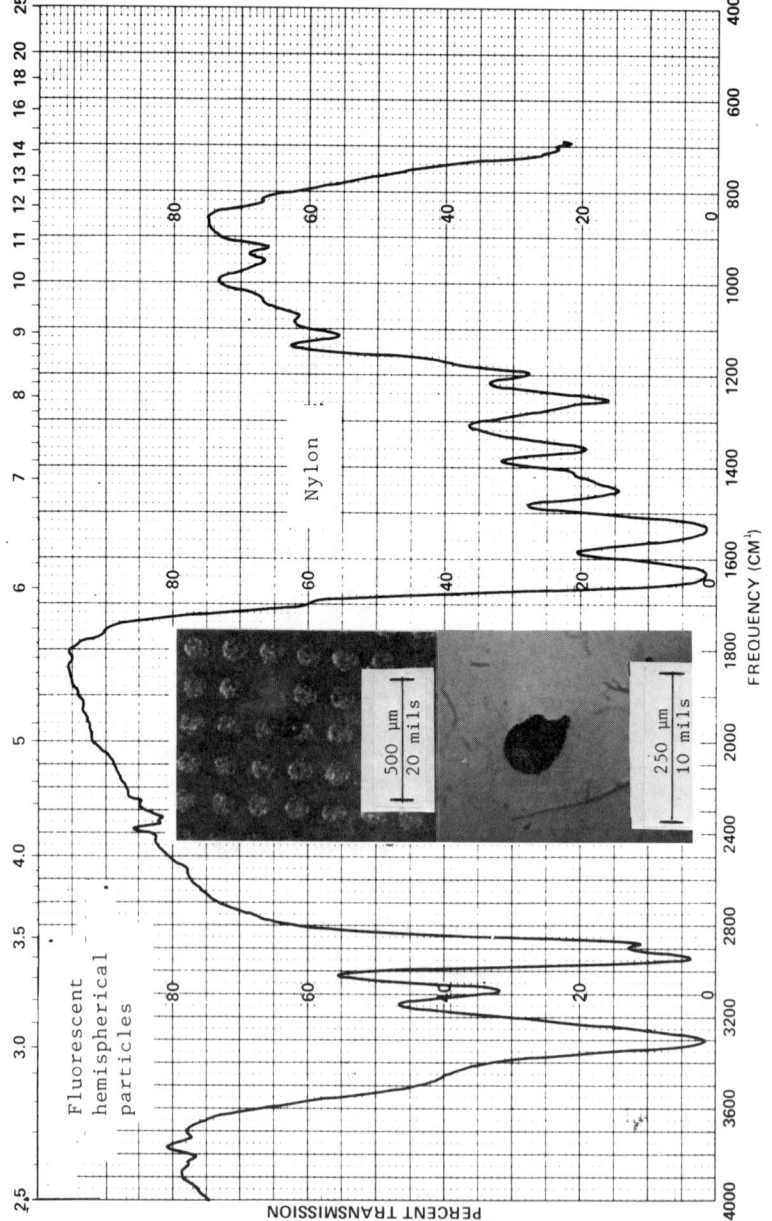

Figure 14. Fluorescent hemispherical particles on module after braze operation—by small area IR.

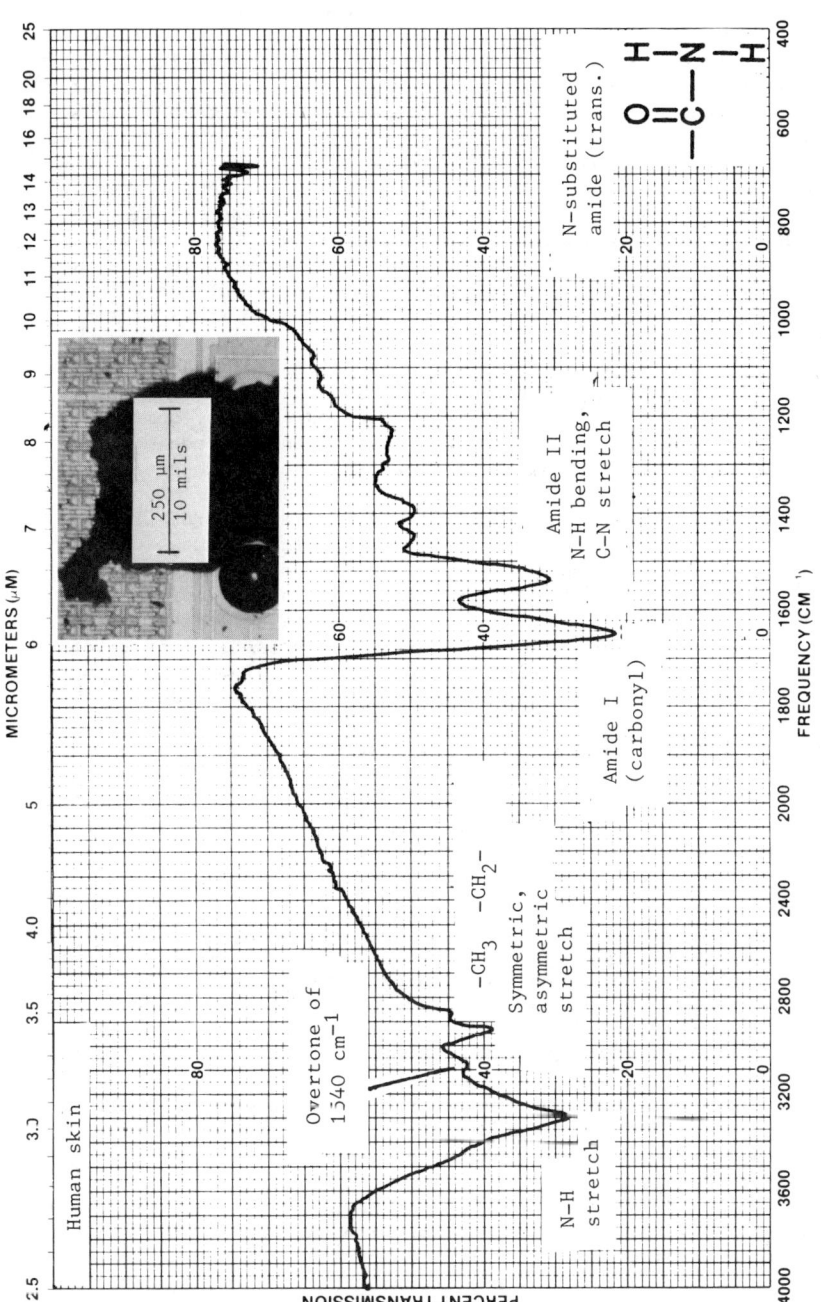

Figure 15. Contamination on wafer--by small area IR.

Microscopy experience, viz. through-focus series, into the LAMMA. Defocussing obviously reduces the energy density of the beam in the sample, allowing one to irradiate the sample without even volatilization and ionization. Then, by progressively improving the focus and pulsing the laser, one can go through initial desorption and soft fragmentation and ionization to the point of extreme fragmentation, where the spectrum is not recognizable. Thus, while the laser power, per se, is constant and stabilized, the power to the specimen is variable and controlled, allowing fragmentation that can be recognized by an experienced organic mass spectroscopist(27). Frank Anderson and Hans Heinen (of Leybold-Heraeus) worked out this procedure which has been discussed elsewhere(28), and have provided some of their early results to known organic materials, which enabled them to identify an unknown material within the scope of this present paper, namely small area molecular analysis applied to micro-electronic device fabrication problems. The attempt was to get LAMMA fragmentation reproducible and to produce spectra recognizable to an "organic" mass spectroscopist by controlling and measuring the low laser power applied to the specimen. Five sample materials, well characterized by Electron Impact/Mass Spectroscopy, were chosen and are given below.

Figure 16 shows the LAMMA 1000 negative spectrum from Benzophenone, an aromatic ketone with a molecular mass of 182. The base peak by LAMMA is 183, probably from the protonation by the abstraction of hydrogen from a second molecule of benzophenone. This process could occur because of the higher vapor pressure of benzophenone in the laser beam. This reaction would be pressure sensitive and possibly even lower laser powers would give 182. EI/MS also gives the 105 and 77 peaks.

Figure 17 shows the LAMMA 1000 negative spectrum of Rhodamine B, an indicator dye. Neither LAMMA or EI "sees" the full molecule (478) because HCl is lost, giving a parent peak of 442. As indicated, all EI mass peaks are found in LAMMA, along with many extraneous peaks that might confuse a novice mass spectroscopist, but experienced organic mass spectroscopists can "see" the pertinent peaks. This points out the high level background required in the early (?) stages of any analytical/characterization technique, especially for organics.

Figure 18 shows the LAMMA 1000 negative spectrum for Poly Alpha Methyl Styrene, an unzippable polymer (meaning that it breaks apart sequentially as monomers). Again, an EI mass spectroscopist would recognize the material.

Figure 19 shows the LAMMA 1000 positive spectra of Irganox 1010 (Ciba Geigy$^{(R)}$), a high molecular weight (1176), multifunctional anti-oxidant and thermal stabilizer. The LAMMA matches the EI spectra well, with the LAMMA giving additional information: the t-butyl, phenyl and benzyl ions suggest an aromatic compound.

Figure 20 shows the LAMMA 1000 negative spectrum of Glycerol monostearate, a mono-ester of glycerine used as an emulsified and lubricant. As stated, the major EI peaks are also in the LAMMA spectrum. Of particular interest are the strong 358 and 340

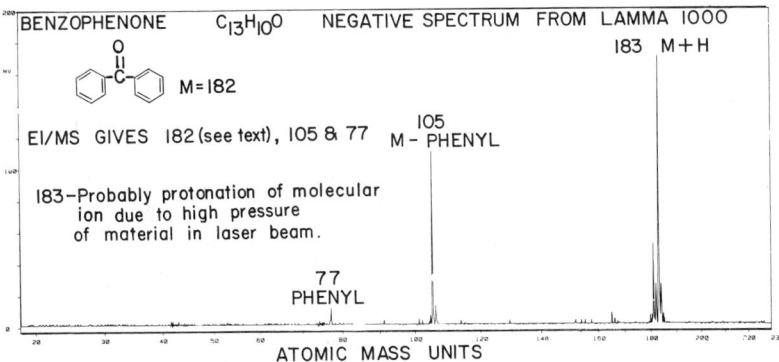

Figure 16. Benzophenone--by LAMMA. (Reproduced with permission from Ref. 27. Copyright 1983 International Conference on X-ray Optics and Microscopy.)

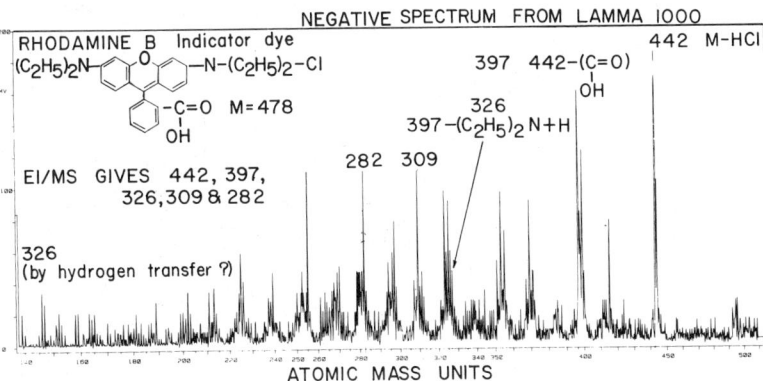

Figure 17. Rhodomine B--by LAMMA. (Reproduced with permission from Ref. 27. Copyright 1983 International Conference on X-ray Optics and Microscopy.)

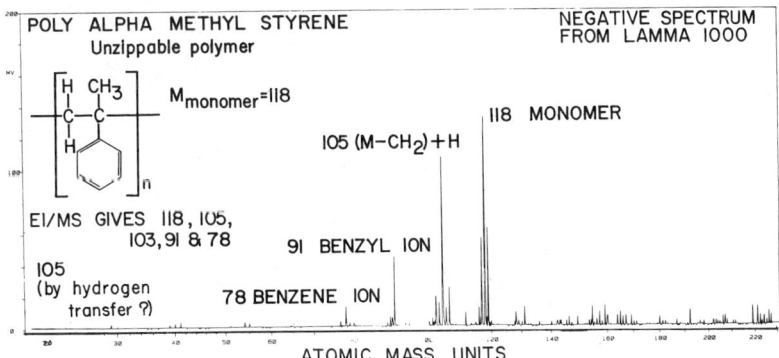

Figure 18. Poly alpha methyl styrene, an unzippable polymer--by LAMMA. (Reproduced with permission from Ref. 27. Copyright 1983 International Conference on X-ray Optics and Microscopy.)

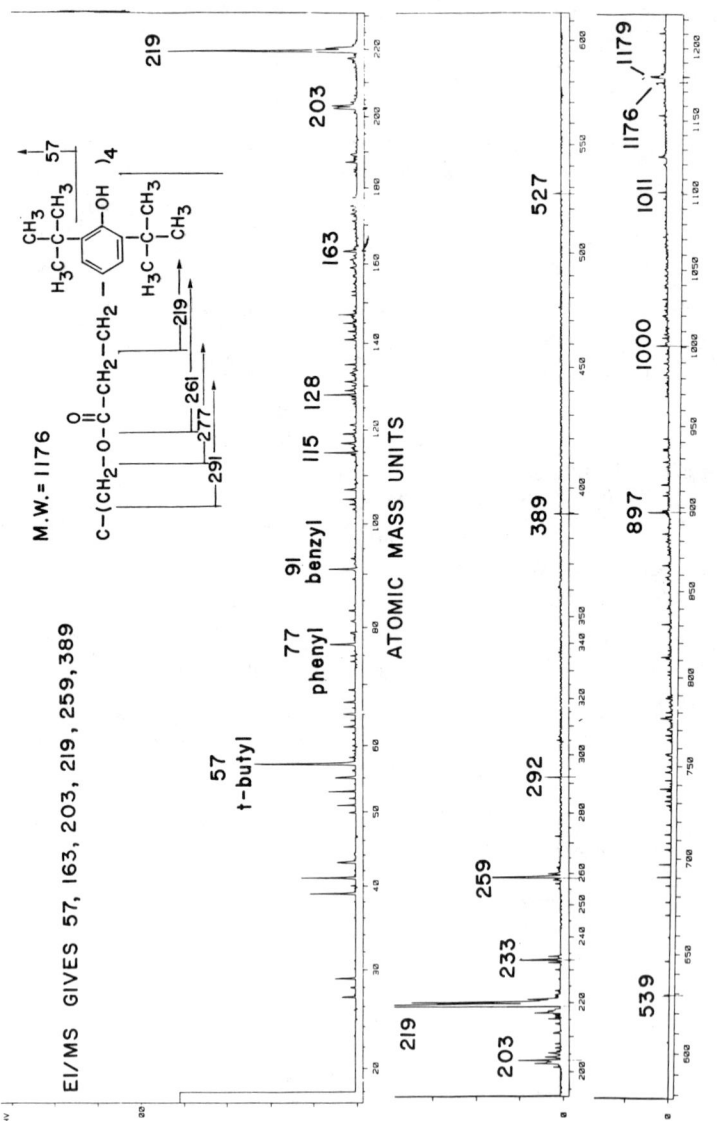

Figure 19. Irganox--by LAMMA.

peaks by LAMMA, representing a loss of H_2O, typical of the rearrangement loss in alcohol or aldehydes: this information is lacking in EI.

These LAMMA "successes" (defined as recognizable to EI/MS) gave confidence to try an unknown, a residue discovered after an etching and cleaning process in device manufacture. Figure 21 showed the material to be nitro benzene sulfonic acid. (Note the overlap of the two mass range spectra, with 157 on both.)

Figure 22 shows the spectra from 1,4 diphenoxy benzene, a polyphenyl ether, taken on the LIMA, recently run with Frank Anderson's method, as outlined above: again, recognizable fragments are produced. This shows that the method is universal, as expected.

While it is evident that the spectra of these unknown organic materials contain recognizable fragments, there are many extraneous peaks beyond EI that add a great deal of uncertainty in analyzing unknown organic materials. Considerable and rather basic studies of the fragmentation paths of this new and different means of ionization must be done to sort out concentration, proton donor and pressure effects. Then the technique should be very useful for analyses of organics.

Conclusions

The problem-solving mode offers great advantages in developing and maintaining the high level of process control required in the microelectronic industry today (and tomorrow). The analytical/characterization group can quickly show that it is a vital part of the entire series of process steps from research and development to manufacturing and into the field. The problem-solving mode of close, mutually-accepted cooperation between the analyst and the person (process engineer/scientist) is essential, with the analyst following up on who ultimately "owns" the problem to see that fixes are put into process - and documented to minimize repeats of the problem(29). To be successful, the group must be equipped with a variety of techniques, used in a complementary manner to obtain the high level of synergism inherent in the analysis of complex problems. There is no _universal_ technique.

Acknowledgments

I am indebted to my colleagues for sharing their data and ideas: F.W. Anderson, LAMMA; R.M. Anderson, TEM; P.W. DeHaven, X-ray diffraction; C.C. Goldsmith, X-ray diffraction; B.W. Griffiths (Cambridge Mass Spect.), LIMA; H. Heinen (Leybold-Heraeus), LAMMA; D.P. Kirby, FTIR and small area infrared; S. Lawhorne, X-ray diffraction; K.P. Madden, small area infrared; C.D. Needham, small area Raman; N.A. O'Neil, specimen preparation; F.M. Ordonez, SEM; R.M. Scott, small area infrared; L.B. Wiggins, electron probe microanalysis; C.E. Wilson, small area infrared.

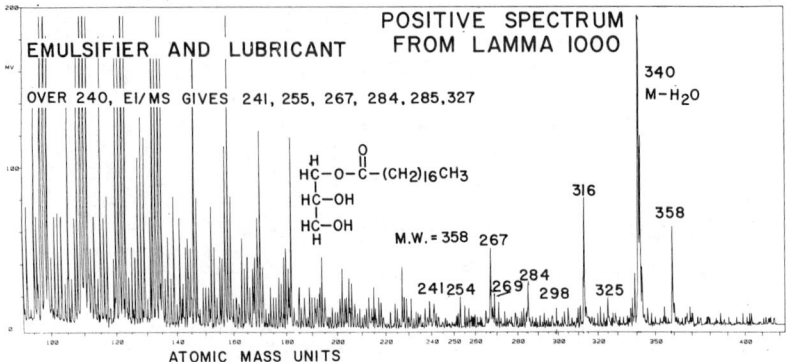

Figure 20. 1-Glycol monostearim--by LAMMA.

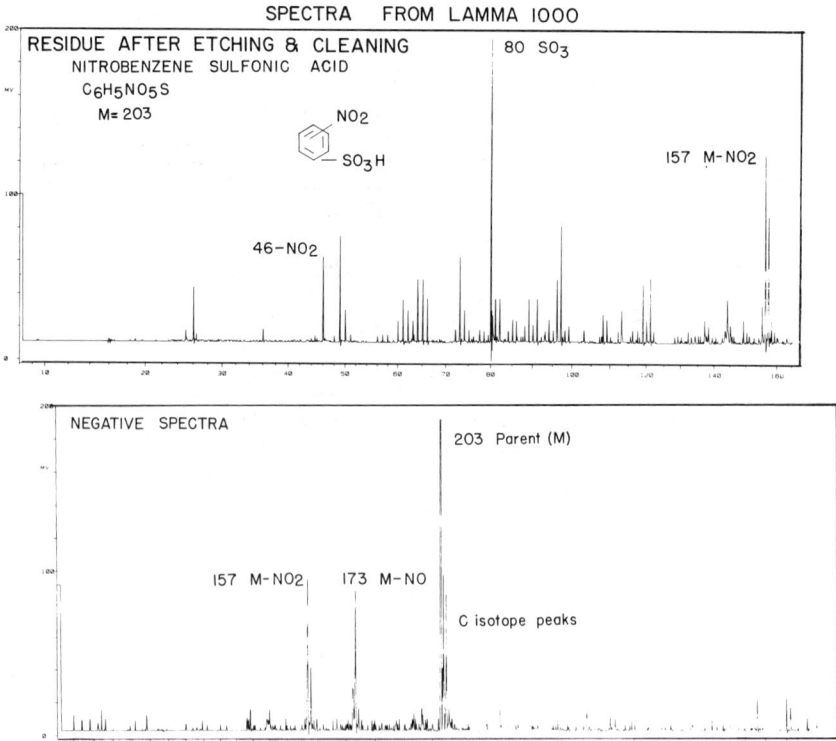

Figure 21. Residue after etching and cleaning--by LAMMA. (Reproduced with permission from Ref. 27. Copyright 1983 International Conference on X-ray Optics and Microscopy.)

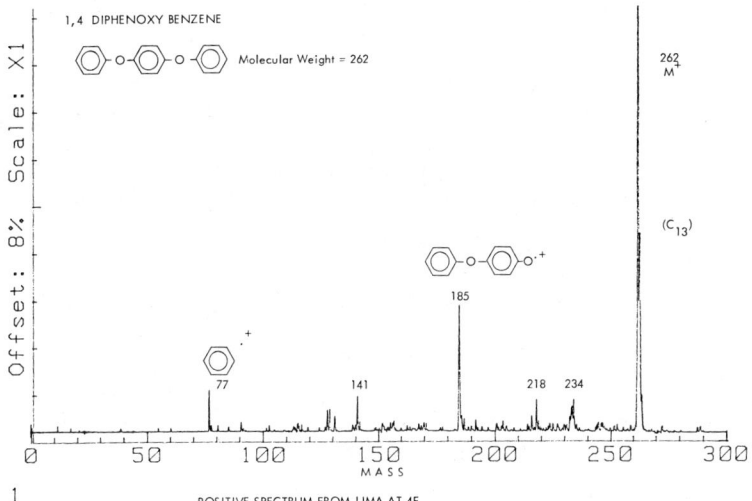

Figure 22. 1,4 Diphenyl benzene--by LIMA. (Reproduced with permission. Copyright Cambridge Mass Spectroscopy Ltd.)

Literature Cited

(1) Wittry, D.B; Axelrod, J.M.; McCaldin, J.O. <u>Proc. Semiconductor Conf. Am. Inst. of Met. Engr.</u>, Boston, August 31 - September 2, 1959.

(2) Ramsey, J.N.; Weinstein, P. "The Electron Microprobe", Editor R. McKinley, Wiley (1966), pg. 715-747.

(3) Ramsey, J.N. <u>J. Vac. Sci. Techn.</u> A1(2) April - June, 1983 p. 721-31.

(4) JCPDS, 1601 Park Lane, Swarthmore, PA. 19081, USA.

(5) Cameron, D.P.; Schneider, F.W.; Ramsey, J.N. <u>Proc. El. Mic. Soc. of Am.</u>, 1969.

(6) McCrone, W.; Delly, J. "The Particle Atlas", Ann Arbor Sci. Publ. Ann Arbor Mich., USA.

(7) Delhaye, M; Dhamenlincourt, P. <u>J. Raman Spectr.</u> 3 (1975) 33.

(8) (a) Jobin Yvon, 91160 Longjumeau, FR.
    (b) Instruments, S.A. 173 Essex Ave. Metuchen, NJ 08840..

(9) L'actualite' chemique, Avril 1980.

(10) Proc. Microbeam Analy. Soc., 1978, 1981-5.

(11) Needham, C.D.; Ramsey, J.N. Semiconductor International 4 (March 1981) 75.

(12) Needham, C.D.; Proc. SPIE (Soc. of Photo-Optical Engineers), Alexandria, VA. Meeting, April, 1982.

(13) (a) Zaring, D.J.; Coates, V.J. <u>Proc. SPIE</u> 276 (1981) 249.
     (b) Nanometrics, 930 W. Maude Ave., Sunnyvale, CA 94086.

(14) Ramsey, J.N.; Hausdorff, H.H. <u>Proc. 16th Meeting, Microbeam Analy. Soc.</u> (1981) 91.

(15) Scott, R.M.; Ramsey, J.N. <u>Proc. 17th Meeting Microbeam Analy. Soc.</u> (1982) 239.

(16) Popek, K.M.; Ramsey, J.N. <u>Proc. SPIE Alexandria, VA. Meeting</u>, April, 1982.

(17) Anderson, M.E.; Muggli, R.Z. Analy. Chem. 53 (October, 1981) 1772.

(18) Leybold-Heraeus, D-5000 Koln 51, FRG.

(19) Kaufman, R. Proc. 17th Meeting Microbeam Analy. Soc. (1982) 341.

(20) Hillencamp, F.; Kaufman, R.; Wechsung, R. Proc. 18th Meeting Microbeam Analy. Soc. (1983).

(21) Cambridge Mass Spectrometry, Ltd., Cambridge CB4 4BH, England.

(22) Evans, C.A.; Griffiths, B.W.; Dingle, T. Proc. of 18th Meeting of Microbeam Analy. Soc. (1983).

(23) Rigaku, Danvers, Massachusetts.

(24) Walker, G.A.; Goldsmith, C.C.; Proc. 16th Reliab. Phys. Symp. IEEE, 1978, p. 56-8.

(25) Eggerding C.L. Proc. Int'l. Conf. on New Trends in Passive Components, Apr. 1982, Paris.

(26) DeHaven, P.W.; Walker, G.A.; O'Neil, N. Accepted for Publication, Adv. in X-Ray Anal., 1984, V27.

(27) Ramsey, J.N. Proc. 10th Int'l. Conf. on X-Ray Optics and Microscopy (ICXOM) Toulouse, Fr., Sept. 1983. Shortened version published Jour. de Phys., Fev. 84 PgC2-881-5.

(28) Anderson, F.W.; Heinen, H.; Ramsey, J.N. Presented as late news paper at Microbeam Analysis Society Annual Meeting, July 1984 and printed in Proc. 1985.

(29) Ramsey, J.N. Proc. 35th Pittsburgh Conf. on Analy. Chem. and Applied Spectr. Atlantic City, NJ, March 1984. Published as Applic of Surf. Sci. 20 (1985), 413-428.

(30) Comprehensive sessions were held at the "Pittsburg" Conference, New Orleans, LA, March 1985 and at Microbeam Analysis Society, Annual Meeting, Louisville, KY, August 1985.

RECEIVED July 23, 1985

INDEXES

Author Index

Adar, Fran, 230
Aspnes, D. E., 192
Baghdadi, Aslan, 208
Balazs, Marjorie K., 320
Becker, D. Scott, 366
Bomben, Kenneth D., 144
Boorn, A., 284
Burkman, Don C., 366
Casper, Lawrence A., 1
Chin, Roland L., 349
Dharmadhikari, Vineet S., 333
Douglas, D. J., 284
Downing, R. G., 163
Fleming, R. F., 163
Guidoboni, Richard J., 308
Gupta, Satish K., 349
Ho, Kim-Khanh N., 320
Houskova, Jana, 320
Jones, Douglas L., 118
Kalin, Ronald V., 49
Kee, Ronald W., 118
Leipziger, Frederic D., 308
Lindfors, Paul A., 118
Lindstrom, Richard M., 294
Maki, J. T., 163
Mazur, Robert G., 34
Nowakowski, C. E., 378
Passoja, D. E., 1
Penzel, W. H., 267
Peterson, Charlie A., 366
Pinillos, J. V. Martinez de, 378
Quan, E. S. K., 284
Ramsey, J. N., 398
Reynolds, D. C., 240
Rosenblatt, G., 284
Rosencwaig, Allan, 181, 253
Rozgonyi, G. A., 75
Ryan-Hotchkiss, Mary, 96
Sadana, D. K., 75
Scharman, A. J., 1
Schmidt, William R., 366
Schroder, Dieter K., 18
Shushan, B., 284
Stickle, William F., 144
Young, Rodney A., 49

Subject Index

A

Absorption bands
  silica films, 357, 358f
  silicon-carbon complexes, 213, 214f
  silicon-oxygen complexes, 209, 211
Absorption coefficient due to an impurity, silicon wafer, 220
Absorption spectrum, silicon wafer with a rough back surface, 222f
Activation analysis
  electronics materials, 294-305
  measurement procedure, 300-303
Activities of semiconductors after neutron irradiation, 299t
Adhesion and planarization characteristics, spin-on glass films, 355, 356f
Air, refractive index, 187
Alloy compositions, characterization by modulated reflectometry, 196-197
Aluminum
  composite ion image, 87f
  metallization
    geometry, 138f
    secondary electron image, 136f
  pad with bonding problems, AES survey spectrum, 129f
  surface image, 89f
Aluminum-zinc alloy
  electron micrograph, 259f
  thermal-wave micrograph, 259f
American Society for Testing and Materials, recommendations for impurity content determinations of silicon wafers, 226
Ammonium hydroxide solutions, silicon-wafer cleaning, 369
Analysis volume for AES, 119-120
Analytical approaches, microelectronic devices characterization, 1-17

Analytical chamber of the FTMS, block
  diagram, 270f
Analytical image space,
  features, 8-9,10f
Analytical problems of microelec-
  tronics, application of Raman
  microprobe, 230-238
Analytical requirements, integrated
  circuit processing, 14-15
Analyzer assembly, FTMS, 269
Angular resolved XPS, 155f
Anisokinetic sampling, effect on
  particle concentration, 388,392f
Annealing
  on the configuration, effects on
    silicon-oxygen complexes, 210t
  SIMS depth profiling, 104-105
Apodization functions, use in FTIR
  instruments, 223,225f,226
Arsenic trichloride, analysis of
  volatile dopants, 318t
Artificial intelligence community,
  definition of knowledge, 5
Atomic contamination on a
  surface, 367-368
Atomic resolution transmission electron
  microscopy, 91-94
Auger depth profile
  spin-on glass film, 360f
  thermal oxide film, 360f
Auger electron(s), energy of, 149
Auger electron emission in XPS, 148-149
Auger electron spectroscopy (AES)
  failure analysis, 124-130
  microelectronics
    applications, 118-140
  process development, 130-140
  silicon wafers, 370,371f
Auger parameter, in XPS, 149-150
Auger spectroscopy-microscopy, 83,85
Auger studies, spin-on glass films, 357
Autodoping, definition, 105

B

Backscatter electron SEM photograph,
  VLSI bipolar transistor, 57f
Beam-specimen interactions, SEM, 52-53
Benzophenone, LAMMA 1000 negative
  spectrum, 419f
Binding energy of photoelectrons, 148
Blistered nickel over fired molybdenum,
  SEM, 402,403f
Boron
  implantation distributions in silicon
    wafers, 169,171,172f
  in silicon and glasses, activation
    analysis, 304

Boron--Continued
  ion implant, profiles, 42-47
  solution, ICP-MS analysis, 290,291f
Boron tribromide, analysis of volatile
  dopants, 318t
Bragg's law, 59
Braze material, gold-tin, 402,404-408
Braze pastes, gold-tin, 404,407f
Brewster's angle for silicon, 219

C

Calibration
  particle counters, 388
  spreading resistance system, 41
Capacitance, definition, 31
Capacitance-voltage profiling,
  semiconductor electrical
  characterization, 23-24
Carbon and oxygen content of silicon
  wafers, measurement by
  FTIR, 208-226
Carrier concentrations
  characterization by modulated
    reflectometry, 196-197
  semiconductor materials and
    devices, 23,25f
Cationic contamination of surfaces,
  reduction, 369
Cellulose, identification on a silicon
  wafer with Raman
  microprobe, 233,235f,236
Channel blocking, study by NDP, 173
Charge neutralization, SIMS
  instrument, 97-98
Chemical sequencing
  comparison by SIMS, 375
  silicon-wafer cleaning, 370,372-375
Chemical shift, description, 146
Chemical state information in AES
  data, 137,139f
Chemical vapor deposition
  low pressure, thin films deposited
    by, 200-202
  silica glass films, 349-350
Chromium
  Auger electron maps, 134f
  composition control chart, thin-film
    resistor fabrication, 342-345f
  effect of composition on temperature
    coefficient of resistance of thin
    films, 339,340f
  X-ray dot map, 63f
  X-ray line profile, 63f
Chromium-nickel process
  summary after process control, 346t
  summary before process control, 336t

INDEX

Circuit boards, AES sputter depth profiling, 131f
Cleaning chemistries, silicon wafers, 368-375
Cleaning operation, device line, contamination problem, 410,412f
Cobalt-chromium alloy
  electron micrograph, 259f
  thermal-wave micrograph, 259f
Colorimetry vs. ion chromatography for weight percent phosphorus determinations, 329t
Complex reflectance coefficients, 194
Conditioning of probes, spreading resistance system, 37
Conductors, thermal, thermal-wave spatial resolution, 254t
Contaminants
  effect on miniaturization in the microelectronics industry, 1,4
  identification by FTMS, 278,281f
  on a surface, 367-368
Contamination layer, on E-beam aperture, FTIR spectra, 409f
Contamination problem
  cleaning operation, device line, 410,412f
  recording surface, 410,413-415
Contamination studies, use of NAA, 303-304
Control charts
  for percent chromium, thin-film resistor fabrication, 342-345f
  in statistical quality control, 342-344
Copper, determination in an orchard leaf digest, 286,289f,290
Copper-tin intermetallics formed on solder reflow, X-ray diffraction, 400-403f
Core levels of atoms, description, 146
Correction factor for multiple reflection, silicon-wafer absorption, 220-221
Critical dimension measurement via digitized SEM signal profile, 71f
Cross-sectional analysis, SEM, 56
Cross-sectional transmission electron microscopy, 88,90f,202,204f
Crucible materials, analysis, 318t
Crystal(s), in fired molybdenum, extraction replication, 402,403f
Crystalline disruptions and variations, imaging with a thermal-wave microscope, 260,262f
Czochralski crystals, oxygen concentration, 209
Czochralski wafer, IR transmission spectra, 217f

D

Data handling, FTMS, 282
Deep-level impurities, semiconductor materials and devices, 26
Deep-level transient spectroscopy, semiconductor electrical characterization, 29-31
Defect(s)
  characterization by AES, 125-127
  effect on miniaturization in the microelectronics industry, 1,4
  in an oxide film, 11,14
  in microelectronic devices, scaling behavior, 4
  on wafers before and after cleaning, 373
  subsurface mechanical, detection with a thermal-wave microscope, 257,258f
Defect stringers following plasma etching, study by AES, 135-138f
Delamination, study by AES, 130,131f
Deposition rate, control in thin-film resistor fabrication, 339
Depth profiling
  SIMS, 103-107
  XPS, 150-152f
Depth resolution, SIMS, 101
Detection, thermal waves, 254
Detection limits, AES, 121-124
Deutsches Institut fur Normung, recommendations for impurity content determinations of silicon wafers, 226
Dielectric properties, spin-on glass films, 353t
Differential interference contrast microscopy for examining semiconductor wafer surfaces, 75-79
Diffusion barriers, studies by AES, 135,136f
Diffusion length, relation to the recombination lifetime of electrons, 27
Digitized image profile, electron-beam lithography features on a device, 71f
Digitizing of ion images, direct ion-imaging microscopy, 85
1,4-Diphenoxybenzene, LIMA spectra, 423f
Direct imaging instrument, SIMS, 98-99,111
Direct ion-imaging microscopy, 85-88
Donor(s)
  identification in doped GaAs, 248,250f,251

Donor(s)—Continued
  magnetic field separation, definition, 248
  neutral states, exciton transitions from, 245-250f
Dopant(s)
  profiles by the spreading resistance technique, 34-48
  SIMS depth profiling, 103-104
  SIMS detection limits, 100t
Doped GaAs, indentification of donors, 248,250f,251
Doped oxide film from a wafer, analysis with and without oxidation, 322f
Doping concentrations, semiconductor materials and devices, 23
Drain current-drain voltage relationship, transistor, 31
Dynamic range, SIMS, 99,101

E

E-beam polymerized film, inadvertent, 408-410
Effective mass approximation, 242
Effective mass model of an impurity state, 243
Effective medium theories, use in optical data analysis, 195
Electrical characterization
  semiconductor materials and devices, 18-33
  spin-on glass films, 353-354
Electrical junction delineation in a doubly implanted n-p-n structure, 79-81
Electrical techniques, SEM, 62-67
Electrically active defects
  silicon-carbon complexes, 213,215
  silicon-oxygen complexes, 211
Electrochemical profiler, 25f
Electron, incident, effect on specimen being analyzed, 124
Electron-beam energy dissipation range, low-atomic-number elements, 52
Electron-beam-induced current
  for analysis of semiconducting materials, 260-261,263f
  photograph, VLSI bipolar transistor, 68f
  signal formation and intensity in p-n diode structure, 64,66f
Electron-beam lithographic features on a device, digitized image profile, 71f
Electron diffraction, general discussion, 398

Electron micrograph
  aluminum-zinc alloy, 259f
  cobalt-chromium alloy, 259f
  GaAs device, 262f
  silicon solar-cell material, 263f
Electron microprobe
  general discussion, 398
  video scans of intermetallic phases, 400,401f
Electron shake-off, description, 148
Electron shake-up, description, 148
Electron spectroscopy, uses of, 145
Electron spectroscopy for chemical analysis (ESCA) studies, spin-on glass films, 357
Electronic control system, FTMS, 271,273f
Electronics, spreading resistance system, 41
Electronics materials, activation analysis, 294-305
Element detection by AES, 119
Elemental analyses of semiconductor materials using ICP-MS, 284-293
Elemental concentration calculation, activation analysis, 302-303
Elemental sensitivities of NDP, 167-170f
Ellipsometers, use in thickness determinations in laminar systems, 198
Ellipsometry, description, 194
Energy dispersive X-ray spectroscopy (EDS)
  detector system, 60f
  performance characteristics, 60f
  X-ray spectra, 61f
Energy-level diagrams, silicon-oxygen complex, 212f,214f
Epoxy-bonded die contaminant, mass spectrum, 281f
Etch-back method for planarizing interlevel dielectric layers, 350
Etch rates, spin-on glass films, 354
Etching, preferential, junction location, 79-83
Ethanol, mass spectrum, 277f
Ethylene diamine tetraacetic acid, identification on a silicon wafer with Raman microprobe, 233
Exciton(s)
  bound to a neutral donor, radiative recombination, 246f
  bound to nonrigid-rotational states, 244-245
  spectra, 241-242
  transitions from neutral donor states, 245-250f

INDEX

Expert systems
  definition, 6
  microelectronic devices
    characterization, 1-17
  problem solving, 5-6
External gettering via misfit
  dislocation, 83,84f
Extraction replication, crystals in
  fired molybdenum, 402,403f
Extrinsic excitons, 241-242

F

Fabry-Perot fringes, elimination in
  FTIR instruments, 215,219
Failure analysis, use of AES, 124-130
Field-effect mobility, definition, 32
Film(s)
  inadvertent E-beam
    polymerized, 408-410
  spin-on glass, composition and
    structure, 355
  thickness
    as a function of spin speed, 352t
    measurements, use of NAA, 304
Flow diagram, laser aerosol
  spectrometer, 379f
Fluorinated hydrocarbon, identification
  on a silicon wafer with Raman
  microprobe, 233,234f
Focused-spot reflectometer,
  characterization of surface and
  sample quality, 195-197f
Fourier transform IR (FTIR) instruments
  characteristics, 193
  elimination of interference
    fringes, 215,219
  use of apodization
    functions, 223,225f,226
Fourier transform IR (FTIR)
  spectroscopy
  contamination layer on E-beam
    aperture, 409f
  isopropyl alcohol extract of new
    resistor coating, 409f
  measurement of oxygen and carbon
    content of silicon
    wafers, 208-226
  use in studies of GaAs, 242-243
Fourier transform mass spectrometry
  (FTMS)
  analytical techniques available, 268
  applications, 271-282
  equipment, 269-271
  in the microelectronics service
    laboratory, 267-282
Fractional error, particle
  concentration in a gas, 386

Free carriers
  absorption in n-type
    silicon, 223,225f
  effects on oxygen content deter-
    minations in silicon
    wafers, 223,224f
Fresnel reflectance expressions, use in
  optical data analysis, 194-195

G

Gallium arsenide (GaAs)
  characterization by magneto-optical
    photoluminescent
    spectroscopy, 240-251
  crystal
    thermal-wave image, 265f
    X-ray topography, 265f
  device
    electron micrograph, 262f
    thermal-wave image, 262f
  materials, study by XPS, 157-158
Gallium metal
  analysis, 317f
  impurities, analysis, 316t
Gas(es)
  detection by FTMS, 278
  particles in, monitoring with a laser
    counter, 377-396
Gas chromatogram
  conformal coating on wiring
    board, 279f
  spin-on fluid solvent package, 276f
Gate oxides, metal-oxide-semiconductor
  technology, 4
Gaussian distribution, small spherical
  particles, 11
Generation lifetime for electrons,
  definition, 27
Geometric elements, microelectronic
  devices, 9,11
Gettering, external, via misfit
  dislocation, 83,84f
Glass films, spin-on, as planarizing
  dielectrics, characterization
  of, 349-364
Glycerol monostearate, LAMMA 1000
  negative spectrum, 422f
Gold, Auger electron maps, 136f
Gold film plated on palladium
  absorbed current image of a void
    in, 127f
  AES sputter depth profile from a void
    in, 127f
Gold-nickel-copper metallization,
  secondary electron image, 136f
Gold-tin braze material, 402,404-408
Gorey-Schneider probe grinder, 37

H

High-frequency thermal-wave imaging in an SEM, experimental configuration, 256,258f
High-resolution photoluminescence studies of GaAs, 243-245
High-resolution thermal-wave imaging in an SEM, applications, 257
Hydrogen chloride solutions, silicon-wafer cleaning, 369
Hydrogen fluoride
  removal of metal ions from silicon wafers, 372
  use in treatment of silicon wafers, 369
Hydrogen peroxide solutions, silicon-wafer cleaning, 369-370

I

Image depth profiling (IDP), SIMS, 112
Image processing, direct-ion imaging microscopy, 85
Image space, analytical, features, 8-10f
Imaging techniques, SEM, 52-56
Implantation distributions, boron, in silicon wafers, study by NDP, 169,171,172f
Impurity analysis in semiconductor matrices, SIMS, 83,85-88
Impurity energy state, definition, 243
Inadvertent E-beam polymerized film, 408-410
Incident electron, effect on specimen being analyzed, 124
Indium phosphide sample, mass spectrum, 288f
Inductively coupled plasma mass spectrometry (ICP-MS)
  elemental and isotopic analyses of semiconductor materials, 284-293
  instrumentation, 284,285f
Infrared absorption spectroscopy, silicon wafers, 215-221
Infrared analysis, general discussion, 398-399
Infrared photodetectors, characteristics, 193
Infrared spectra, spin-on glass films, 355,357,358f
Infrared transmission spectra
  Czochralski wafer, 217f
  silicon wafer, 217f
Inlet systems, 269,271,272f
Instrumental NAA, 300
Instrumentation for XPS, 150,152f,153
Insulators, thermal, thermal-wave spatial resolution, 254t
Integrated circuit processing, analytical requirements, 14-15
Integrated circuit technology, science and engineering disciplines, 2f
Intensity of light transmitted through a plane parallel plate, 219,220
Interfaces
  characterization by optical reflectance and ellipsometric techniques, 192-206
  study by XPS, 158
Interfacial profiling, use of NDP, 171-173
Interference fringes, elimination in FTIR instruments, 215,219
Interferogram, silicon wafer, 218f
Intermetallics, molecular analysis, 400-403f
Intrinsic excitons, 241
Intrinsic gettering of impurities in semiconductor fabrication, 213
Ion(s), stability of, 268-269
Ion beams, effect on specimen being analyzed, 124
Ion chromatogram
  conformal coating on wiring board, 279f
  spin-on fluid solvent package, 276f
Ion chromatograph, configuration, 325f
Ion chromatography
  study of phosphite ions, 326-328f
  study of plasma phosphosilicate glass films, 327f
  vs. colorimetry for weight-percent phosphorus determinations, 329t
Ion cyclotron resonance, 267
Ion imaging, SIMS, 111-113
Ion-imaging microscopy, direct, 85-88
Ion microprobe, SIMS, 98,111
Ionic contamination on a surface, 367-368
Irgonax 1010, LAMMA 1000 negative spectrum, 420f
Irradiation of samples during activation analysis, 301
Isokinetic sample probe, 388,389f
Isokineticity requirements, monitoring of particles in gases using a laser counter, 383,386-396
Isopropyl alcohol extract of new resistor coating, FTIR spectra, 409f
Isotopic analyses of semiconductor materials using ICP-MS, 284-293
Isotopic dilution, use for copper determination, 286,289f,290
Isotopic studies, SIMS, 108-110

INDEX 435

J

Japan Electronics Development Association, recommendations for impurity content determinations of silicon wafers, 226
Junction delineation by spreading resistance probes, 81-83
Junction depth, semiconductor materials and devices, 26
Junction location, preferential etching, 79-83

K

Kinetic energy of photoelectrons, 148
Kinetics of activation analysis, 298-300
Knowledge, definition of the artificial intelligence community, 5
Knowledge representation, major themes, 5-6

L

Laboratory manifold used to test different sampling techniques, 388,389f
Laminar samples, optical data analysis, 194-195
Laminar systems, thickness determinations, 198-199
LAMMA 1000, analyses with, 414,418-423
Laser, for generating and detecting thermal waves, 182
Laser aerosol spectrometer, 377-380f
Laser-annealed silicon on silicon oxide, study with Raman microprobe, 236-238f
Laser-annealing processes, 105
Laser-beam deflection signal
 relative amplitude as a function of film thickness, 190f
 relative amplitude as a function of thermal-wave frequency, 190f
Laser-beam deflection technique
 schematic, 184f
 used to measure thin-film thickness, 183-191
Laser counter, monitoring of particles in gases, 377-396
Laser desorption-ionization and mass analysis, 399-400
Laser-probe beam, schematic depiction of physical processes affecting, 185f,189f
Leaching, study by NDP, 173-175f

Lead bonding, study by AES, 128-131f
Lead samples, ICP-MS analysis, 292f,293
Least-squares regression analysis, use in optical data analysis, 195
Light transmitted through a plane parallel plate, intensity, 219,220
LIMA spectra, 1,4-diphenoxybenzene, 423
Locked-in interface, nondestructive analysis, 202-204f
Low-beam-energy SEM image, photoresist pattern, 70
Low-pressure chemical vapor deposition, thin films deposited by, 200-202

M

Magnetic field separation of donors, definition, 248
Magneto-optical photoluminescent spectroscopy, characterization of GaAs, 240-251
Magnification
 analytical image space, 8-9
 determination of optimal range, 12f
Manifold, laboratory, used to test different sampling techniques, 388,389f
Mass analyzer, SIMS instrument, 97
Mass spectra
 epoxy-bonded die contaminant, 281f
 ethanol, 277f
 indium phosphide sample, 288f
 silicon, 314,317f
 SIMS, 107-108
 transition metals, 287f
 trichlorotrifluoroethane, 280f
Mass spectrometry, inductively coupled plasma (ICP-MS), analyses of semiconductor materials, 284-293
Material(s)
 analysis, 200-204f
 characterization by optical reflectance and ellipsometric techniques, 192-206
 control in semiconductor manufacture, use of FTMS, 275
 information provided by SEM images, 54t
Metal-can integrated circuits, analyses, 274,275t
Metal ions, removal from silicon wafers with hydrogen fluoride, 372
Metal-oxide-semiconductor field-effect transistor, characteristics, 19-21
Metal-oxide-semiconductor technology, gate oxides, 4
Metallic contamination on a surface, 367-368

Metallization pattern on the surface of a transistor, 86f
Metallography, use of thermal-wave imaging, 257,259f,260
Microanalysis
  break in wire connect, use of Raman microprobe, 236
  organic contamination, use of Raman microprobe, 231,233-236
  problems, rules used by experts in solving, 6-8
Microelectronics
  applications
    of AES, 118-140
    of Raman microprobe to analytical problems, 230-238
  processing problem solving, synergism of complementary techniques, 398-423
  service laboratory, FTMS, 267-282
Microelectronics devices
  building blocks, 10f
  characterization
    analytical approaches, 1-17
    expert systems, 1-17
Microelectronics materials
  applications, XPS, 144-159
  characterization, SIMS, 96-114
  processing, application of NDP, 163-177
Microstructurally inhomogeneous samples, optical data analysis, 194-195
Microwave reflection measurement of the recombination lifetime of electrons, 27,28f
Miniaturization in the microelectronics industry
  effect of contaminants, 1,4
  effect of defects, 1,4
  impact, 1,4
Misfit dislocation, external gettering, 83,84f
Mobility
  definition, 32
  semiconductor materials and devices, 31-32
Modulated reflectometry
  characterization of alloy compositions, 196-197
  characterization of carrier concentrations, 196-197
  description, 194
Moisture analysis, use of FTMS, 271,273-277
Molecular analysis, problem solving by, examples, 400-423
Molecular contamination on a surface, 367

Molybdate blue reaction
  oxoacids of phosphorus, time study, 322f
  phosphosilicate glass films, time study, 324f
Molybdenum, fired, blistered nickel over, SEM, 402,403f
Molybdenum nitride, SIMS depth profiles, 106
Molybdenum silicide, SIMS depth profiles, 106
Monitoring of particles in gases using a laser counter, 377-396
Multilayer corrections, spreading resistance system, 41-42
Multiplet splitting, occurrence of, 148

N

N-channel metal-oxide-silicon transistor
  cross-sectional diagram, 58f
  secondary electron SEM photograph, 58f
Negative-positive ion mode, SIMS, 113
Neutral donor states, exciton transitions from, 245-250f
Neutron activation analysis (NAA)
  basic features, 295-303
  consequences of using nuclear reactions, 297-298
  electronics materials, 294-305
  histogram of detection limits, 297f
Neutron depth profiling (NDP), 45
  advantages, 177
  applications to microelectronic materials processing, 163-177
  boron implants, 45
  foundations, 164-169
Neutron irradiation, activities of semiconductors after, 299t
New donors, silicon-oxygen complexes, 211
Nickel
  Auger electron maps, 136f
  blistered, over fired molybdenum, SEM, 402,403f
Nickel-chromium, vacuum deposited, process control for the fabrication of reproducible thin-film resistors, 333-347
Noncontacting measurements, semiconductor materials and devices, 21
Nonrigid-rotational states, excitons bound to, 244-245
Nonrigid-rotator model, excited states in InP and GaAs, 244

Nonspecularly reflected light, information obtained from, 193
Nuclear analytical techniques, 304-305
Nuclear magnetic resonance, plasma phosphosilicate glass films, 330,331f
Nylon, IR spectrum, 416f

O

Optical characterization techniques, real-time applications, 202,205-206
Optical effects, thermal-wave measurement of thin-film thickness, 188,189f
Optical fibers, characterization with Raman microprobe, 237-238
Optical microscopy, for examining semiconductor wafer surfaces, 75-79
Optical photograph, of VLSI bipolar transistor, 55f
Optical reflectance and ellipsometric techniques, characterization of materials, thin films, and interfaces, 192-206
Optics diagram, laser aerosol spectrometer, 379f
Organic cleaning procedures, silicon wafers, 368
Oscillator strength of bound and intrinsic excitons, 242
Oxidation
effect on sheet resistance of thin films, 339,341f
effect on temperature coefficient of resistance of thin films, 339,341f
reaction, phosphorus trioxide, 321
Oxide film, defects in, 11,14
Oxide layers
on a silicon wafer, removal, 368-369
on a stepped silicon substrate, high-resolution TEM micrographs, 92f
Oxide removal, influence on cleaning of silicon wafers, 372
Oxygen
and carbon content of silicon wafers, measurement by FTIR, 208-226
Auger electron maps, 136f

P

Packaging of samples prior to activation analysis, 301
Partial pressure of gases in deposition chamber, control in thin-film resistor fabrication, 339
Particle(s)
characterization by AES, 125-127
counter calibration, 388
diameter range, laser aerosol spectrometer counter, 378,382t
high-pressure generator, 388,389f
in gases, monitoring with a laser counter, 377-396
Particulate analysis, nylon vs. skin, 414,416-417f
Peroxide solutions, silicon-wafer cleaning, 369
Phonon frequency, Raman, laser-annealed silicon on silicon oxide, 237,238f
Phosphate ions
analysis with and without oxidation, 322f
ion chromatographic study, 326-328f
Phosphine, in doped oxide films, 320-321
Phosphorus
determinations, colorimetry vs. ion chromatography, 329t
suboxide determinations, 321
weight percent, plasma doped oxides, 330t
Phosphorus compounds that give positive Mo ion tests without a reducing agent, 324f
Phosphorus-doped oxide(s), plasma, characterization of components, 320-332
Phosphorus-doped oxide films, quantitative analysis by EDS, 61f
Phosphorus molybdate blue reaction, oxoacids of, time study, 322f
Phosphorus oxides and acids, 323f
Phosphorus trioxide, oxidation reaction, 321
Phosphosilicate glass films molybdate blue reaction, time study, 324f
Photodetectors, UV-visible, characteristics, 193
Photoelectric effect, 144
Photoelectrons
binding energy, 148
kinetic energy, 148
Photoluminescence studies, high resolution, of GaAs, 243-245
Photoluminescent spectroscopy, magneto-optical, characterization of GaAs, 240-251
Photoresist pattern, low-beam-energy SEM image, 70

Photoresist stripping methods, 368
Physical characteristics, spin-on glass films, 352
Physical constraints, microelectronic devices, 9,11
Physics of NDP, 164-165
Piping bends, effect on particle concentration, 393,395f
Planarization and adhesion characteristics, spin-on glass films, 355,356f
Planarizing dielectrics, spin-on glass films, characterization of, 349-364
Plasma etching
 AES survey spectra from aluminum surfaces exposed to, 138f
 spin-on glass films, 354-355
Plasma phosphorus-doped oxides, characterization of components, 320-332
Plasma phosphosilicate glass, suboxides of phosphorus in, 321
Plasma phosphosilicate glass films
 ion chromatographic study, 327f
 NMR, 330,331f
Platinum-silicon, AES sputter depth profiles, 133f
Polarimetry, description, 194
Polarized light microscopy, general discussion, 398
Poly(ethylene terephthalate), identification on a silicon wafer with Raman microprobe, 233
Poly(α-methylstyrene), LAMMA 1000 negative spectrum, 419f
Positive-negative ion mode, SIMS, 113
Postirradiation chemistry, activation analysis, 301-302
Preferential etching, junction location, 79-83
Pressure-reduction diffuser, laser aerosol spectrometer, 378,383-385f
Pretreatment of samples for activation analysis, 300
Primary ion source, SIMS instrument, 97
Probability distributions, microelectronics devices, 11
Probes, spreading resistance technique, 35-41
Problem solving
 by molecular analysis, examples, 400-423
 expert systems, 5-6
 guidelines, 7-8
 microelectronics processing, the synergism of complementary techniques, 398-423

Process background and requirements, thin-film resistor fabrication, 334-338
Process control of vacuum-deposited nickel-chromium for the fabrication of reproducible thin-film resistors, 333-347
Process development
 thin-film resistor fabrication, 335,338f-341
 use of AES, 130-140
Process history, thin-film resistor fabrication, 335,336t
Process influence and improvements, thin-film resistor fabrication, 344,346-347
Prompt γ-ray activation analysis, histogram of detection limits, 297f

Q

Qualitative aspects of AES, 121
Quantitative aspects of AES, 121

R

Radiative recombination, exciton bound to a neutral donor, 246f
Radioactivity assay, activation analysis, 302
Radioactivity of a radionuclide at the end of an irradiation interval, 298
Radio-frequency spark source, 313f
Radiotracer methods, evaluation of metallic and ionic cleaning procedures, 370
Raleigh distribution, small spherical particles, 11
Raman microprobe
 application to analytical problems of microelectronics, 230-238
 characterization of optical fibers, 237-238
 characterization of semiconductors, 237-238
 general instrument design, 231,232f
 study of laser-annealed silicon on silicon oxide, 236-238f
 use in microanalysis of break in wire connect, 236
 use in microanalysis of organic contamination, 231,233-236
Raman phonon frequency, laser-annealed silicon on silicon oxide, 237,238f
Raman scattering, description, 230

INDEX 439

Raman spectroscopy, general discussion, 398-399
Reaction ion etching, spin-on glass films, 354
Real-time applications of optical characterization techniques, 202,205-206
Recombination-generation lifetime of electrons, 26-29
Recording surface, contamination problem, 410,413-415
Reflectance-based optical techniques
 characterization of materials, thin films, and interfaces, 192-206
 general principles, 193-195
Reflection electron diffraction, Au-Sn preforms, 404,406f
Reflectometers, use in thickness determinations in laminar systems, 198
Reflectometry, description, 194
Reflow glass process for planarizing interlevel dielectric layers, 350
Refractive index of air, 187
Relaxation time for a particle, definition, 386
Reproducibility of IR absorption spectroscopy, silicon wafer, 221,222f
Residual donors in GaAs, 243-244
Residual gas analyzer traces, use in thin-film resistor fabrication, 339
Residues, determinations using TMS, 275,278-280f
Resistance, effect on drain and transconductance, 32
Resistance plots using wafer mapping, 22f
Resistivity
 of contacts, study by AES, 128,129f
 of semiconductor materials and devices, 21-22
Resistor performance, effect of major process variables, 335,338f-341
Resolution of NDP, 167
Rhodamine B, LAMMA 1000 negative spectrum, 419f
Rigid-rotator model, excited states in CdTe, 244
Rutherford backscattering
 thin-film analysis, 83
 use with NDP, 176
Rydberg, effective, definition, 243

S

Sampling through a tee, effect on particle concentration, 393,394f
Satellite structure, description, 146,148
Scanning electron microscopy (SEM)
 advantages, 49-50
 beam-specimen interactions, 52-53
 blistered nickel over fired molybdenum, 402,403f
 cross-sectional analysis, 56
 electrical techniques, 62-67
 imaging techniques, 52-56
 role in VLSI processing, 67,69-72
 semiconductor materials characterization, 49-72
 thermal-wave imaging, 253-265
 wafer inspection, 67,70,72
 working principle, 50-52
 X-ray microanalysis, 56,59-62
Schottky barrier heights, study by XPS, 157
Secondary electron SEM photograph
 shorted metal lines, 65f
 VLSI bipolar transistor, 55f,68f
Secondary ion detector, SIMS instrument, 97
Secondary ion mass spectroscopy (SIMS)
 characteristics, 99-102
 chemical sequence comparison, 375
 comparison to spreading resistance technique, 45-47
 description, 96-97
 detection limits for dopants, 100t
 images
  factors affecting, 111-112
  uses, 112
 impurity analysis in semiconductor matrices, 83,85-88
 instruments, 97-100
 low levels of contamination on silicon surfaces, 370-375
 microelectronic materials characterization, 96-114
 modes of analysis, 102-113
 semiconductor electrical characterization, 24
 spin-on glass films, 357
 use with NDP, 174,176
Semiconductor(s)
 after neutron irradiation, activities, 299t
 characterization
  electrical, 18-33
  elemental and isotopic, using ICP-MS, 284-293
  using SEM techniques, 49-72
  with Raman microprobe, 238
 defect diagnostics for submicrometer VLSI technology, 75-94
 impurities, vibrational modes, 216f

Semiconductor(s)--Continued
  matrices, impurity analysis,
    SIMS, 83,85-88
  technology, application of
    SSMS, 315-319
  XPS studies, 157-158
Sensitivity, analytical image
    space, 8-9
Sequencing, chemical, silicon-wafer
    cleaning, 370,372-375
Sheet resistance of thin films, effect
    of oxidation, 339,341f
Shewhart control charts in statistical
    quality control, 342-344
Shorted metal lines
  secondary electron SEM
    photograph, 65f
  voltage contrast SEM photograph, 65f
Silica films
  absorption bands, 357,358f
  chemical vapor deposition, 349-350
Silicides
  formation, study by AES, 132
  study by XPS, 154,157
Silicon
  AES spectrum, 139f
  Auger electron maps, 134f
  Brewster's angle, 219
  bulk, size distribution of
    defects, 13f
  composite ion image, 87f
  crystal, striation, 76,77f
  device, subsurface mechanical
    defects, 258f
  donors, in GaAs, 245-251
  integrated circuit processing,
    recombination-generation lifetime
    of electrons, 29
  laser annealed, on silicon oxide,
    study with Raman
    microprobe, 236-238f
  mass spectra, 314,317f
  oxygen and carbon in, 209-215
  2p photoelectron
    region, 147f,152f,156f
  sample impurities, comparison, 316t
  secondary electron images, 139f
  solar cell
    electron micrograph, 263f
    thermal-wave images, 262-263f
  spin-on glass interface
    studies, 357,359-361f
  surface image, 89f
  trace element characterization, 303
  XPS studies, 153-156f
Silicon-carbon complexes
  absorption bands, 213,214f
  carbon distribution, 213
  electrically active defects, 213,215

Silicon-chromium thin-film resistors,
    light optical photograph, 134f
Silicon-oxygen complexes
  absorption bands, 209,211
  effects of annealing on the
    configuration, 210t
  electrically active defects, 211
  energy-level diagrams, 212f,214f
  new donors, 211
Silicon wafer(s)
  cleaning, effects of various
    chemistries, 366-375
  interferogram, 218f
  IR transmission spectra, 217f
  oxygen and carbon content,
    measurement by FTIR, 208-226
  with a rough back surface, absorption
    spectrum, 222f
Silicone, identification on a silicon
    wafer with Raman microprobe, 233
Skin, IR spectrum, 416f
Small-area molecular spectroscopies,
    general discussion, 398-399
Soft XPS, 153
Solder pad joining, 410,411f
Solder reflow, Cu-Sn intermetallics
    formed on, X-ray
    diffraction, 400-403f
Solids, thermal conductivity of,
    temperature dependence, 187
Spark source mass spectrometry (SSMS)
  application to semiconductor
    technology, 315-319
  common interferences, 314-315
  data interpretation, 309,315
  history, 309-310
  instrument design, 309,311-313f
  mass spectra, 314-315,317f
  sample requirements, 311,314
  trace element survey analyses, 308-31
Specimen preparation, AES, 124-125
Spectra obtained from NDP, 165-167
Spectroscopic ellipsometry,
    description, 194
Specularly reflected light, information
    obtained from, 193
Spin-on and cure procedure, spin-on
    glass films, 352
Spin-on dielectrics, 350-352
Spin-on fluid solvent package
  composition, 275t
  gas chromatogram, 276f
  ion chromatogram, 276f
Spin-on glass films
  as interlevel dielectrics, 350-352
  as planarizing dielectrics,
    characterization of, 349-364
  Auger depth profile, 361f
  Auger studies, 357

INDEX

Spin-on glass films—Continued
  dielectric properties, 353t
  electrical characteristics, 353-354
  ESCA studies, 357
  etch rates, 354
  film composition and structure, 355
  film thickness as a function of spin speed, 352t
  general characteristics, 352
  IR spectra, 355,357,358f
  physical characteristics, 352-353
  planarization and adhesion characteristics, 355,356f
  SIMS studies, 357
  spin-on and cure procedure, 352
  stress, 359,362-364
Spin-on glass-silicon interface studies, 357,359-361f
Spreading resistance probes, junction delineation, 81-83
Spreading resistance technique
  accuracy, 42,45
  applications, 47-48
  comparison to SIMS, 45-47
  doping profiles, 34-48
  general discussion, 35-42
  semiconductor electrical characterization, 24
  use with NDP, 176
Sputter-etching technique for planarizing interlevel dielectric layers, 350
Static SIMS
  general discussion, 110
  instrument, 99
Statistical analysis, thin-film resistor fabrication, 335,337-338f
Statistical quality control charts, 342-344
Statistical significance, number of samples needed, 378,381t
Stokes number for a sampling tube inlet, definition, 383
Stress
  laser-annealed silicon on silicon oxide, study with Raman microprobe, 236-238f
  spin-on glass films, 359,362-364
Submicrometer VLSI technology, semiconductor materials defect diagnostics, 75-94
Suboxide determinations, phosphorus, 321
Subsurface mechanical defects, detection with a thermal-wave microscope, 257,258f
Success, definition, 6

Sulfur donors, in GaAs, 245-251
Superconducting magnet
  instrumentation using, 268
  of the FTMS, block diagram, 270f
Surface
  and sample quality, characterization with a focused-spot reflectometer, 195-197f
  contaminants on, 367-368

T

Tantalum 4f photoelectron region, 151f
Tantalum silicide on silicon, depth profile, 151f
Temperature coefficient of resistance of thin films
  effect of chromium concentration, 339,340f
  effect of oxidation, 339,341f
Temperature dependence, thermal conductivity of solids, 187
Texels
  analytical image space, 9
  definition, 1
Thermal conductivity of solids, temperature dependence, 187
Thermal-cooled module, 400
Thermal lens effects, thermal-wave measurement of thin-film thickness, 183,185f,187
Thermal oxide film, Auger depth profile, 360f
Thermal wave(s)
  detection, 182,254
  generation and description, 253
  images
    GaAs crystal, 265f
    GaAs device, 262f
    silicon solar cell, 262-263f
  imaging
    in an SEM, 253-265
    physical processes, 255f
  measurement of thin-film thickness, 181-191
  micrographs
    aluminum-zinc alloy, 259f
    cobalt-chromium alloy, 259f
  microscope system as part of an SEM, 258f
Thermoacoustic probe for detecting thermal waves, 182
Thermoacoustic waves, description, 254
Thickness determination in laminar systems, 198-199

Thin film(s)
 characterization
  by AES, 130-135
  by optical reflectance and ellipsometric techniques, 192-206
 deposited by low-pressure chemical deposition, 200-202
 deposition procedure, 334-336t
 NDP study, 173-175f
 SIMS depth profiling, 105-107
 thickness, thermal-wave measurement, 181-191
Thin-film resistor(s)
 AES study, 132,134f
 requirements, 336t
Thin-film resistor fabrication
 control chart for percent chromium, 342-345f
 process background and requirements, 334-338
 process control of vacuum-deposited nickel-chromium, 333-347
 process development and characterization, 335,338f-341
 process influence and improvements, 344,346-347
Tin-copper intermetallics formed on solder reflow, X-ray diffraction, 400-403f
Tin-gold braze material, 402,404-408
Titanium metal, photoelectron spectrum, 147f
Titanium-silicon, AES sputter depth profiles, 133f
Trace element survey analyses by SSMS, 308-319
Trace methods of analysis, 310f
Transfer of samples during activation analysis, 301
Transistor, metallization pattern on the surface, 86f
Transition metals, mass spectrum, 287f
Transitions from neutral donor states, excitons, 245-250f
Transmission electron microscopy (TEM), 88,90-92
Trichloroethane sample, analysis of impurities, 319t
Trichlorotrifluoroethane, mass spectrum, 280f

V

Vacuum-deposited nickel-chromium, process control, for the fabrication of reproducible thin-film resistors, 333-347

Very large scale integration (VLSI)
 bipolar transistor
  backscatter electron SEM photograph, 57f
  electron-beam-induced current photograph, 68f
  optical photograph, 55f
  secondary electron SEM photograph, 55f,68f
  Y-modulation SEM photograph, 57f
 processing, role of SEM, 67,69-72
 technology
  sensitivity to defects and contamination, 4
  submicrometer, semiconductor materials defect diagnostics, 75-94
Vibrational modes, semiconductor impurities, 216f
Visible-UV photodetectors, characteristics, 193
Voltage contrast
 mechanism, 63f
 SEM photograph, shorted metal lines, 65f

W

Wafer(s)
 inspection, use of SEM, 67,70,72
 mapping, recombination lifetime of electrons, 27,28f
 silicon, effects of various chemistries on cleaning, 366-375
Wavelength dispersive X-ray spectroscopy, performance characteristics, 60f
Wiring board, conformal coating on
 gas chromatogram, 279f
 ion chromatogram, 279f

X

X-ray diffraction
 Cu-Sn intermetallics formed on solder reflow, 400-403f
 general discussion, 398
 poorly wetting Au-Sn preform, 404,405f
X-ray microanalysis, SEM, 56,59-62
X-ray photoelectron spectroscopy (XPS)
 general discussion, 146-153
 microelectronic materials applications, 144-159
X-ray photon, generation, 56,59

X-ray topography
  GaAs crystal, 265f
  imaging of crystalline
    samples, 261, 264, 265f

Y

Y-modulation SEM photograph, VLSI
  bipolar transistor, 57f

Production by Meg Marshall
Indexing by Karen McCeney
Jacket design by Pamela Lewis

Elements typeset by Hot Type Ltd., Washington, DC
Printed and bound by Maple Press Co., York, PA

RECENT ACS BOOKS

"Environmental Applications of Chemometrics"
Edited by Joseph J. Breen and Philip E. Robinson
ACS SYMPOSIUM SERIES 292; 286 pp; ISBN 0-8412-0945-6

"Desorption Mass Spectrometry: Are SIMS and FAB the Same?"
Edited by Philip A. Lyon
ACS SYMPOSIUM SERIES 291; 248 pp; ISBN 0-8412-0942-1

"Catalyst Characterization Science:
Surface and Solid State Chemistry"
Edited by Marvin L. Deviney and John L. Gland
ACS SYMPOSIUM SERIES 288; 616 pp; ISBN 0-8412-0937-5

"Polymer Wear and Its Control"
Edited by Lieng-Huang Lee
ACS SYMPOSIUM SERIES 287; 421 PP; ISBN 0-8412-0932-4

"Ring-Opening Polymerization:
Kinetics, Mechanisms, and Synthesis"
Edited by James E. McGrath
ACS SYMPOSIUM SERIES 286; 398 pp; ISBN 0-8412-0926-X

"Trace Residue Analysis: Chemometric Estimations
of Sampling, Amount, and Error"
Edited by David A. Kurtz
ACS SYMPOSIUM SERIES 284; 284 pp; ISBN 0-8412-0925-1

"Polycyclic Hydrocarbons and Carcinogenesis"
Edited by Ronald G. Harvey
ACS SYMPOSIUM SERIES 283; 406 pp; ISBN 0-8412-0924-3

"Reactive Oligomers"
Edited by Frank W. Harris and Harry J. Spinelli
ACS SYMPOSIUM SERIES 282; 261 pp; ISBN 0-8412-0922-7

"Reverse Osmosis and Ultrafiltration"
Edited by S. Sourirajan and Takeshi Matsuura
ACS SYMPOSIUM SERIES 281; 508 pp; ISBN 0-8412-0921-9

"Multicomponent Polymer Materials"
Edited by D. R. Paul and L. H. Sperling
ADVANCES IN CHEMISTRY SERIES 211; 354 pp; ISBN 0-8412-0899-9

"Formaldehyde: Analytical Chemistry and Toxicology"
Edited by Victor Turoski
ADVANCES IN CHEMISTRY SERIES 210; 393 pp; ISBN 0-8412-0903-0

For further information contact:
American Chemical Society, Sales Office
1155 16th Street NW, Washington, DC 20036
Telephone 800-424-6747